ACS SYMPOSIUM SERIES 627

Hydrogels and Biodegradable Polymers for Bioapplications

Raphael M. Ottenbrite, EDITOR
Virginia Commonwealth University

Samuel J. Huang, EDITOR
University of Connecticut

Kinam Park, EDITOR
Purdue University

Developed from a symposium sponsored
by the Division of Polymer Chemistry, Inc.,
at the 208th National Meeting
of the American Chemical Society,
Washington, DC
August 21–26, 1994

American Chemical Society, Washington, DC 1996

Library of Congress Cataloging-in-Publication Data

Hydrogels and biodegradable polymers for bioapplications / Raphael M. Ottenbrite, Samuel J. Huang, Kinam Park, editors.

p. cm.—(ACS symposium series, ISSN 0097–6156; 627)

"Developed from a symposium sponsored by the Division of Polymer Chemistry, Inc., at the 208th National Meeting of the American Chemical Society, Washington, DC, August 21–26, 1994."

Includes bibliographical references and indexes.

ISBN 0–8412–3400–0

1. Polymers in medicine—Congresses. 2. Polymeric drugs—Congresses. 3. Colloids in medicine—Congresses. 4. Gels (Pharmacy)—Congresses. 5. Biodegradation—Congresses.

I. Ottenbrite, Raphael M. II. Huang, Samuel J., 1937– . III. Park, Kinam. IV. American Chemical Society. Meeting (208th: 1994: Washington, D.C.) IV. American Chemical Society. Division of Polymer Chemistry. (Washington, D.C.) VI. Series.

R857.P6H96 1996
610'.28—dc20 96–33785
CIP

This book is printed on acid-free, recycled paper.

PRINTED IN THE UNITED STATES OF AMERICA

Advisory Board

ACS Symposium Series

Foreword

THE ACS SYMPOSIUM SERIES was first published in 1974 to provide a mechanism for publishing symposia quickly in book form. The purpose of this series is to publish comprehensive books developed from symposia, which are usually "snapshots in time" of the current research being done on a topic, plus some review material on the topic. For this reason, it is necessary that the papers be published as quickly as possible.

Before a symposium-based book is put under contract, the proposed table of contents is reviewed for appropriateness to the topic and for comprehensiveness of the collection. Some papers are excluded at this point, and others are added to round out the scope of the volume. In addition, a draft of each paper is peer-reviewed prior to final acceptance or rejection. This anonymous review process is supervised by the organizer(s) of the symposium, who become the editor(s) of the book. The authors then revise their papers according to the recommendations of both the reviewers and the editors, prepare camera-ready copy, and submit the final papers to the editors, who check that all necessary revisions have been made.

As a rule, only original research papers and original review papers are included in the volumes. Verbatim reproductions of previously published papers are not accepted.

ACS BOOKS DEPARTMENT

Contents

INDEXES

Preface

THE PROVINCE FOR BIORELATED POLYMERS has evolved significantly over the past few years with remarkable advances in many areas such as polymeric drugs and drug delivery, self-assembly systems, implant materials, and controlled release of excipients. These areas have now become well established in the realm of multidisciplinary technology and science. The blending of advanced biochemistry, biophysics, and physicotechnology with polymer science has led to the design and preparation of many new and viable systems. In tandem with the merging of these disciplines has been the development of ancillary science and technology such as modes of encapsulation and degradations of materials used in bioapplications.

The theme of the symposium upon which this book is based was hydrogel biodegradation and bioapplications. Understanding hydrogel applications has become important in the development of hydrogel technology. The study and evaluation of hydrogel systems under clinical conditions has now become a reality. Hydrogels are formed by adding a small amount of cross-linked macromolecular material to a large amount of water, which produces an apparent solid. This book addresses reversible hydrogels, stimuli-sensitive hydrogels, and some in vivo applications of hydrogels.

Biodegradation has become a very important technology in many respects such as controlled release of excipients from degradable matrices, degradation of synthetic materials in marine and soil environments, and time-related degradation of temporary support devices for limbs and other applications.

This book comprises 20 chapters related to these topics, based on work being widely pursued. The contents should be of interest to those who need crossdisciplinary solutions to problems in the area of biorelated polymers. Therefore this volume is recommended to organic, physical, polymer, and biochemists as well as to biologists, materials scientists, and bioengineers.

The editors thank all the contributors to this publication and wish them well in their continued quests to make this world a better place to live.

RAPHAEL M. OTTENBRITE
Department of Chemistry
Virginia Commonwealth University
1001 West Main Street
Richmond, VA 23284–2006

SAMUEL J. HUANG
Polymer Science Program
University of Connecticut
97 North Eagleville Road
Storrs, CT 06269

KINAM PARK
School of Pharmacy
Purdue University
Robert E. Heine Pharmacy Building
West Lafayette, IN 47907–1336

December 28, 1995

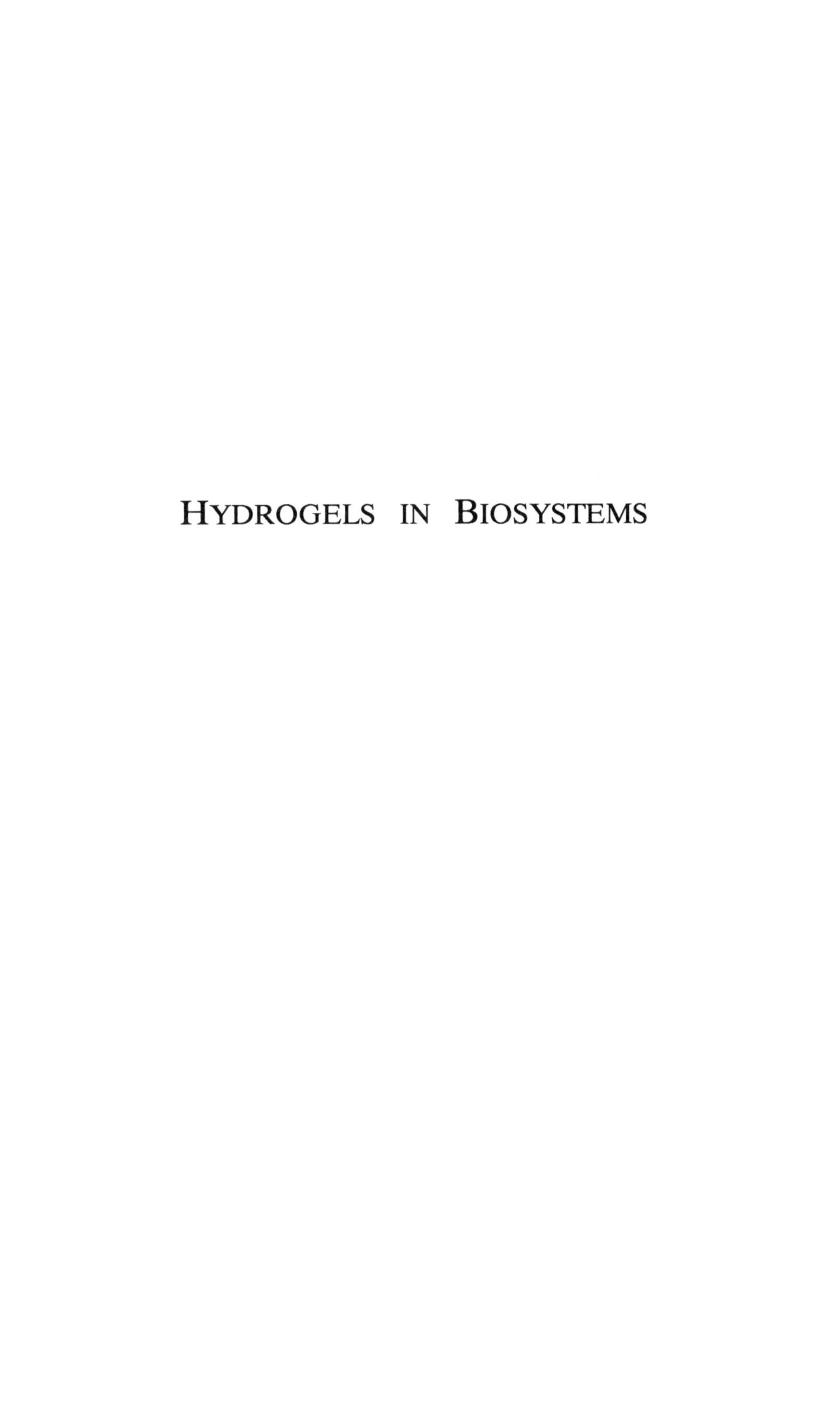

HYDROGELS IN BIOSYSTEMS

Chapter 1

Hydrogels in Bioapplications

Haesun Park and Kinam Park[1]

School of Pharmacy, Purdue University, Robert E. Heine Pharmacy Building, West Lafayette, IN 47907–1336

Research on hydrogels has been geared toward biomedical applications from the beginning due to their relatively high biocompatibility. Initially only the hydrophilic nature and the large swelling properties of hydrogels was explored. Continued research on hydrogels has resulted in the development of new types of hydrogels, such as environment-sensitive hydrogels, thermoplastic hydrogels, hydrogel foams, and sol-gel phase-reversible hydrogels. Application of hydrogels ranges from biomedical devices to solute separation. Examples of hydrogel applications in pharmaceutics, biomaterials, and biotechnology are briefly described.

Hydrogel is a three-dimensional network of hydrophilic polymers in which a large amount of water is present. In general, the amount of water is at least 20% of the total weight. If water is composed of more than 95% of the total weight, then the hydrogel is called superabsorbent. The most characteristic property of hydrogel is that it swells in the presence of water and shrink in the absence of water. The extent of swelling is determined by the nature (mainly hydrophilicity) of polymer chains and the crosslinking density. If hydrogel is dried, the swollen network of the hydrogel is collapsed during drying due to the high surface tension of water. Thus, the dried hydrogel (or xerogel) becomes much smaller in size than the hydrogel swollen in water. During swelling and shrinking process, hydrogels can preserve its overall shape.

To maintain the three-dimensional structures, polymer chains of hydrogels are usually crosslinked either chemically or physically. In chemical gels polymer chains are connected by covalent bonds, and thus it is difficult to change the shape of chemical gels. On the other hand, polymer chains of physical gels are connected through non-covalent bonds, such as van der Waals interactions, ionic interactions, hydrogen bonding, or hydrophobic interactions (1). Since the bonding between polymer chains are reversible, physical gels possess sol-gel reversibility. For example, sodium alginate becomes a gel in the presence of calcium ions, but the gel becomes sol if the divalent cations are removed.

Strong interest in biomedical applications of hydrogels was caused by the landmark paper by Wichterle and Lim on poly(2-hydroxyethyl methacrylate) or p(HEMA)(2). Since then the research on hydrogel has been steadily increased. It is

[1]Corresponding author

0097–6156/96/0627–0002$15.00/0

not until the end of 1970's, however, when the research on hydrogels began to take off. Fig. 1 shows the number of articles on gels published each year from 1907 to 1994. As shown in Fig. 1, the number of publications on gel started to increase in the late 70's and continued to increase during 80's and early 90's. The number increased dramatically in 1994. The involvement of a large number of scientists resulted in more understanding on the physicochemical properties of hydrogels and development of new types of hydrogels (3,4).

New Types of Hydrogels

Environment-Sensitive Hydrogels

One of the inherent properties of hydrogels is their ability to swell in the presence of water and to shrink in the absence of water. This property is common to all hydrogels. Thus, by merely having the swelling-shrinking properties does not make any particular hydrogel of great interest. Recently, many investigators have prepared hydrogels with additional functions, such as the ability to swell or shrink in response to a signal. These hydrogels with additional functions are called "smart (or intelligent) hydrogels" (5).

The most widely known smart hydrogels are those which respond (i.e., either swell, shrink, bend, or degrade) to changes in environmental conditions. For this reason, they are usually known as environment-sensitive hydrogels. One of the unique properties of environment-sensitive hydrogels is that they change their swelling ratio (which is the volume of the swollen hydrogel divided by the volume of the dried hydrogel) rather abruptly upon small changes in environmental factors. Table 1 lists the environmental factors which are known to cause such an abrupt volume change. Fig. 2 shows dramatic volume changes as a result of only a minute change in the environmental condition. Volume change is so dramatic it is often called volume collapse (or volume phase transition) (6). The volume collapse phenomena have been applied in a variety of areas ranging from pharmaceutics to biotechnology (see below).

Table 1. Factors which cause volume collapse of hydrogels

Factor	Reference
1. pH	(7,8)
2. Temperature	(9-11)
3. Electric field	(12-15)
4. Ionic strength	(16)
5. Salt type	(17,18)
6. Solvent	(19,20)
7. Stress	(21,22)
8. Light	(23,24)
9. Pressure	(25)

Thermoplastic Hydrogels

One of the physical gels is thermoplastic hydrogels (26). Thermoplastic hydrogels are based on linear copolymers of hydrophilic and hydrophobic monomers. Physical gel is formed by hydrophobic interactions between hydrophobic chains of the copolymer. They dissolve in organic solvents, while only swell without dissolving in water, and

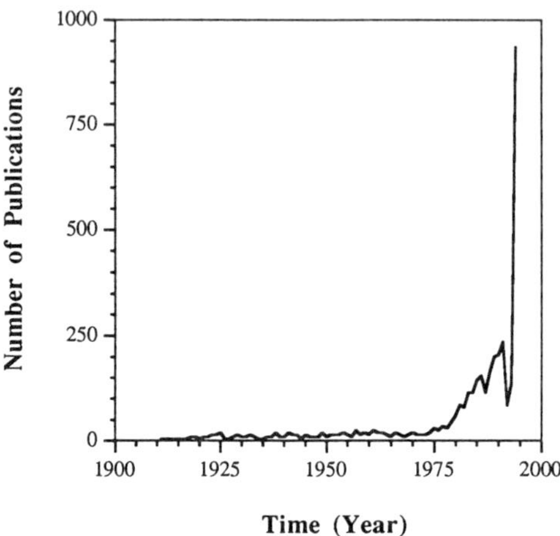

Figure 1. The number of articles on gels published between 1907 and 1994. The articles appeared on Chemical Abstract were counted. For the years earlier than 1982, keywords of "gels/hydrophilic" and "hydrogels" were searched under the subject heading of "colloid". For later years articles under the heading of "gels/hydrophilic & hydrogels" were counted.

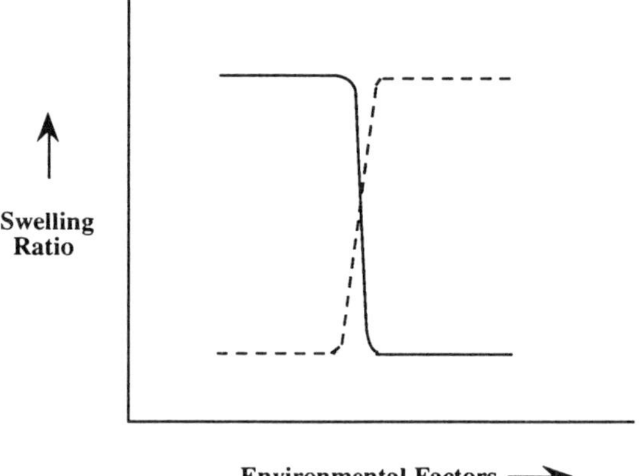

Figure 2. Volume collapse (or volume phase transition) of smart hydrogels in response to a small change in environmental factors. Hydrogels may undergo dramatic increase (dotted line) or decrease (solid line) in the swelling ratio.

this property provides an advantage of easy processibility. Copolymers of N-vinyl-2-pyrrolidone and methyl methacrylate are known to form thermoplastic hydrogels which have useful properties such as film-forming capacity and melt-processibility (27).

Hydrogel Foams

While hydrogels swell to a large extent in water, the equilibrium swelling usually takes a long time from hours to days depending on the size and shape of hydrogels. To overcome this slow swelling of hydrogels, hydrogel foam was recently developed (28,29). Hydrogel foams were made by synthesizing hydrogels in the presence of gas bubbles. The hydrogel foams prepared with macroscopic gas cells are different from hydrogel sponges (30) or macroporous hydrogels (31). The size of pores in the hydrogel foams is orders of magnitude larger than the pore size (which is typically a few micrometers) in hydrogel sponges or macroporous hydrogels. In addition, the kinetics and the extent of swelling of hydrogel foams are much faster and larger than others.

The kinetics of hydrogel swelling is limited by the diffusion of water through the glassy layer of dried hydrogels. Thus, the swelling is actually quite slow for many applications. On the other hand, hydrogel foams swell much faster since water is absorbed into hydrogel foams by capillary reaction by the pores in dried hydrogels rather than the diffusion of water through the glassy layer. Recently developed comb-type grafted hydrogels (11) showed faster deswelling but the swelling was still quite slow. The hydrogel foams made of poly(acrylic acid) can swell more than thousand times its original size. The property of fast swelling with a very large swelling ratio of hydrogel foams should be useful in many bioapplications.

Ligand-Specific Sol-Gel Phase-Reversible Hydrogels

Physical gels are capable of undergoing sol-gel phase transition due to non-covalent crosslinking of polymer chains. There are only a few hydrogels which undergo such a phase change in response to interaction with specific molecules. For example, hydrogels which become sol in the presence of glucose were developed. Hydrogels made of boronic acid-containing polymers and poly(vinyl alcohol) are known to degrade in the presence of glucose in the environment (32-34). Recently, glucose-containing copolymer and concanavalin-A molecules were used to prepare hydrogels which are more specific to glucose than boronic acid gels (35,36). The hydrogels become sol in the presence of glucose in the medium, and the sol becomes gel again if the free glucose molecules are removed by dialysis. Since there are many ligand specific interactions, such as antigen-antibody, avidin-biotin, and carbohydrate-lectin, various types of ligand-specific phase-reversible hydrogels can be made for different applications.

Hydrogels in Bioapplications

Pharmaceutical applications

Much of the research on hydrogels have been focused on the application in controlled drug delivery (37,38). While zero-order drug release is important for most drugs, there are many drugs that need to be delivered in a pulsatile fashion (39). The most widely used example is the delivery of insulin. Temporal control of insulin delivery can be achieved by utilization of smart hydrogels which release more insulin in response to increase in glucose level. Most glucose-responsive hydrogel systems are made of pH-sensitive polymers such as poly(diethylaminoethyl methacrylate) (PDEAEMA) (40-42) and glucose oxidase, which transforms glucose into gluconic acid. In addition to the pH-responsive hydrogels, the glucose-sensitive dissolvable hydrogels have been used to control the insulin release (34,43). Microspherical hydrogels such as alginate microparticles have been used to encapsulate insulin-producing cells for the delivery of insulin (44).

Pulsatile delivery of drugs can also be achieved by temperature-responsive hydrogels. Thermo-sensitive hydrogels are usually made of polyacrylamide derivatives with hydrophobic groups which promote hydrophobic interactions necessary for shrinking at elevated temperatures. The volume collapse temperature can be adjusted by varying the hydrophobic groups. By altering the temperature around the thermosensitive hydrogels, the release of drug from the gel can be turned on and off at will (10,45).

Biomedical Applications

The applications of hydrogels in biomedical fields are diverse ranging from diagnostic device (46) to artificial muscle (47). Application of hydrogels as contact lenses and intraocular lenses has a rather long history compared with other applications. Soft contact lenses made of hydrogels possess desirable properties such as high oxygen permeability, although they have problems of protein deposits and lens spoilation (48). Soft intraocular lenses have advantages over rigid types. Their ability to be folded allows surgeon to use a much smaller surgical incision (49). The hydrogel contact and intraocular lenses can be sterilized by autoclaving, which is more convenient than the sterilization by ethylene oxide needed for rigid lenses made of poly(methyl methacrylate).

Hydrogels are commonly used as wound dressing materials, since they are flexible, durable, non-antigenic, and permeable to water vapor and metabolites, while securely covering the wound to prevent bacterial infection (50). Methylcellulose hydrogel has been used to deliver allergens in skin testing. When test allergens are delivered in the hydrogel vehicle, less skin irritation was observed (51). Hydrogels are also often coated on the urinary catheter surface to improve its biocompatibility (52). The hydrogel layer not only provide smooth, slippery surface, but also it can prevent bacterial colonization on the surface (53).

Hydrogel layers formed on the inner surface of injured arteries are known to reduce thrombosis and intimal thickening in animal models (54). Intimal thickening was prevented by inhibiting contact between blood and subendothelial tissue with a hydrogel layer. The swelling pressure of p(HEMA) hydrogel was used to stabilize the bone implants (55). With improved design of the implant, such hydrogels are expected to be effectively used as a stabilizing interface.

The potential applications of hydrogels in sterilization and cervical dilatation have been explored (56). When a hydrogel which forms an in situ plug was placed into fallopian tubes of rabbits by transcervical catheterization, conception was prevented. With more structurally rigid and biocompatible hydrogels the tubular sterilization system can be developed (57). Hydrogel rods were developed to deliver hormones such as prostaglandin analogs as well as to mechanically dilate the cervix. Dilatation of the cervical canal is necessary for the first trimester-induced abortion by suction curettage (56).

One of the advanced applications of hydrogels is in the development of artificial muscles. Smart hydrogels which can transform electrochemical stimuli into mechanical work (i.e., contraction) can function like the human muscle tissue (58). Polymeric gels capable of reversible contraction and expansion under physicochemical stimuli are essential in the development of advanced robotics with electrically driven muscle-like actuators (12) Smart materials that emulate the contractions and secretions of human organs in response to changes in environmental conditions such as temperature, pH, or electric field may soon find a use in medical implants, prosthetic muscles or organs, and robotic grippers (59).

Applications in Biotechnology

Hydrogels have been used as reactive matrix membranes in sensors. Hydrogels possess many advantageous properties, such as rapid and selective diffusion of the analyte, necessary for effective sensing. Hydrogels can also be made tough and flexible with desirable refractive indices (60).

Hydrogels made of p(HEMA) in an electrolyte solution were used as a salt bridge which separate the metallic electrodes from the biological system to prevent contamination by electrolysis products (61). The p(HEMA) salt bridges are inexpensive, easy to make, easy to sterilize, very durable, and nontoxic to cell systems. They provides a viable and effective alternative to the widely used agar salt bridge.

The ability of smart hydrogels in solutions to reversibly swell and shrink with small changes in the environmental conditions can be used to prepare purification devices (62). Smart hydrogels, especially thermo- and pH-sensitive hydrogels, have been used to concentrate dilute aqueous solutions of macromolecular solutes including proteins and enzymes, with no adverse effect on the activity of the enzyme (63). Water is absorbed into the hydrogel while macromolecules are excluded from the hydrogel network primarily by size and net charge (64). The absorbed water can be released from the hydrogel by altering temperature or pH of the environment, and thus the hydrogels can be reused repeatedly. Separation of bioactive proteins produced by recombinant DNA technology in a cost effective manner remains as one of the major huddles for the wide use of the technology. Separation of products by direct adsorption to adsorbents is attractive, but the adsorbents become fouled by colloidal contaminants and large macromolecules. This problem can be overcome by immobilizing adsorbents into hydrogels such as agarose and calcium alginate gel (52). Since the immobilized adsorbents do not contact with contaminants, separation becomes easier and more effective.

The smart hydrogels can also be used to control the reactions of substrates with immobilized enzymes by controlling the substrate diffusivity via swelling changes (65). Park and Hoffman immobilized Arthrobacter simplex in a thermally reversible hydrogel and examined the effect of temperature cycling on steroid conversion (66). The steroid conversion was higher in more hydrophobic gels due to the high partitioning of water-insoluble steroids into the hydrophobic regions and the reduced product inhibition within the hydrophobic gel matrices (66).

Future of Hydrogels

In the age of nanofabrication (67), the size of hydrogels in various applications is also expected to shrink. Gel electrophoresis is widely used for the separation of proteins and DNA. Recent report showed that miniaturized electrophoresis gel instrument in the size of 25 mm long and 50 µm wide was constructed (68). The gels that can be used in the instrument is orders of magnitude smaller than the gels used in conventional instruments. Miniaturization of hydrogels is also important in all the areas mentioned above. For example, the ability to reproducibly prepare hydrogels in microscale is essential in the preparation of glucose microsensors (69).

Hydrogels are generally biocompatible, but they are not perfect biomaterials, i.e., they still cause undesirable body reactions. Further improvement in biocompatibility will be critical in the wider applications of hydrogels in biomedical and pharmaceutical areas. Hydrogels have rather poor mechanical strength and durability for some applications. Enhancing these properties will make hydrogels more acceptable for many applications to come.

References

(1) Guenet, J.-M. *Thermoreversible Gelation of Polymers and Biopolymers*, Academic Press, New York, **1992**.
(2) Wichterle, O.; Lim, D. *Nature* **1960**, 185, 117-118.
(3) Aharoni, S. M.*Synthesis, Characterization, and Theory of Polymeric Network and Gels*, Plenum Press: New York, NY., 1992.
(4) DeRossi, D.; Kajiwara, K.; Osada, Y.; Yamauchi, A.*Polymer Gels. Fundamentals and Biomedical Applications*, Plenum Press: New York, NY., 1991.
(5) Takagi, T.; Takahashi, K.; Aizawa, M.; Miyata, S.*Proceedings of the First International Conference on Intelligent Materials*, Technomic Publishing Co., Inc.: Lanaster, PA, 1993.
(6) Tanaka, T. *Sci. Amer.* **1981**, 244, 124-138.
(7) Siegel, R. A. in *Pulsed and Self-Regulated Drug Delivery*, CRC Press, Boca Raton, FL., **1990**, Chapter 8.
(8) Brannon-Peppas, L.; Peppas, N. A. *Chem. Eng. Sci.* **1991**, 46, 715-722.
(9) Feil, H.; Bae, Y. H.; Feijen, J.; Kim, S. W. *Journal Of Membrane Science* **1991**, 64, 283-294.
(10) Dong, L. C.; Hoffman, A. S. *Journal Of Controlled Release* **1990**, 13, 21-32.
(11) Yoshida, R.; Uchida, K.; Kaneko, Y.; Sakal, K.; Kikuchi, A.; Sakurai, Y.; Okano, T. *Nature* **1995**, 374, 240-242.
(12) De Rossi, D.; Suzuki, M.; Osada, Y.; Morasso, P. *J. Intell. Mater. Syst. Struct.* **1992**, 3, 75-95.
(13) Kaetsu, I.; Uchida, K.; Morita, Y.; Okubo, M. *Radiat. Phys. Chem.* **1992**, 40, 157-160.
(14) Tanaka, T.; Nishio, I.; Sun, S.-T.; Ueno-Nishio, S. *Science* **1982**, 218, 467-469.
(15) Kwon, I. C.; Bae, Y. H.; Kim, S. W. *Nature (Lond).* **1991**, 354, 291-293.
(16) Hooper, H. H.; Baker, J. P.; Blanch, H. W.; Prausnitz, J. M. *Macromolecules* **1990**, 23, 1096-1104.
(17) Ohmine, I.; Tanaka, T. *J. Chem. Phys.* **1982**, 77, 5725-5729.
(18) Inomata, H.; Goto, S.; Otake, K.; Saito, S. *Macromolecules* **1992**, 8, 687-690.
(19) Hu, Y.; Horie, K.; Ushiki, H.; Yamashita, T.; Tsunomori, F. *Macromolecules* **1993**, 26, 1761-1766.
(20) Amiya, T.; Tanaka, T. *Macromolecules* **1987**, 20, 1162-1164.
(21) Okuzaki, H.; Osada, Y. in *Proceedings of the First International Conference on Intelligent Materials*, Technomic Publishing Co., Inc., Lanaster, PA, **1993**, 273-278.
(22) Sawahata, K.; Hara, M.; Yasunaga, H.; Osada, Y. *J. Controlled Rel.* **1990**, 14, 253-262.
(23) Suzuki, A. in *Proceedings of the First International Conference on Intelligent Materials*, Technomic Publishing Co., Inc., Lanaster, PA, **1993**, 297-300.
(24) Mamada, A.; Tanaka, T.; Kungwatchakun, D.; Irie, M. *Macromolecules* **1990**, 23, 1517-1519.
(25) Lee, K. K.; Cussler, E. L.; Marchetti, M.; McHugh, M. A. *Chem. Eng. Sci.* **1990**, 45, 766-767.
(26) Capozza, R. C.; Meyers, W. E.; Neidlinger, H. H.; Stoy, V. A. *Polymer Preprints* **1990**, 31, 57.
(27) Liu, Y.; Huglin, M. B.; Davis, T. P. *Eur. Polym. J.* **1994**, 30, 457-463.
(28) Park, H.; Park, K. *The 20th Annual Meeting of the Society for Biomaterials* **1994**, Abstract #158.
(29) Park, H.; Park, K. *Proc. Intern. Symp. Control. Rel. Bioact. Mater.* **1994**, 21, 21-22.

(30) Chirila, T. V.; Constable, I. J.; Crawford, G. J.; Vijayasekaran, S.; Thompson, D. E.; Chen, Y. C.; Fletcher, W. A.; Griffin, B. J. *Biomaterials* **1993**, 14, 26-38.
(31) Oxley, H. R.; Corkhill, P. H.; Fitton, J. H.; Tighe, B. J. *Biomaterials* **1993**, 14, 1065-1072.
(32) Miyazaki, H.; Kikuchi, A.; Kitano, S.; Kataoka, K.; Koyama, Y.; Okano, T.; Sakurai, Y. in *Proceedings of the First International Conference on Intelligent Materials*, Technomic Publishing Co., Inc., Lanaster, PA, **1993**, 481-484.
(33) Kitano, S.; Hisamitsu, I.; Koyama, Y.; Kataoka, K.; Okano, T.; Yokoyama, M.; Sakurai, Y. in *Proceedings of the First International Conference on Intelligent Materials*, Technomic Publishing Co., Inc., Lanaster, PA, **1993**, 383-388.
(34) Shino, D.; Kataoka, K.; Koyama, Y.; Yokoyama, M.; Okano, T.; Sakurai, Y. in *Proceedings of the First International Conference on Intelligent Materials*, Technomic Publishing Co., Inc., Lanaster, PA, **1993**, 301-304.
(35) Lee, S. J.; Park, K. *Polymer Preprints* **1994**, 35, 391-392.
(36) Lee, S. J.; Park, K. *J. Mol. Rec.* submitted for publication.
(37) Peppas, N. A. *Hydrogels in Medicine and Pharmacy. Volumes I-III.*, CRC Press, Boca Raton, FL, **1987**.
(38) Park, K.; Shalaby, S. W. S.; Park, H. *Biodegradable Hydrogels for Drug Delivery*, Technomic Publishing Co., Lancaster, **1993**.
(39) Kost, J.*Pulsed and Self-Regulated Drug Delivery*, CRC Press: Boca Raton, FL., **1990**.
(40) Klumb, L. A.; Horbett, T. A. *Journal Of Controlled Release* **1993**, 27, 95-114.
(41) Albin, G.; Horbett, T. A.; Ratner, B. D. in *Pulsed and Self-Regulated Drug Delivery*, CRC Press, Boca Raton, FL., **1990**, 159-185.
(42) Ishihara, K.; Kobayashi, M.; Shionohara, I. *Makromol. Chem. Rapid Commun.* **1983**, 4, 327-331.
(43) Kitano, S.; Koyama, Y.; Kataoka, K.; Okano, T.; Sakurai, Y. *J. Controlled Rel.* **1992**, 19, 162-170.
(44) Lacy, P. E. *Sci. Amer.* **1995**, 273, 50-58.
(45) Bae, Y. H.; Okano, T.; Kim, S. W. *Pharm. Res.* **1991**, 8, 624-628.
(46) Hoffman, A. S. *J. Controlled Rel.* **1987**, 6, 297-305.
(47) Suzuki, M. *Kobunshi Ronbunshu* **1989**, 46, 603-611.
(48) Myers, R. I.; Larsen, D. W.; Taso, M.; Castellano, C.; Becherer, L. D.; Fontana, F.; Ghormley, N. R.; Meier, G. *Optometry And Vision Science* **1991**, 68, 776-782.
(49) Carlson, K. H.; Cameron, J. D.; Lindstrom, R. L. *Journal Of Cataract And Refractive Surgery* **1993**, 19, 9-15.
(50) Corkhill, P. H.; Hamilton, C. J.; Tighe, B. J. *Biomaterials* **1989**, 10, 3-10.
(51) Darsow, U.; Vieluf, D.; Ring, J. *Journal Of Allergy And Clinical Immunology* **1995**, 95, 677-684.
(52) Nigam, S. C.; Sakoda, A.; Wang, H. Y. *Biotechnology Progress* **1988**, 4, 166-172.
(53) Graiver, D.; Durall, R. L.; Okada, T. *Biomaterials* **1993**, 14, 465-469.
(54) Hill-West, J. L.; Chowdhury, S. M.; Slepian, M. J.; Hubbell, J. A. *Proceedings Of The National Academy Of Sciences Of The United States Of America* **1994**, 91, 5967-5971.
(55) Netti, P. A.; Shelton, J. C.; Revell, P. A.; Pirie, C.; Smith, S.; Ambrosio, L.; Nicolais, L.; Bonfield *Biomaterials* **1993**, 14, 1098-1104.
(56) Molin, A.; Brundin, J. *Gynecologic And Obstetric Investigation* **1992**, 34, 12-14.
(57) Maubon, A. J.; Thurmond, A. S.; Laurent, A.; Honiger, J. E.; Scanlan, R. M.; Rouanet, J. P. *Radiology* **1994**, 193, 721-723.

(58) Studt, T. *R&D Magazine* **1992**, April, 55-60.
(59) Constance, J. *Mech. Eng.* **1991**, 113, 51-53.
(60) Davies, M. L.; Murphy, S. M.; Hamilton, C. J.; Tighe, B. J. *Biomaterials* **1992**, 13, 991-999.
(61) Kindler, D. D.; Bergethon, P. R. *Journal of Applied Physiology* **1990**, 69, 373-375.
(62) Marchetti, M.; Cussler, E. L. *Sep. Purif. Methods* **1989**, 18, 177-192.
(63) Gehrke, S. H.; Andrews, G. P.; Cussler, E. L. *Chem. Eng. Sci.* **1986**, 41, 2153-2169.
(64) Vasheghani-Farahani, E.; Cooper, D. G.; Vera, J. H.; Weber, M. E. *Chem. Eng. Sci.* **1992**, 47, 31-40.
(65) Park, T. G.; Hoffman, A. S. *Biotechnology Progress* **1994**, 10, 82-86.
(66) Park, T. G.; Hoffman, A. S. *Journal Of Biomedical Materials Research* **1990**, 24, 21-38.
(67) Drexler, K. E. *Nanosystems. Molecular Machinery, Manufacturing, and Computation*, John Wiley & Sons, New York, NY, **1992**.
(68) Studt, T. *R&D Magazine* **1995**, June Issue, 22-26.
(69) Pishko, M. V.; Michael, A. C.; Heller, A. *Analytical Chemistry* **1991**, 63, 2268-2272.

RECEIVED July 11, 1995

Chapter 2

Glucose-Sensitive Phase-Reversible Hydrogels

Samuel J. Lee and Kinam Park[1]

School of Pharmacy, Purdue University, Robert E. Heine Pharmacy Building, West Lafayette, IN 47907–1336

A novel hydrogel which undergoes sol-gel phase transformation by changes in glucose concentration of the surrounding medium was synthesized. The specific interactions between glucose and Concanavalin A (Con-A) was utilized to provide glucose sensitivity and reversible crosslinking. Glucose-containing copolymers were synthesized using allyl glucose and N-vinyl-2-pyrrolidinone (VP). Mixing of the copolymer and Con-A solutions resulted in the hydrogel formation. The efficiency of hydrogel formation increased as the relative concentration of Con-A to copolymer was increased. Upon addition of free glucose, the gel became a sol as a result of replacement of the polymer-attached glucose by free glucose molecules. The removal of the free glucose from the sol by dialysis caused the formation of hydrogel again. The sensitivity of the hydrogel to glucose can be adjusted by controlling the concentrations of copolymer and Con-A as well as the glucose concentration in the copolymer.

Hydrogels have been used in various applications ranging from controlled drug delivery to biotechnology. Many hydrogels have the ability to respond (i.e., swell, shrink, or degrade) to the changes in environmental stimuli, such as pH and temperature. Such a response has been limited to changes in the volume of hydrogels in the same gel phase. Currently, there are no hydrogels which undergo reversible sol-gel (i.e., liquid-solid) phase transformation in the presence of a specific molecule such as glucose. Recently, a few hydrogel systems which dissolve to the liquid state in the presence of excess glucose were developed (1, 2). Although they respond to changes in the environmental glucose concentration, they are not specific only to glucose and their sol-gel reversibility was not tested.

Here, we report the synthesis of hydrogels which can undergo phase transformation from sol to gel and vice versa depending on the presence of a specific biomolecule in the environment. In this particular study, we synthesized sol-gel phase-reversible hydrogels which are sensitive to glucose. The glucose sensitivity was chosen, since one of the goals of our study was to design self-regulating insulin delivery systems based on phase-reversible hydrogels.

[1]Corresponding author

0097–6156/96/0627–0011$15.00/0

Sol-Gel Phase-Reversible Hydrogels Sensitive to Glucose

We used specific interactions between glucose and Concanavalin-A (Con-A) to form physical crosslinks between polymer chains. Con-A is a glucose-binding protein extracted from plants. Con-A consists of four polypeptide subunits, which is also called protomers. The molecular weight of each protomer is 26,000. We synthesized polymer chains containing glucose molecules. The glucose molecules attached to the polymer backbone react with Con-A. Since Con-A exists as a tetramer at physiological pH and each subunit has a glucose binding site, Con-A can function as a crosslinking agent for glucose-containing polymer chains. Because of the non-covalent interaction between glucose and Con-A, the formed crosslinks are reversible. Individual glucose molecules can compete with the polymer-attached glucose molecules. Thus, the maintenance of the crosslinks depends on the relative concentration of free glucose in the environment. This concept is described in Fig. 1.

The gel is formed by mixing a solution of glucose-containing polymers with a Con-A solution. Upon addition of free glucose molecules, the hydrogel becomes a sol (i.e., the hydrogel dissolves) due to the detachment of polymer chains from Con-A as a result of competitive binding of free glucose to Con-A. The sol can become a gel again upon removal of free glucose by dialysis. In Fig. 1, Con-A can also be attached to the polymer backbone. It is important to note that the sol-gel phase transition in our study is totally different from volume phase-transition, which is not really a phase transition, since the volume change occurs in the same gel state.

Materials and Methods

Synthesis of Allyl Glucose. Dry HCl (Aldrich) was dissolved in allyl alcohol (Aldrich). The allyl alcohol/HCl mixture was added to a-D-Glucose (MW 180.16; Aldrich) and reacted at 80°C for 4 h with reflux. After the reaction, unreacted allyl alcohol was evaporated under reduced pressure and heat. The prepared allyl glucose (MW 220.22) was then extracted with dry acetone. Allyl glucose was crystallized out from the concentrated extract.

Synthesis of Glucose-Containing Copolymers. Copolymers of allyl glucose and 1-vinyl-2-pyrrolidinone (VP; MW 111.14; Aldrich) were synthesized by free radical polymerization. The copolymers were synthesized from a solution of VP, allyl glucose, 2,2'-azobis[2-(2-imidazolin-2-yl)propane] dihydrochloride (Wako Pure Chemical Industries), and N,N,N',N'-tetramethylethylenediamine (Bio-Rad). Synthesized copolymers were dialyzed extensively against double distilled water. The molecular weight cutoff of the dialysis membrane was 6,000-8,000. Copolymers were then separated by precipitation with acetone and dried. A representative structure of the prepared copolymer is shown in Fig. 2.

Hydrogel Formation. Concanavalin-A (Con-A; MW 110,000; Sigma) was dissolved in 2X phosphate-buffered saline (PBS) solution containing 1mM $CaCl_2$, 1mM $MnCl_2$. The concentration of Con-A in the stock solution was 30% (w/v). The copolymer was also dissolved in PBS. The copolymer concentration was varied from 22.5 mg/ml to 180 mg/ml before the addition of Con-A solution. Equal volumes of Con-A and copolymer solutions were mixed to form hydrogels.

Determination of Glucose Concentration. The concentration of glucose molecules incorporated into the copolymer was measured by the phenol-sulfuric acid assay (3). A standard curve was constructed using various amounts of a-D-

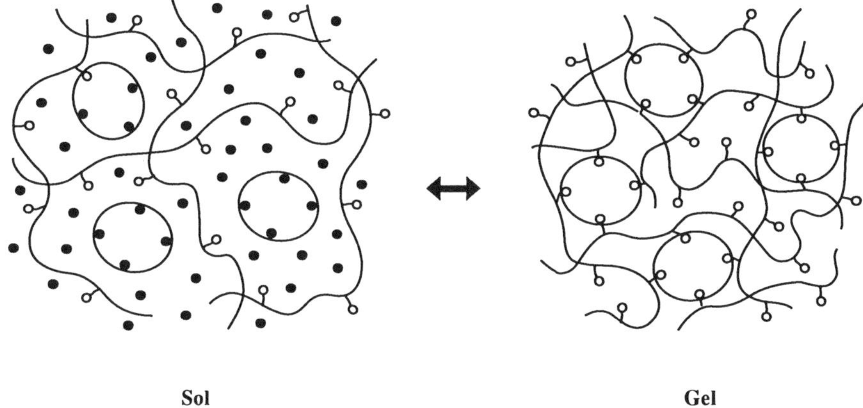

Figure 1. Pictorial representation of the sol-gel phase transition. Large circles represent Con-A molecules. Small open circles represent glucose attached to the polymer chain and small closed circles represent free glucose molecules.

Figure 2. Representation of a copolymer made from allyl glucose and vinylpyrrolidinone monomers.

glucose. To glucose standard solutions and to copolymer solutions (2 ml each) was added 0.1 ml of phenol and 5 ml of concentrated sulfuric acid. After 30 min, the absorbance of the solutions at 490 nm was read on a Beckman DU-7 Spectrophotometer to determine the glucose concentrations.

Determination of Reactivity Ratios. For the measurement of the reactivity ratios, two monomers with various molar ratios in the initial feeds were copolymerized only for 10 min to produce copolymers at low conversion. Polymerization was stopped by pouring the reaction mixture into an excess amount of acetone, which resulted in dilution of more than 200 times by volume. From the measured concentrations of glucose and VP, the reactivity ratio was determined using the graphical method of Fineman and Ross (4).

Turbidity Measurements. Gelation of the mixture of Con-A and copolymers was determined by measuring the turbidity of the solution. To each well of a 96 well micro test plate were added 100 ml of copolymer solution and 100 ml of Con-A solution. The final concentration of copolymer solution was varied from 4.5 to 90.4 mg/ml, and that of Con-A solution was varied from 50 mg/ml to 150 mg/ml. The absorption was measured at 630 nm on EL311 Microplate Autoreader (Bio-Tek Instruments).

Results and Discussions

Polymerization of allyl glucose and VP produced copolymers containing various concentration of glucose on the backbone chain. Table I shows the compositions of the synthesized copolymers along with the molar ratios in the initial feed solution. The composition of the synthesized copolymer was determined by phenol-sulfuric assay. As shown in Table I, the incorporation of glucose into the copolymer increased with increase in initial feed composition of allyl glucose. The reactivity ratios obtained were 1.095 and -0.085 for r_1 (VP) and r_2 (allyl glucose), respectively. The reactivity ratio r_2 is essentially zero. The result indicates that the copolymer is composed of small blocks of VP with allyl glucose monomer placed in between the blocks.

Copolymer solutions were mixed with Con-A solution in test tubes at various concentrations to find out the condition for the gel formation. The final concentrations varied from 11 mg/ml to 90 mg/ml for copolymers and 50 mg/ml to 150 mg/ml for Con-A. The mixing resulted in either a viscous solution or a gel depending on the concentrations of the copolymer and Con-A. When the gel was formed, it formed immediately after the mixing. The sample became turbid and was not transparent any more. At that time, the gel could be taken out of the container as one solid piece. The results showed that the gel was formed more easily as the copolymer concentration became lower or the Con-A concentration became higher. The easier gel formation at low copolymer concentrations may be due to the less competition for Con-A among polymer chains. At higher Con-A concentrations, the competition among polymer chains is expected to be reduced and all the polymer chains can be crosslinked to form a gel.

The effect of copolymer concentration on the formation of gel was examined by measuring the turbidity of the mixture of Con-A and copolymer. The microturbidity plate wells have a capacity of 0.2 ml. The formation of a gel resulted in higher opacity. Thus, the increase in absorbance value or the increase in turbidity can be followed for the progress of gel formation. Fig. 3 shows the effect of copolymer and Con-A concentration on the turbidity values. On the x-axis, the first number denotes the [AG]/[VP] ratio in the copolymer and the second number denotes the Con-A concentration (w/v). The figure clearly shows that the turbidity increases with decrease in copolymer concentration and/or increase in Con-A concentration.

Table 1. Copolymerization of AG and VP

Concentration of [VP] in feed	Molar ratio of [Ag]/[VP] in feed	Copolymer composition	
		[AG] mol%	[VP] mol%
8%	0.19	11.2	88.9
	0.36	19.2	80.8
	0.51	26.9	73.1
	0.72	35.6	63.4
12%	0.13	10.3	89.7
	0.16	15.3	84.7
	0.32	23.6	76.4
	0.48	31.5	68.5
16%	0.10	9.1	90.9
	0.15	15.4	84.6
	0.30	22.1	77.9
	0.45	25.2	74.8

In a study to examine the glucose sensitivity, we prepared the gel composed of 150 mg/ml of Con-A and 22.5 mg/ml of copolymer in dialysis tubes. The weight ratio of [AG]/[VP] was 0.61. From this information, it was calculated that the polymer-attached glucose concentration in dialysis tube was 8.6 mg/ml. The gel-containing dialysis tubes were placed in glucose solutions of various concentrations. The results showed that the gel remained in the gel state as long as the concentration of free glucose was less than 38 mg/ml. If the solution concentration of free glucose, however, was greater than 38 mg/ml, the gel was converted into a sol. The sol became a gel again when the free glucose was removed by placing the dialysis tube in PBS solution. The concentration of free glucose necessary to dissolve the gel was about 4 times that of the glucose concentration in copolymer. When copolymers containing different amount of glucose was used, the result was about the same, i.e., about 4 times larger free glucose concentration was necessary to dissolve the gel. This study has demonstrated that our hydrogels are able to undergo phase changes between the sol and the gel states, and such phase changes are sensitive to the glucose concentration in the environment.

The glucose-sensitive phase-reversible hydrogels are useful in the design of self-regulating insulin delivery systems as well as glucose sensors. The phase transition and the resulting changes in permeability of insulin through the gel layer can be utilized to control the delivery of insulin. The sol state resulting from high free glucose levels is expected to allow faster insulin diffusion, and thus more insulin release. The gel state present at low free glucose levels, however, is expected to slow down or inhibit the release of insulin. The ability to sense glucose levels using our hydrogel instead of enzymes such as glucose oxidase (5) is that the Con-A system is independent of the oxygen level which often affect the result of the enzymatic reactions.

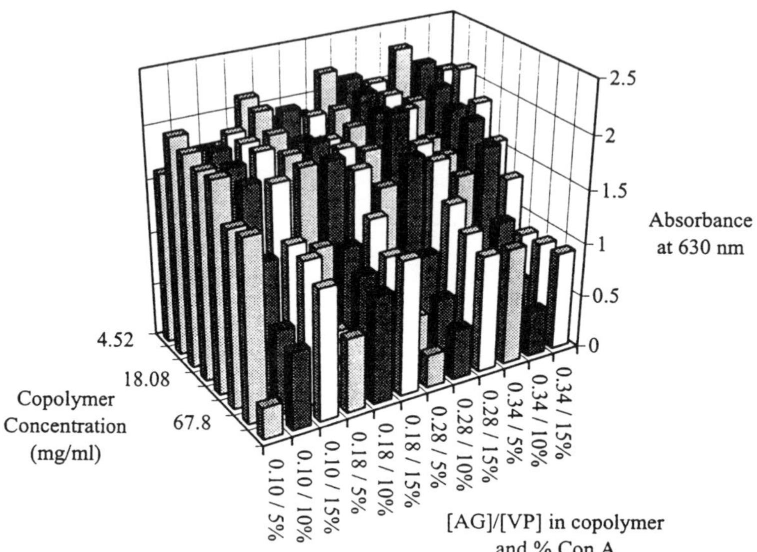

Figure 3. Turbidity of the mixtures of Con-A and copolymer solutions as a function of the copolymer concentration. Absorbance was measured at 630 nm. The left numbers (0.10, 0.18, and 0.28) in the x-axis indicate the molar ratios of [AG]/[VP] in the copolymer. The right numbers (5%, 10%, 15%) indicate the final concentrations of Con-A.

In this study we have focused on the synthesis of glucose-sensitive phase-reversible systems. The optimization of our hydrogel system can certainly lead to the development of better self-regulating insulin delivery systems and better glucose sensors. The existence of many specific interactions in nature provides opportunity for other new sol-gel phase-reversible hydrogels. The presence of antibody to almost any molecule makes it possible to synthesize hydrogels sensitive to any biomolecule.

Acknowledgments

This study was supported by the American Diabetes Association, Indiana Affiliate, Inc.

References

1. Kitano, S.; Koyama, Y.; Kataoka, K.; Okano, T.Sakurai, Y. *J. Controlled Release* **1992**, 19, 162-170.
2. Kataoka, K.; Miyazaki, H.; Okano, T.Sakurai, Y. *Macromolecules* **1994**, 27, 1061-1062.
3. Dubois, M.; Gilles, K. A.; Hamilton, J. K.; Rebers, P. A.Smith, F. *Anal. Chem.* **1956**, 28, 350-356.
4. Fineman, M.Ross, S. D. *Journal of Polymer Science* **1950**, 2, 259-262.
5. Chung, D.-J.; Ito, Y.Imanishi, Y. *J. Controlled Release* **1992**, 18, 45-54.

RECEIVED October 19, 1995

Chapter 3

New Bioartificial Hydrogels: Characterization and Physical Properties

Jean-Charles Gayet and Guy Fortier[1]

Laboratoire d'Enzymologie Appliquée, Groupe de Recherche en Biothérapeutique Moléculaire, Département de Chimie-Biochimie, Université du Québec à Montréal, C.P. 8888, Succursale Centre-Ville, Montréal, Québec H3C 3P8, Canada

Physical properties of a bioartificial hydrogel prepared by cross-linking activated poly(ethylene glycol) of molecular mass of 8000 with bovine serum albumin (BSA) are described. This hydrogel shows a great swelling capacity and a high equilibrium water content (EWC > 95%). The volume expansion factor of a BSA-PEG 8000 hydrogel during swelling was a function of its initial thickness. This hydrogel shows a high elasticity since it was reversibly deformable until a level of compression of 60% and it broke only when this level reached 80%. DSC heating scans demonstrated that differentiation of two kinds of water in the hydrogel is possible : the free (bulk) and the bound water. There was evidence that bound water is in the form of a trihydrate complex with ethylene oxide units of the PEG in the network. This hydrogel was suitable for controlled drug release systems, as demonstrated with release experiments.

Hydrogels have attracted significant attention during the last decade for their potential use in the biomedical field (*1-2*). They are a hydrophilic polymeric network which is glassy in the dehydrated state and which swells in the presence of water to form an elastic gel. Whether they are of natural or synthetic origin, they share a high water content, a permeability to small (even large) molecules, a cross-linked structure and a potentially good biocompatibility. Biomedical applications of hydrogels are extending into areas such as materials for blood contact, contact lenses, artificial tendons, controlled release devices, bioadhesive materials, and so forth. Since synthetic polymers are available with a wide variety of compositions, properties and forms, and exhibit good mechanical behaviour, they often lack biocompatibility in living systems, whereas natural polymers have the opposite properties. New bioartificial polymeric materials which combine properties of synthetic and natural polymers may overcome the deficiencies of synthetic polymers and enhance the mechanical properties of biopolymers, leading to a better composite material (*3*). In agreement with this point of view, we have recently introduced a new family of bioartificial hydrogels prepared by cross-linking activated poly(ethylene glycol) (PEG) of various molecular masses with bovine serum albumin (BSA) (*4*). We present herein the characterization and some physical properties of a hydrogel made with PEG of molecular mass of 8000 (BSA-PEG 8000).

[1]Corresponding author

0097–6156/96/0627–0017$15.00/0

Materials and Methods

PEG of molecular mass of 8000, methylene blue and phosphate bufferred saline (PBS) were obtained from Sigma Chemical Co. (St Louis, MO, USA). All chemical and biochemical reagents were of the purest grade available.

Hydrogel Preparation. The hydrogel BSA-PEG was prepared with BSA and activated PEG. Activation of PEG with p-nitrophenyl-chloroformate was described previously and yielded di(p-nitrophenylcarbonate)PEG (PEGa) (*5*). To a solution of BSA (50 mg/ml) in 400 mM sodium borate buffer was added the desired amount of PEGa to achieve a molar ratio of OH/NH_2 of 1.4 (*6*). The mixed and degassed solution was poured between two glass plates mounted in a MiniProtean II electrophoresis gel maker and allowed to polymerize. When polymerization was achieved, the hydrogel film was cut into small pastilles (diameter : 8 mm; thickness : 2 mm) and washed with 100 mM borate buffer, pH 9.4, until no more p-nitrophenol or unreacted PEG or BSA was released. The hydrogel was then fully swollen and had reached its final shape and thickness. The equilibrium water content (EWC) was determined by weighing the pastilles before and after vigorous drying. The hydrogel pastilles were stored in borate buffer at 4°C.

Mechanical Tests. Compression tests were performed on hydrogel pastilles with a LiveCo Vitrodyne V1000 attached to a 5000 g strength cell and interfaced with a 486 microcomputer. The hydrogel pastilles were compressed between two parallel steel plates at a compression rate of 20 µm/sec until the hydrogel was broken. The force applied was recorded along with variations in pastille thickness. For the calculations, the variation of the surface of the pastille was evaluated from the instant thickness assuming a constant volume for the hydrogel pastille.

Differential Scanning Calorimetry. A Dupont 912 Dual Sample DSC provided with a Dupont Mechanical Cooling Accessory connected to a Dupont 2100 Thermal Analyser was used for the determination of thermal properties. The 912 DSC cells were calibrated in accordance to the instrument manufacturer's specifications, with certified indium standard samples at the experiment scanning rate, under an atmosphere of nitrogen. Five to ten mg of hydrogel were transferred into an aluminium sample pan and hermetically sealed. The pan was cooled to -40°C, left to equilibrate and then was heated to ambient temperature at a rate of 2°C/min.

Release Experiments. Pastilles of BSA-PEG 8000 hydrogel were equilibrated in a solution of methylene blue at 250 µg/ml in PBS for two days. After loading, each pastille was mounted on a glass slide and its edge covered with paint in order to leave only one face available for diffusion in bulk medium. Each glass slide was immersed in a beaker containing 30 ml of PBS under stirring. The bulk solution was continuously pumped into the flow cell of a 8452A Hewlett Packard spectrophotometer to monitor the release of methylene blue at 664 nm.

Results and Discussion

Synthesis and Swelling Properties. The gelation process occurred by cross-linking between the two activated hydroxyl groups of the PEGa and the free amino groups of the protein. The parameters affecting the reticulation process such as the molar ratio of OH/NH_2, the pH of the reaction and the reaction time, have been optimized previously (*6*). It was demonstrated that during the rinsing process, the hydrogel underwent a swelling phase along with a large expansion in volume until an equilibrium state was reached. The magnitude of the swelling depended on the molecular mass of the PEG used during the synthesis and on the ionic strength of the rinsing solution (*4,6*). The equilibrium water content (EWC) for the BSA-PEG 8000 was 95.5%.

We have evaluated the effect of the pastille's shape on the expansion volume of the hydrogel. Table I highlights the final thickness, diameter and volume of the swollen pastilles of BSA-PEG 8000 hydrogels made with an initial diameter of 8 mm and with an initial thickness of 1, 1.5 and 2 mm.

Table I. Effect of swelling on the dimensional characteristics of BSA-PEG 8000 hydrogel pastilles [a]

e_o (mm)	e_s (mm)	$\varnothing_s$ (mm)[b]	V_o (µl)	V_s (µl)	e_s/e_o	V_s/V_o
1.0	1.65	13	50	219	1.65	4.38
1.5	2.40	12	75	271	1.60	3.61
2.0	3.00	12	101	339	1.50	3.36

[a] Ø, e and V are diameter, thickness and volume of the hydrogel pastille. o and s indices refer respectively to initial (before swelling) and final (after swelling) dimensions of the hydrogel pastille.

[b] $\varnothing_o$ = 8 mm.

Although the initial thickness had little effect on the diameter after swelling, it greatly affected the final thickness of the pastille. The radial expansion was constant and reached a value of 12-13 mm, independently of the initial thickness (1.0, 1.5 or 2 mm) of the hydrogel pastille. The variations were more important on the thickness. Pastilles made at an initial thickness of 1 and 2 mm reached a thickness after swelling of 1.65 and 3.0 mm, which corresponded to a thickness expansion factor (e_s/e_o) of 1.65 and 1.50, respectively. This led to a volume expansion factor (V_s/V_o) of 4.38 and 3.36, respectively. The same phenomenon was also observed with larger pastilles made with an initial diameter of 18 mm and an initial thickness of 7 and 8 mm. In these cases, the final thicknesses observed after swelling were 11 and 10 mm, respectively, and corresponded to volume expansion factors of 3.03 and 2.41 (results not shown). It can be concluded that the final volume of the hydrogel depends on the initial size and shape of the sample.

Mechanical Tests. The mechanical behaviour of the BSA-PEG 8000 hydrogel was evaluated under a constant compression rate by measuring the compressive stress and the deformation until the breaking point of the pastille (Figure 1). The compressive stress increased slowly as the force was applied to the hydrogel. Between 60 and 70% of compression, the force increased quickly until breaking of the pastille occurred at 80% compression. Due to the shape of the stress-strain curve which lacked a linear portion, it is impossible to evaluate a Young's modulus to represent the hydrogel strength. This modulus characterizes the pure elastic domain of a material where the deformation is reversible and linear. Although the BSA-PEG 8000 hydrogel shows visco-elastic behaviour, the applied deformation was totally reversible up to 60% of compression. So, we have used the compressive stress at 60% compression as a measure of the hydrogel's strength. The BSA-PEG 8000 hydrogel has a strength of 122 g/cm^2. After this point, the applied deformation was not reversible and the breaking of the hydrogel occurred at 80% of compression with a compressive stress of 389 g/cm^2. This "breaking force" represents the resistance of the hydrogel to stress.

This mechanical behaviour is good considering the high EWC (95.5%) of this hydrogel. However, it is difficult to compare with published data on other hydrogels,

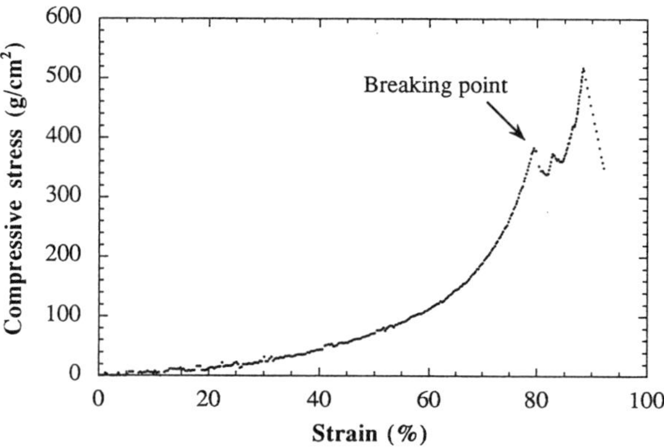

Figure 1. Typical compressive test of BSA-PEG 8000 hydrogel.

since tensile tests are usually reported and they are not suitable due to the high water content of the hydrogel.

Differential Scanning Calorimetry. Previous DSC studies on aqueous PEG solutions have shown the presence of two melting endotherms, both below the freezing point of water (*7*). These fusion peaks have been identified as the melting of two types of crystalline water : the bound and the free water. The peak for bound water was found at -10°C and corresponded to the melting of a polymer hydrate of fixed composition (3 molecules of water per ethylene oxide unit). The peak for free water near 0°C was caused by the melting of ice. Other DSC studies on swollen, chemically reticulated poly(ethylene oxide) hydrogels have also shown the presence of bound and free water, and the existence of the trihydrate complex has been confirmed by other physico-chemical experiments (*8*).

Two fusion endotherms are visible on the DSC heating scans of BSA-PEG 8000 hydrogels (Figure 2) : a weak peak observed around -20°C and a strong peak observed just below 0°C. The temperatures and enthalpies of fusion are summarized in Table II for the two scans. Scan A and scan B were obtained with BSA-PEG 8000 hydrogels at EWC of 80% (before swelling) and 95.5% (after swelling), respectively.

Table II. Temperatures and enthalpies of fusion of BSA-PEG 8000 hydrogels [a]

DSC scan	*Temperature of fusion (°C)*		*Enthalpy of fusion (J/g)*	
	T_1	T_2	ΔH_1	ΔH_2
before swelling (a)	-20.7	-2.99	9.24	185.6
after swelling (b)	-17.9	-0.60	2.37	291.0

[a] 1 and 2 indices refer respectively to the low and high temperature peaks of fusion on Figure 2.

The high temperature endotherms, around 0°C, were assumed to correspond to the melting of ice (free water) in the hydrogel. The area under this peak was greater for the hydrogel after swelling (scan B - ΔH_2 = 291.0 J/g) than before swelling (scan A - ΔH_2 = 185.6 J/g), due to its higher water content. The fusion of the free water is confirmed by the temperature T_2 in the swollen hydrogel which was -0.60°C, near that of pure water. T_2 reaches a value of -2.99°C for the free water in the hydrogel before swelling, due to the antifreeze effect of the polymer network in this more concentrated phase. The low temperature endotherm was -20.7°C for the hydrogel before swelling (scan A) and -17.9°C for the hydrogel after swelling (scan B). Such low temperatures of fusion were assumed to represent the fusion of bound water, in agreement with the literature (*7-8*). Similarly to what was observed for free water, the fusion of bound water in the hydrogel before swelling was 3°C lower than after swelling. The number of molecules of water bound on each ethylene oxide unit in the BSA-PEG 8000 swollen hydrogel was evaluated with the method used by Antonsen and Hoffman (*9*). We found that 2.7 molecules of water were bound, a value which is close to that of 3 molecules commonly found for PEG. The measurement of the enthalpy of fusion with this kind of hydrogel is sometimes difficult, since a certain amount of the water inside the hydrogel is not able to freeze. This can be seen when the enthalpies of fusion of the low and the high melting endotherms are added. The value obtained does not represent all of the water present inside the hydrogel. As an example, the sum of the enthalpies of fusion for the hydrogel before swelling was 194.8 J/g which represents 60% of the enthalpy of fusion of pure water (330 J/g). This can lead to uncertainty and underestimation of the final results. Despite this fact, free and bound water appear to be present in this type of hydrogel, similar to a solution of PEG or hydrogels of pure poly(ethylene oxide).

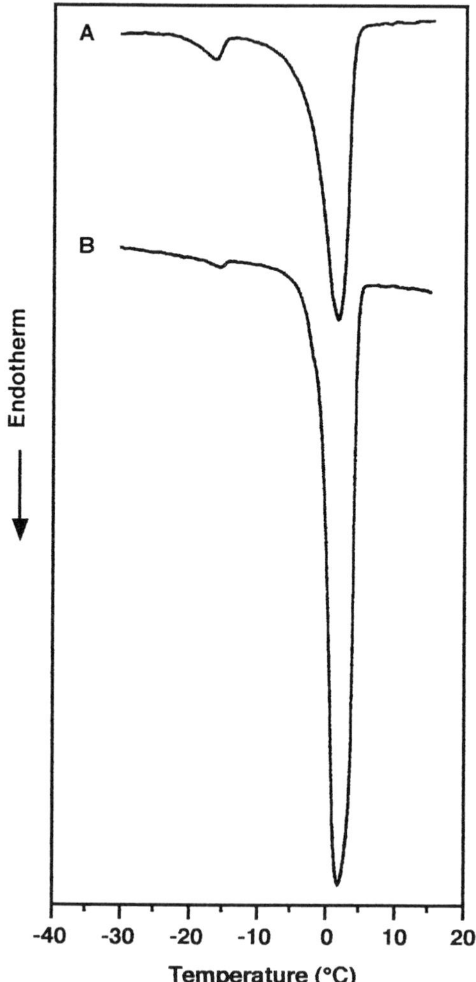

Figure 2. DSC heating scans of BSA-PEG 8000 hydrogel at EWC of 80% (A) and 95.5% (B).

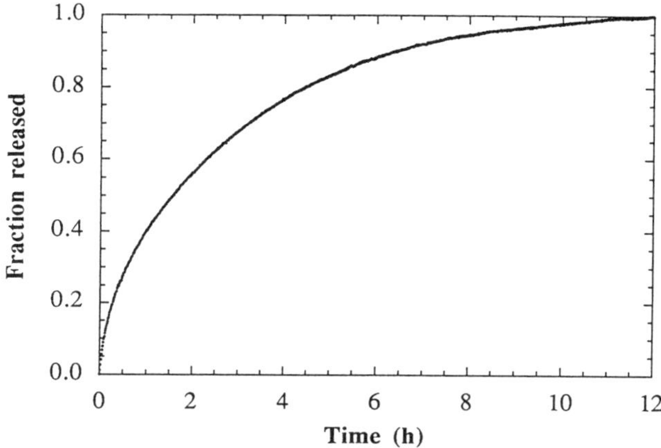

Figure 3. Release of methylene blue from BSA-PEG 8000 hydrogel in PBS as a function of time.

Release Experiments. As an example application, Figure 3 shows the release of methylene blue from a BSA-PEG 8000 hydrogel pastille with a thickness of 1.5 mm, expressed as the fraction released M_t/M_∞ as a function of time. The complete release was obtained after 12 hours and methylene blue has an half-life time of release of 90 min. This release was apparently diffusion controlled and followed a Fickian mechanism (*10*). This allows the calculation of the diffusion coefficient D of methylene blue (equation 1) from the BSA-PEG 8000 hydrogel pastille under these conditions, which is 8.3 x 10^{-7} cm^2/sec.

$$M_t/M_\infty = \frac{4}{e} \cdot \sqrt{\frac{D}{\pi}} \cdot \sqrt{t} \qquad \text{for } 0 \leq M_t/M_\infty \geq 0.6 \qquad (1)$$

A mechanism of Fickian diffusion is consistent with the high water content of this type of hydrogel where most of the water is in the form of bulk water, as concluded from DSC experiments.

Conclusion

We have studied some physical properties of a biopolymeric hydrogel obtained by cross-linking of bovine serum albumin and activated poly(ethylene glycol) M.W. 8000. This hydrogel has a high water content (>95%) and a large volume of expansion during swelling. In spite of its high water content, the BSA-PEG hydrogel has good mechanical properties. It is highly elastic and withstands a level of compression of at least 80% before breaking. This mechanical behaviour is not common among materials containing so much water and is favorised for applications such as wound dressing or supporting biopolymers. We have also shown with DSC experiments that a part of the water present in the hydrogel was bound to the PEG moiety in the form of a trihydrate complex, similar to what was observed in a solution of free PEG or with swollen polymers of pure poly(ethylene oxide). The remaining water is free inside the polymer network and is responsible for the bulk properties of the hydrogel, such as the diffusion of species through it, as demonstrated with the release of methylene blue. Therefore, such a hydrogel is a good candidate for controlled release devices.

Acknowledgments

We would like to thank Mme. Dorina Banu (Center for Building Studies, Concordia University, Montréal) for her very kind help with DSC experiments and M. Pierre Limoge (École Polytechnique, Montréal) for his assistance in mechanical testing. We are grateful to the Conseil de Recherches en Sciences Naturelles et en Génie (subventions stratégiques) and the Université du Québec à Montréal for their financial support.

References

1. Andrade, J. D., Ed. *Hydrogels for Medical and Related Applications*; ACS Symp. Ser. 31; ACS: Washington, D.C., 1976.
2. Peppas, N. A., Ed. *Hydrogels in Medicine and Pharmacy*; CRC Press: Boca Raton, FL, 1987; Vol. 1-3.
3. Gusti, P.; Lazzari, L.; Lelli, L. *TRIP* **1993**, *9*, 261-267.
4. D'Urso, E. M.; Fortier, G. *Biotechnol.Tech.* **1994**, *8*, 71-76.
5. Fortier, G.; Laliberté, M. *Biotechnol. Appl. Biochem.* **1993**, *17*, 115-130.
6. D'Urso, E. M.; Fortier, G. *J. Bioact. Comp. Polym.* **1994**, *9*, 367-387.
7. Graham, N. B.; Zulfiqar, M.; Nwachuku, N. E.; Rashid, A. *Polymer* **1989**, *30*, 528-533.
8. Graham, N. B.; Zulfiqar, M.; Nwachuku, N. E.; Rashid, A. *Polymer* **1990**, *31*, 909-916.
9. Antonsen, K. P.; Hoffman, A. S. In *Poly(Ethylene Glycol) Chemistry : Biotechnical and Biomedical Applications*; Harris, M. J., Ed.; Plenum Press: New York, NY, 1992, Chapter 2, pp 15-27.
10. Baker, R. *Controlled Release of Biologically Active Agents*; John Wiley & Sons: New York, NY, 1987; Chapter 3, pp 50-56.

RECEIVED January 3, 1996

Chapter 4

Bioapplication of Poly(ethylene glycol)–Albumin Hydrogels: Matrix for Enzyme Immobilization

Edith M. D'Urso, Jacques Jean-François, and Guy Fortier[1]

Laboratoire d'Enzymologie Appliquée, Groupe de Recherche en Biothérapeutique Moléculaire, Département de Chimie-Biochimie, Université du Québec à Montréal, C.P. 8888, Succursale Centre-Ville, Montréal, Québec H3C 3P8, Canada

The feasibility of enzyme immobilization into PEG-albumin hydrogels is demonstrated with acid phosphatase and asparaginase. After immobilization, the apparent Km of acid phosphatase was increased by about 11 times when compared with the soluble enzyme. To the contrary, the apparent Km of asparaginase dramatically increased by 150 times, after immobilization. A denaturing effect of the hydrogel swelling on the enzyme activity is hypothesised. For both enzymes, the operational stability at 37 °C was markedly improved by immobilization into the PEG-albumin hydrogels. The *in vitro* biodegradability of these hydrogels is also demonstrated through protease digestion.

Enzymes have long been used in the biomedical field as diagnostic tools or disease markers *(1)*. In the last decade however, they have found new applications as therapeutic agents *(2-3)*. The therapeutic action of an enzyme consists mainly in degrading a substance which is toxic for the organism, due to a pathologic accumulation, an inherited deficiency or an intoxication. A list of enzymes already involved or potentially useful in therapy has been published recently *(4)*. There are various domains of application such as replacement therapy *(5-6)*, specific detoxification, scavengers *(7)*, antineoplastic agents *(8-11)*, and prevention of clot formation. Even though there were numerous possibilities of enzyme therapy, two major problems were rapidly identified

[1]Corresponding author

0097–6156/96/0627–0025$15.00/0

which limited their development. First, enzymes are quite unstable biopolymers which denature quickly. Second, they are eliminated in few hours after injection into the body by different mechanisms such as kidney clearance, reticulo-endothelial adsorption and other non-specific interactions. Repeated injections are then necessary to provide sufficient enzyme activity in the organism. When the enzyme used is from a foreign source, it generally induces an immunological response which can lead to severe anaphylactic shock. To overcome these problems, it has been decided to modify the biocatalyst by coupling chemically polymer chains, like dextran or PEG, to their surface. These surface modifications showed several benefits : the antigenic site can be masked by the polymer chain, the digestibility of the enzyme by protease is restricted (this step plays an important role in the process of antigen presentation), the size of the modified biocatalyst is increased and the plasma clearance is reduced. In some cases, immobilization has also lead to important variations in the kinetic parameters of the enzyme, and particularly if the coupling is achieved through amino acids involved in or close to the catalytic site.

Another approach to increase the life span of an enzyme in the organism while limiting the immunological drawbacks is to immobilize it in a biocompatible matrix. Recently, we have described a new family of PEG-BSA hydrogels *(12-13)* which seems to have interesting characteristics for the purpose of enzyme immobilization for biomedical applications. In fact, the biocompatibility of this family of hydrogels is good, due to the presence of the PEG in the structure (unpublished results). The stabilization of enzyme by surface modification with PEG has been well demonstrated *(14-17)*. It has been observed that coating a surface with PEG reduces the thrombogenecity and the protein adsorption due to the large hydrodynamic volume of this polymer *(18)*. Also, albumin has been used to stabilize enzymes *(19)*, creating a "biological-like" environment. The presence of the protein makes the PEG-albumin hydrogels biodegradable.

This family of PEG-albumin hydrogels is characterized by a very high water content of about 96 % in saline buffer *(12, 13)*, which is favourable to a rapid diffusion of species through the matrix. This point is very important when considering enzyme immobilization because the catalysis efficiency would depend on the ability of the substrate to diffuse into the matrix to reach the catalytic site of the enzyme and on the ability of the product to be expulsed. It is also important to consider the mechanical properties of the hydrogels since they would have to resist manipulation during implantation, or to resist the flow of substrate in a bioreactor. In order to improve the biocompatibility, various sources of serum albumin can be used (bovine, rat, mice, human), depending on the animal model studied. In these hydrogels, the proteins (the albumin and the enzyme) play the role of crosslinkers for the PEG and the reticulation degree depends partly on the number of accessible NH_2 groups of lysyl available at the proteins surface. It was demonstrated in a previous paper that the porosity of hydrogels depends on the chain length of the PEG *(20)*.

The feasibility of enzyme immobilization in such matrices has been studied with two enzymes, acid phosphatase and asparaginase, immobilized in pastille-shaped hydrogels. The acid phosphatase (AP) was chosen as a general model. Asparaginase (ASNase) was selected as a model for enzymotherapy. This enzyme has been extensively used in the treatment of acute lymphoblastic leukaemia and acute myeloblastic leukaemia *(21)*. The apparent kinetic parameters of both immobilized enzymes (Km, Vmax, stability) were determined and compared with those of the soluble enzymes. The biodegradability of the PEG-albumin hydrogels, which is an important characteristic for *in-vivo* applications, was evaluated *in vitro* by protease digestion.

Materials and Method

Acid phosphatase (AP, EC 3.1.3.2), grade II, from potato, and Pronase, from Streptomyces griseus) were purchased from Boehringer. Asparaginase (ASNase, EC 3.5.1.1) from E.coli, poly(ethylene glycol) of molecular mass 10,000 and bovine serum albumin (BSA) were purchased from Sigma Chemical Co (St. Louis MO, USA).

The protein content of the solutions was determined according to the micro-BCA assay *(22)*. The PEG released during proteolysis was assayed using the iodine procedure *(23)*.

Synthesis of the AP-PEG-BSA hydrogels. Mixtures of acid phosphatase, AP (5-20 mg/ml - 2 U/mg - protein content 60 %), bovine serum albumin, BSA (47-38 mg/ml) and activated poly(ethylene glycol) p-nitrophenyl di-carbonate of molecular mass 10,000 (PEG 10000) *(24)* were prepared in borate buffer 200 mM pH 8.8. In every cases, the total protein concentration was fixed at 50 mg/ml. The concentration of activated PEG was kept at 131 mg/ml according to the optimum molar reagent ratio determined previously *(13)*. The AP-PEG-BSA solution was then cast between two glass slides in order to obtain a slab of 1.5 mm thickness. After polymerization, which occurred within two hours, the hydrogel was cut in pastilles of 7 mm in diameter. The pastilles were then washed extensively, first in borate buffer at pH 8.8, then in citrate buffer at pH 5.7. After completion of washing, the hydrogels were stored at 4°C in citrate buffer pH 5.6 containing 0.02 % NaN_3. After polymerization, the hydrogels contain about 60 % of water. During washings, they are subjected to swelling till an equilibrium water content of about 96% (13).

The AP activity was assayed by following the appearance at 405 nm of the p-nitrophenol (pnP) enzymatically generated from the p-nitrophenyl phosphate (pnPP), after addition of NaOH 0.5 N in the reaction solution [reaction [1], *(20)*].

$$\underset{\text{(pnPP)}}{\text{p-nitrophenyl phosphate}} + H_2O \xrightarrow{\textbf{\textit{acid phosphatase}}} \underset{\text{(pnP)}}{\text{p-nitrophenol}} + Pi \qquad [1]$$

Each AP-PEG-BSA hydrogel is incubated in 4 ml of 50 mM citrate buffer pH 5.7.

Aliquots are taken over a 20 minutes time-frame and assayed for their content of pnP. The initial velocity (linear part of the curve) is calculated.

Synthesis of the ASNase-PEG-BSA hydrogels. Asparaginase was immobilized in pastille-shaped hydrogels. Two mg (336 U) of *E.coli* asparaginase were mixed with 100 mg of BSA and 204 mg of activated PEG 10000 in 2 ml of 200 mM borate buffer pH 9. The reaction mixture was then cast between 2 glass slides in order to obtain an enzymatic hydrogel slab of 1.5 mm thickness. Pastilles of 8 mm in diameter were then cut and washed extensively in borate buffer.

The asparaginase activity was assayed by measuring the appearance of aspartic acid produced enzymatically from the hydrolysis of asparagine (asn) (reaction [2]).

$$\underset{\text{(asn)}}{\text{asparagine}} + H_2O \xrightarrow{\textbf{\textit{asparaginase}}} \underset{\text{(asp)}}{\text{aspartic acid}} + NH_3 \qquad [2]$$

Aspartic acid (asp) and asparagine (asn) were quantitatively analysed by high performance liquid chromatography (HPLC) on a pre-packed Ultrasphere XL-ODS column (Beckman). After pre-column derivatization with ophthalaldehyde *(25)*, the fluorescent amino acids were detected using a spectrofluorometer (Shimadzu RF-551).

RESULTS AND DISCUSSION

Effect of the Enzyme Loading on the Hydrogel Activity. The enzyme retention ability of our hydrogels has been evaluated with acid phosphatase. The overall protein concentration in the initial mixture was kept constant at 50 mg/ml. The concentration of AP (protein content 60 %) was then increased from 5 to 20 mg/ml while the BSA one was decreased from 47 to 38 mg/ml. The AP activity of the pastille-shaped hydrogels (7 mm in diameter, 1.5 mm thickness, before swelling) was measured in 50 mM citrate buffer at pH 5.7. As depicted in Table I, the enzymatic activity of the hydrogel (apparent Vmax) increases from 108 to 263 mU for AP loadings increasing from 5 to 20 mg/ml, as described previously for AP immobilized in BSA-PEG microbeads hydrogels *(20)*.

The yield of immobilization, in terms of enzyme activity retention, varied from 22 to 13 %, suggesting that a large amount of enzyme has been denatured during the hydrogel synthesis. Similar results have been obtained previously for AP microbeads [20].

Surface modification of the AP with methoxy-PEG 5000 (mPEG) has been done to limit the crosslinking of the enzyme during the gel formation. Different mPEG/AP weight ratios (5, 15, 50) were studied. Pastille-shape hydrogels (8 mm in diameter, 1.5 mm in thickness, before swelling) were prepared with modified AP (enzyme loading 75 mU/pastille) and their kinetic parameters were then determined. The results are depicted in Table II. The immobilization yield of the modified AP increases slightly with the mPEG/AP ratio, suggesting a slight protection of the enzyme by the mPEG. As the mPEG/AP ratio increases, the number of available NH_2 groups at the enzyme surface decreases. The degree of crosslinking of the enzyme in the hydrogel then decreases which should limit its denaturation, if it is related to a structure deformation during the hydrogel swelling. It has been described previously that soluble AP was not denatured by modification with mPEG. The enzyme activity could be limited due to diffusion restriction inside the hydrogel. Denaturation of the enzyme due to deformation during the cross linking can also be hypothesised to explain the low values of immobilization yields obtained.

Table I : Effect of the enzyme loading on the characteristics of the immobilized acid phosphatase into PEG-BSA hydrogels

	Enzyme loading (U/ml)		
Apparent kinetic parameters	10	20	40
appVmax (mU)[(a)]	108 ± 1.5	191 ± 1	263 ± 13
appKm (mM)[(a)]	4.49 ± 0.33	6.62 ± 0.06	7.52 ± 0.69
Immobilization yield[(b)]	21.6% (38.8%)	19.1 % (43.9%)	13.2% (44.2%)

The AP was immobilized in pastille-shaped hydrogels of 7 mm in diameter and 1.5 mm thickness before swelling. After swelling, the pastille dimensions were 13 mm in diameter and 2.65 mm thickness.
(a) The appKm and Vmax values were determined in optimal stirring conditions in a batch system. One unit of enzyme activity corresponds to one µmole of pnPP generated per minute in our assay conditions.
(b) The immobilization yields were calculated in terms of enzyme activity taking into account for each pastille the activity loaded and the apparent Vmax. The values in parenthesis were calculated from the AP activities recovered after protease digestion.

Table II : Effect of surface modification by mPEG 5000 on kinetic parameters and immobilization yield of acid phosphatase

mPEG 5000/AP		0[(a)]	5	15	50
app Km (mU)		1.95	1.85	1.82	2.49
appVmax (mU)		35.3	20.1	25.7	24.3
Immobilization yield	(%) (relative)	24 1	28 1.2	32 1.3	33 1.4

The AP was modified by mPEG 5000 according to a procedure described previously (24). Pastilles (8 mm in diameter, 1.5 mm thickness) were prepared with the modified enzyme using about 75 mU/pastille.
(a)Control pastilles with unmodified AP were also prepared.
The appKm and appVmax values were calculated using a non linear regression of the Michaelis-Menten equation.

Effect of Immobilization on the Apparent Kinetic Parameters of Enzymes.

Effect of Immobilization on the Apparent Km of Acid Phosphatase. After optimization of the stirring conditions (Figure 1), Pastille-shaped hydrogels (7 mm in diameter, 1.5 mm thickness) were tested for their AP activity in 4 ml of 50 mM citrate buffer pH 5.7 at 400 rpm (this stirring speed has been found to minimize the external diffusion restriction in the assay solution). The apparent Km values were determined using a non linear regression of the experimental data, according to the Michaelis-Menten equation (Figure 1). As depicted in Table 1, the apparent Km of the immobilized AP increased when the enzyme loading was increased. This result can be explained by an increasing effect of internal diffusion restriction for increasing amounts of enzyme retained in the pastille. Most of the substrate is transformed by the enzyme located at the surface of the pastille. In the micro-environment of the enzyme located deeper in the hydrogel, the substrate concentration is smaller due to diffusion restriction *(26)*, and the activity is lower. When comparing with the Km value determined for the native enzyme, i.e. 0.42 mM *(20)*, the immmobilization induces an increase of the apparent Km of 11 to 18 times.

Effect of Immobilization on the Apparent Km of Asparaginase. The effect of substrate concentration on the ASNase activity has been studied for both soluble and immobilized enzymes. Two asparaginase pastilles were tested under optimum stirring. The ASNase activity was calculated by measuring the increase of the peak height of aspartic acid (reaction product) as a function of the reaction time (Figure 2). The experimental results were fitted with the Michaelis-Menten equation, according to a non linear regression. The apparent Km of asparaginase was dramatically increased after immobilization in the hydrogel, with value increasing from 15 μM for the native enzyme (Figure 3a) to 2.2 mM for the immobilized one (Figure 3b). A similar increase of the apparent Km has been reported in the literature for ASNase immobilized onto a Dacron vascular prosthesis *(27)*, onto a nylon tube *(28)*, or into micro-capsules *(29)*.

The yield of immobilization, in terms of enzyme activity recovering, has been calculated taking into account the enzyme loading per pastille and the value of apparent Vmax calculated from Figure 3b. The yield of ASNase immobilization in the PEG-BSA hydrogels was very low, *i.e.* 6 % comparing with the value obtained for acid phosphatase. This result could be explained

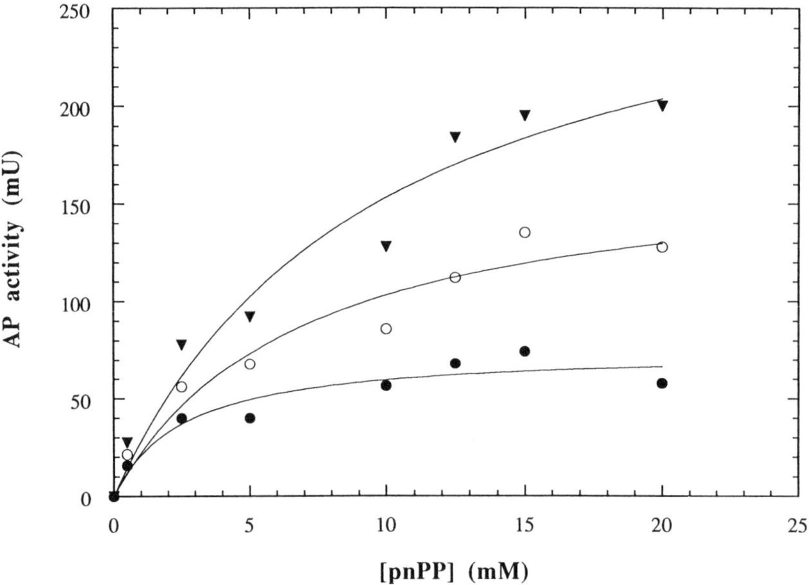

Figure 1. Effect of the enzyme loading on the apparent Km of AP for pnPP. The apparent Km values were determined at ambient temperature in 50 mM citrate buffer pH 5.7 using a non linear regression of the Michaelis-Menten equation. Each hydrogel (1.5 mm thickness, 7 mm in diameter, before swelling) has been tested in duplicate. Full circle : AP loading 5 mg/ml; open circle : AP loading 10 mg/ml; full triangle : AP loading 20 mg/ml in the initial mixture (AP-PEG-BSA).

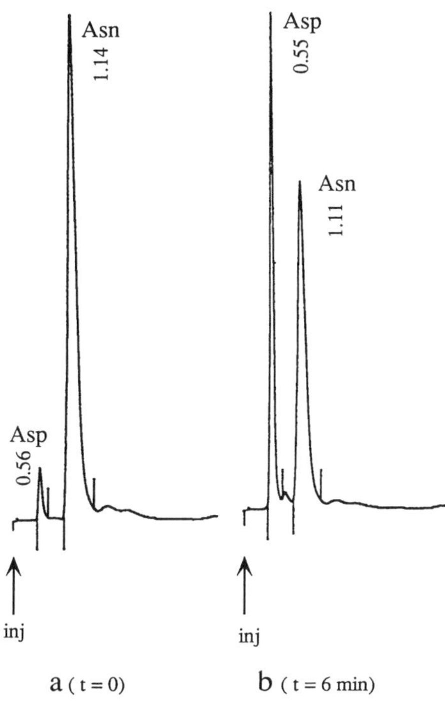

Figure 2. HPLC Chromatogram of aspartic acid and asparagine. ASNase pastille-shaped hydrogel is incubated in optimal conditions with asparagine. The incubation solution of ASNase is diluted and derivatized with o-pthalhaldehyde before injection at a, t = 0 and b, t = 5 min of incubation. It can be noted that the asparagine solution was contaminated by a small amount of aspartic acid (a).

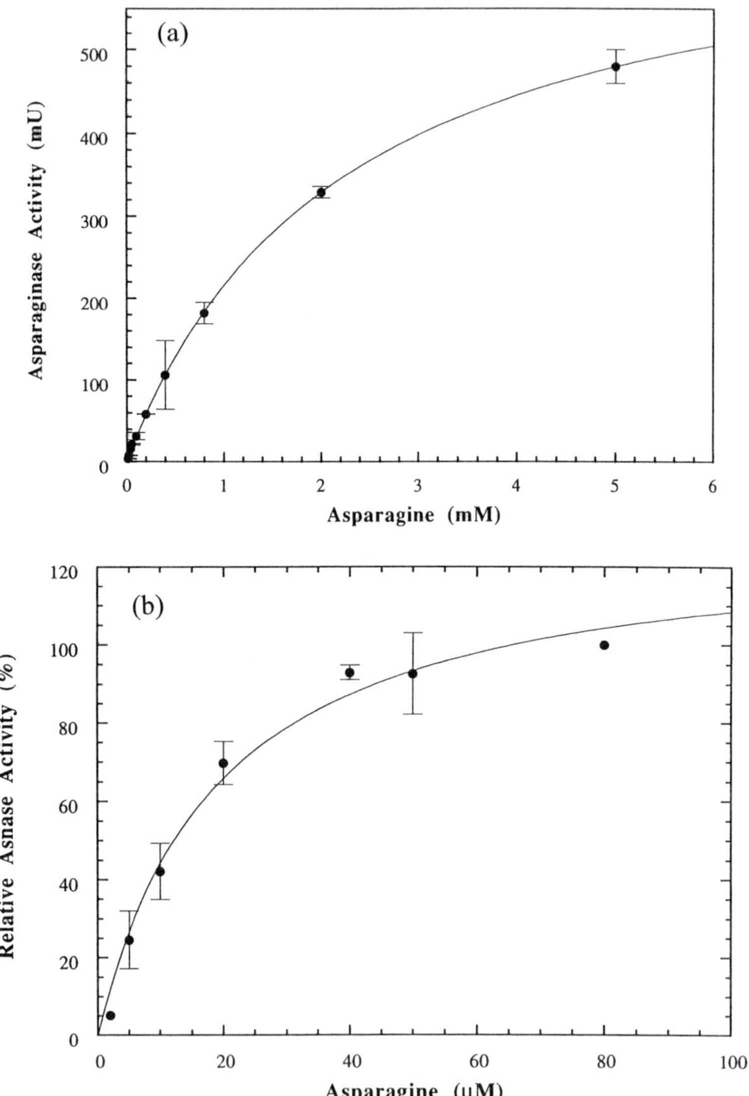

Figure 3. Kinetic behaviour of soluble and immobilized asparaginase. All activity measurements were performed at 37 °C in 50 mM borate buffer pH 8.6 under optimal stirring (400 rpm). The mean values were calculated from triplicates (n=3). (a) soluble ASNase; (b) immobilized ASNase.

either by internal diffusion limitations, or by denaturation of most of the ASNase during the hydrogel synthesis.

The influence of diffusion limitations into the hydrogel was evaluated. ASNase microparticles (300 µm before swelling) were prepared by crunching the pastilles through a calibrated mesh sieve of 300 µm pore size and tested in a plug flow reactor. Increasing the flow rate from 0.32 to 0.6 ml/min induced a decrease of the appKm of the ASNase (Figure 4), revealing diffusion limitations. However, increasing the flow rate over 0.6 ml/min had no more effect on the appKm, demonstrating that this dramatic Km increase was not a consequence of diffusion restrictions alone. Then, the strong apparent Km modification should be explained by a loss of affinity of the ASNase for its substrate, due to a conformational change upon immobilization. During the crosslinking of the ASNase, an NH_2 group close to or in the active site of the enzyme could possibly have reacted with PEG, leading to a modification of its catalytic site.

The swelling of the hydrogel after polymerization (see Material and Methods) could have a denaturing effect which could be responsible for Km modification and enzyme denaturation. This effect could be more pronounced for a tetrameric enzyme as ASNase than for a dimeric one, as AP. These hypothesis are currently under evaluation in our laboratory.

Operational Thermostability of Soluble and Immobilized Enzymes at 37°C. Two pastille-shaped asparaginase hydrogels (700 mU) were continuously incubated at 37 °C in 0.1 mM borate buffer pH 8.6 containing asparagine at low concentration (100µM). The residual activity was evaluated regularly at a saturating asparagine concentration of 5 mM. The residual activity of the soluble ASNase (580 mU initial) incubated in the same conditions but without asparagine was measured at 100 µM of substrate (which is a saturating concentration for the soluble enzyme). The results are depicted in Figure 5. A half-life of 54 hours has been determined for the soluble ASNase. After immobilization, the operational stability is markedly increased (at least 27 times) since less than 10 % of the initial activity was lost after 61 days (>1460 hours) of continuous catalysis.

With acid phosphatase, the operational stability at 37°C was also dramatically increased after immobilization with a half-life of 80 hours when compared with 72 minutes for the soluble AP tested in the same conditions, which corresponds to a 67 time increase of the half-life *(20)*.

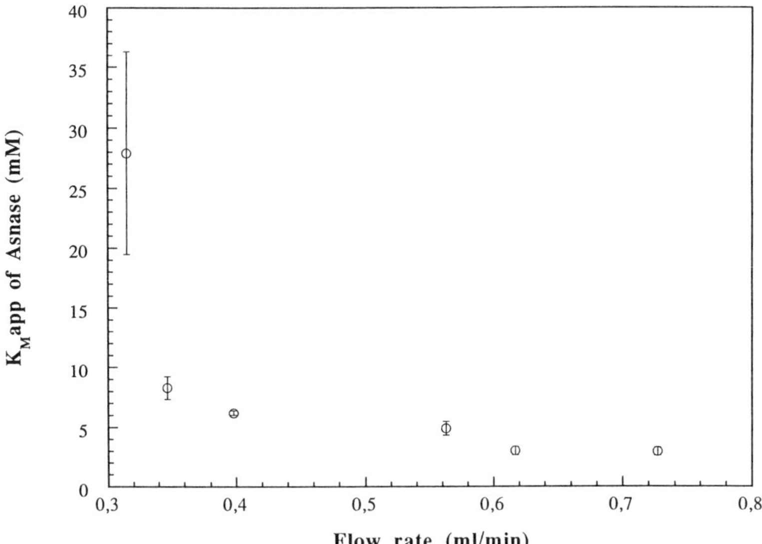

Figure 4. Effect of the flow rate on the activity of the immobilized ASNase. Pastille-shape hydrogel of PEG-BSA have been crunched onto a 300 μm sieve to obtain microparticles. The apparent Km values were determined in 50 mM borate buffer pH 8.6 at ambient temperature using a non linear regression of the Michaelis-Menten equation. The mean values were calculated from triplicates.

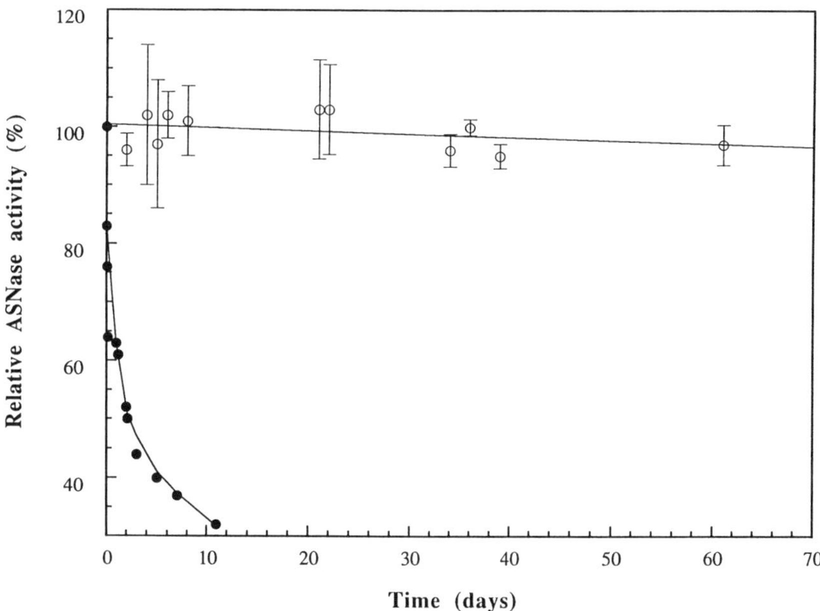

Figure 5. Thermostability of the soluble and immobilized asparaginase. Soluble (10 IU) and immobilized ASNase (700 mU) were incubated at 37 °C in 50 mM borate buffer pH 8.6. The residual activities were measured periodically at 37 °C in the presence of 100 μM and 5 mM of asparagine for soluble and immobilized ASNase, respectively. The ASNase pastille-shaped hydrogels were 12 mm in diameter and 2.5 mm thickness. Full circle : native ASNase; open circle : immobilized ASNase. In both cases, the mean values were obtained from duplicates.

Biodegradability of PEG-BSA Hydrogels. The biodegradability of our PEG-BSA hydrogels has been evaluated with acid phosphatase pastille-shaped hydrogels (7 mm in diameter, 1.5 mm thickness before swelling) incubated at ambient temperature with pronase at a final concentration of 0.1 mg/ml in phosphate buffer saline pH 7.6. Controls have been incubated in the same conditions but without protease. Aliquots were collected regularly from the incubation mixtures and assayed for their content in PEG. Results are depicted in Figure 6 for two hydrogels loaded with 0, 5, 10 and 20 mg/ml of AP (total protein content : 50 mg/ml, as described previously). The digestion of the hydrogels proceeds from the exterior of the disc, which was shrinking gradually upon incubation time with the pronase. The rate of PEG release was constant for 390 to 600 minutes and increased from 5.2 to 10.6 μg/min of PEG for enzyme loading increasing from 0 to 20 mg/ml. Initially, we thought that the rate of protease digestion could be useful to determine whether the presence of the enzyme would modify the degree of crosslinking of the hydrogel. However, this hypothesis could not be demonstrated since the rate of hydrogel digestion by pronase is reduced in the presence of acid phosphatase. To understand this result, the native AP was incubated in the presence of pronase. After 4 days of incubation, the enzyme activity was totally conserved. This resistance of AP to proteolysis can be explained by the fact that this enzyme is a highly glycosylated enzyme.

Moreover, after complete hydrolysis of the hydrogels by pronase, the overall AP activities released were 209 to 390 % of the appVmax determined previously (Table I). This result demonstrates that the hydrogel content of active AP was greater than the apparent Vmax experimentally determined, confirming the hypothesis of diffusion resistance into the hydrogel.

Conclusion

The feasibility of covalent enzyme immobilization in PEG-albumin hydrogels has been demonstrated. The diffusion restrictions, as revealed by the Km increase after immobilization, were limited, especially at low enzyme loading. For both enzymes, *i.e.* acid phosphatase and asparaginase, the *in vitro* operational stability was dramatically increased after immobilization. The biodegradability of these bioactive hydrogels has also been demonstrated *in vitro*. However, it has been shown through *in vivo* studies that the hydrogels could persist in mice and rats

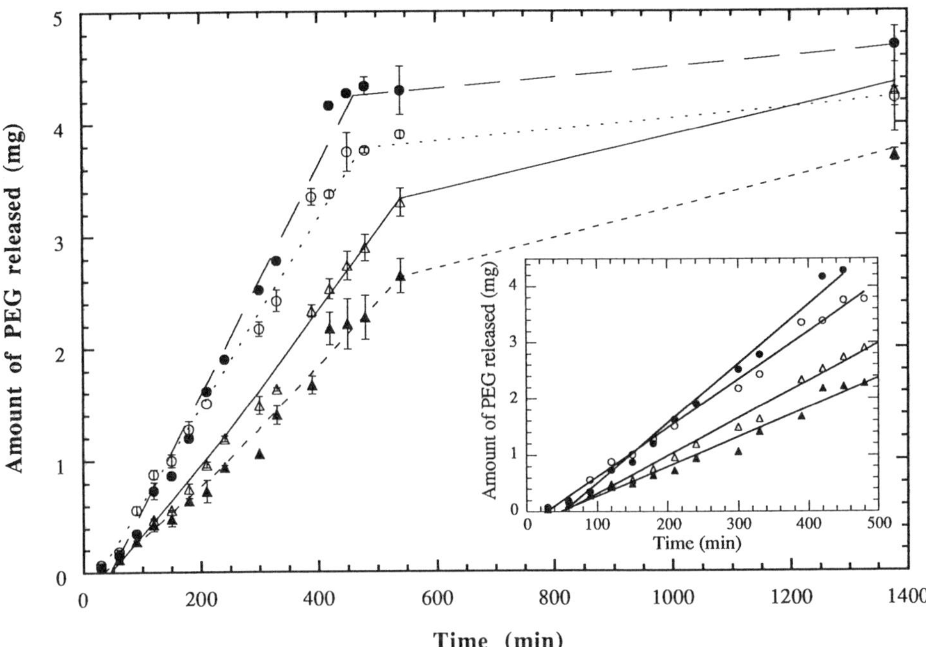

Figure 6. Protease digestion of AP-PEG-albumin hydrogels. Pastille-shape hydrogels (1.5 mm in thickness, 7 mm in diameter, before swelling) of AP-PEG-albumin were incubated at ambient temperature in 50 mM PBS buffer pH 7.6 containing pronase at 0.1 mg/ml. The PEG released from the hydrogels was detected colorimetrically according to the iodine assay (23). Full circle : no enzyme; open circle : AP loading 5 mg/ml; open triangle : AP loading 10 mg/ml; full triangle : AP loading 20 mg/ml. The mean values were obtained from duplicates.

for more than one month when implanted in an intra-peritoneal or subcutaneous position (unpublished results).

The immobilization efficiency seems to depend greatly on the enzyme studied. While satisfying results were obtained with acid phosphatase, the immobilization was less successful with asparaginase, since a marked increase (150 times) of the apparent Km was observed after immobilization. We assumed that this was a consequence of either the hydrogel swelling after polymerization, or a modification of the active site of the enzyme by crosslinking with PEG. It is hypothesised that the negative effect of the hydrogel swelling (i.e., the enzyme stretching) would depend on the oligomeric structure of the enzyme involved. As asparaginase is a tetrameric enzyme, it could be more sensitive to stretching than the AP which is dimeric. To overcome this denaturing effect of swelling, it should be possible to modify the enzyme with PEG in a first step and then to entrap it in the PEG-albumin hydrogel. The enzyme molecule would be then less exposed to mechanical deformation occurring upon swelling but would benefit from the stabilizing and protective effect of the matrix. The active site of the ASNase could also be protected during the crosslinking by introducing a substrate or an inhibitor of the enzyme in the polymerization mixture. These alternatives are currently under evaluation in our laboratory.

ACKNOWLEDGEMENTS

We would like to thank the Conseil de Recherches en Sciences Naturelles et en Génie and the Université du Québec à Montréal for their financial supports. E.M.D. is grateful to the Direction régionale de la région Rhône-Alpes for her post-doctoral fellowship.

Literature Cited :

1 - M.K. Schwartz, *Clin. Chim. Acta,* **1992,** *206,* pp. 77-82.

2 - D.M. Goldberg, *Clin. Chim. Acta,* **1992,** *206,* pp. 45-76.

3 - M.D. Klein and R. Langer, *Trends Biotechnol.,* **1986,** *4,* pp. 179-186.

4 - G. Fortier, *Biotechnol. Gen. Engin. Rev.,* **1994,** *12,* pp. 331-356.

5 - N.W. Barton, R.O. Brady, J.M. Dambrosia, A.M. Di Bisceglie, S.H. Doppelt, S.C. Hill, H.J. Mankin, G.J. Murray, R.I. Parker, C.E. Argoff, R.P. Grewal, K-T Yu, *New Eng. J. Med.,* **1991,** *324,* pp. 1464-1470.

6 - M.S. Hershfield, S. Chaffee and R. U. Sorensen, *Pediatric Res.,* **1993,** *33,* pp. S42-S48.

7 - R. A. Greenwald, *Free Rad. Biol. Med.,* **1990,** *8,* pp. 201-209.

8 - H. Takaku, S. Misawa, H. Hayashi and K. Miyazaki, *Jpn. J. Cancer Res.,* **1993,** *84,* pp.1195-1200.

9 - S.L. Berg, F.M. Balis, C.L. McCully, K.S. Godwin and D.G. Poplack, *Cancer Chemother. Pharmacol.,* **1993,** *32,* pp. 310-314.

10 - A. Abuchowski, G. M. Kazo, C.R. Verhoest Jr., T. Van Es, D. Kafkewitz, M.L. Nucci, A.T. Viau and F.F. Davis, *Cancer Biochem. Biophys.,* **1984,** *7,* pp. 175-186.

11 - R.L. Capizzi and Y-C. Cheng., **1981,** *Enzymes as drugs,* J.S. Holcenberg and J. Roberts eds., Wiley Intersciences, N.Y., p.1-60.

12 - E.M. D'Urso and G. Fortier., *Biotechnol. Techn.,* **1994,** *8* , pp.71-76.

13 - E.M. D'Urso and G. Fortier, *J. Bioact. Comp. Polym.,* **1994,** *9,* pp. 367-387.

14 - N.V. Katre, *Adv. Drug Deliv.Rev.,* **1993,** *10* , pp. 91-114.

15 - A. Abuchowski, J.R. McCoy, T. Von Es, N.C. Palczuk and F.F. Davis., *J. Biol. Chem.*, **1977**, *252* , pp. 3578-3581.

16 - F.M. Veronese, R. Largajolli, E. Boccu, C.A. Bennasi and O. Schiavon, *Appl. Biochem. Biotech.*, **1985**, *11*, pp.141-152.

17 - H. Wada, I. Imamura, M. Sako, S. Katagiri, S. Tarui, H. Nishimura and Y. Inada, *Ann. N.Y. Acad. Sci.*, **1990**, *619*, pp. 95-108.

18 - J.M. Harris, *Poly(ethylene glycol) Chemistry : Biotechnical and Biomedical Applications*, J.M. Harris, Ed. CRC Press, **1992**, p.1-17.

19 - M. Cantarella, M-H. Remy, V. Scardi, F. Alfani, G. Ioro and G. Greco, *Biochem. J.*, **1979**, *179*, pp. 15-20.

20 - E.M. D'Urso and G. Fortier, *Enz. Micr. Tech.,* **1994**. (submitted)

21 - B.L. Asselin, D. Ryan, C.N. Frantz, S.D. Bernal, P. Leavitt, S.E. Sallan and H.J. Cohen, *Cancer Res.*, **1989**, *49*, 4363-4368.

22 - E.B. Rhoderick , K. L. Jarvis and D. J. Hyland, *Anal. Biochem.*, **1989**, *180*, pp.136-139.

23 - G. E. C. Sims and T. J. Snape, *Anal. Biochem.*, **1980**, *107*, 60-63.

24 - G. Fortier and M. Laliberté, *Biotechnol. Appl. Biochem.*, **1993**, *17* , pp.115-130.

25 - R. Schuster, *J. Chromatogr.*, **1988**, *431*, pp. 271-284.

26 - J.M. Engasser & Cs. Horwath, *Applied Biochemistry and Bioengineering*, vol.1, L.B. Wingard, E. Katchalski-Katzir and L. Golstein eds., Academic Press, N.Y., **1976**, pp.127-219.

27 - D.A. Cooney, H.H. Weetall and E. Long, *Biochem. Pharma.*, **1975**, *24*, pp. 503-515.

28 - J.P. Allison, L. Davidson, A. Gutierrez-Hartman and G.B. Kitto, *Biochem. Biophys. Res. Com.*, **1972**, *47*, pp. 66-73.

29 - T. Mori, T. Tosa and Chibata, *Biochim. Biophys. Acta*, **1973**, *321*, pp. 653-661.

RECEIVED January 3, 1996

Chapter 5

Sulfonylurea-Grafted Polymers for Langerhans Islet Stimulation

You Han Bae[1,2], Sung Wan Kim[1], Akihiko Kikuchi[1,3], Suk-Ja Chong[1], and Soo Chang Song[1]

[1]Center for Controlled Chemical Delivery, University of Utah, 570 Biomedical Polymers Research Building, Room 205, Salt Lake City, UT 84112

[2]Department of Materials Science and Engineering, Kwangju Institute of Science and Technology, Kwangju, Korea

The preliminary results of the *in vitro* bioactivities of sulfonylurea (a glyburide analogue) grafted copolymers, evaluated by adding the copolymers into normal rat islet cultures and measuring insulin concentration in the supernatant, was presented. A hydrophobic polymerizable sulfonylurea monomer showed the stimulatory effect on rat islets at glucose concentrations of 50 mg/dL and 100 mg/dL with equivalent stimulation indices to glyburide, a potent sulfonylurea drug, while less effect was observed at a glucose concentration of 200 mg/dL. The water solubilities and bioactivities of sulfonylurea polymers may be influenced by the polymer backbone structure. Since it is suggested that the sulfonylurea receptors are located in the lipid phase on the ATP-sensitive K^+ channels, solubilization of sulfonylurea into and its diffusion through the lipid phase in the cell membrane may affect the interactions of sulfonylurea with its receptors. With sulfonylurea polymer, sulfonylurea unit in the polymer can interact with its receptor, but this interaction would be affected to some extent by the polymer structure.

Many efforts have been devoted to develop a biohybrid artificial pancreas for the treatment of insulin dependent diabetes mellitus (IDDM) (for reviews, 1-3). Islet macroencapsulation into a chamber with vascular grafts (4,5) and microencapsulation of islets within permselective membranes (6-8) have been major approaches for the development of a biohybrid artificial pancreas. Until now, there appear several unsolved problems in developing a bioartifical pancreas, such as biocompatibility of

[3]Current address: Institute of Biomedical Engineering, Tokyo Women's Medical College, 8–1 Kawadacho, Shinjuku-ku, Tokyo 162, Japan

0097–6156/96/0627–0042$15.00/0

immuno-protective membranes, necrosis or malfunction of the implanted islets due to the local hypoxia and reduced nutrient supply which is caused by the lack of a microvascular system around the islets, large implant volume, and removal of implant or reseeding islets. Assuming that one will be able to replenish implanted islets after a certain implant period as research advances, it would be beneficial that one could reduce the number of implant islets by maximizing the insulin secretion function of islets because of limited islet resources, low isolation yield, preservation problems, as well as implant recipients comfort. It was reported that the minimal requirement of islets implanted for correction of hyperglycemia is more than 5,000 islets per kg body weight (human islet size equivalent), indicating a large number (i.e. volume) of islets would be required for a biohybrid artificial pancreas (9).

Sulfonylureas have been used for the treatment of non-insulin dependent diabetes mellitus (NIDDM) for decades (10). A NIDDM patient is characterized by low response in insulin secretion toward increased blood glucose level. Sulfonylureas directly interact with β-cells of islets of Langerhans which results in increased insulin secretion (11). Sulfonylureas interact with sulfonylurea receptors, which exists on the ATP-sensitive K^+ channel on the pancreatic β-cell membrane surface (12 - 26). This interaction inhibits K^+ efflux causing membrane depolarization followed by increase in Ca^{2+} influx through voltage-dependent Ca^{2+}-channels (14, 19, 23, 27, 28). The increase in cytoplasmic Ca^{2+} concentration eventually triggers insulin secretion from pancreatic islets (14, 19, 28, 29). Since sulfonylureas react with cell surface membrane receptors (12, 16, 18), these may not necessarily be internalized into cells for insulin secretion (12, 16).

Orally taken sulfonylureas may, although not confirmed, interact with implanted islets for enhanced insulin secretion. It would be more beneficial for the implant recipient to avoid taking oral sulfonylureas. Sulfonylureas covalently attached to soluble polymer chains may interact with islets and enhance insulin secretion from the islets when these conjugates are entrapped with islets within the immunoprotective membranes, thus residing inside the membranes until the whole content is replaced. This may reduce the required number of islets for glucose homeostasis. In addition, this conjugate concept may offer more means for detailed investigation of drug unit interactions with corresponding receptor cells.

Pioneer work of sulfonylurea conjugation to water soluble polymer was performed by Obereigner et al. (30). They synthesized several types of polymerizable derivatives from carbutamide, and then copolymerized with 2-hydroxypropyl methacrylamide. These polymers exhibited the lowering effect of blood glucose level by 70 % of basal glucose concentration *in vivo* after administration of polymer solution (20 mg/kg body weight corresponding to sulfonylurea unit) to rat intravenously. Although they showed the reduced blood glucose level *in vivo*, there is no detailed information of enhanced insulin secretion from Langerhans islets. Furthermore, carbutamide is a less potent sulfonylurea drug compared to a second generation sulfonylurea drug, glyburide (see Figure 1 for its structure), which has two-orders of higher potency (70 mg/kg in human) than carbutamide.

The primary aims of this study are to synthesize sulfonylurea compounds (glyburide analogues) for the grafting onto water soluble polymers and to test the bioactivities of the compounds. Two approaches were involved to synthesize sulfonylurea grafted polymers: i) random copolymerization of a water soluble monomer such as N,N-dimethylacrylamide with monomeric sulfonylurea and ii) direct coupling a sulfonylurea compound with amino group onto water soluble polymers having carboxylic groups.

Materials

Terephthalic acid, ethylenediamine, isobutyl chloroformate, tributylamine, 4-(2-aminoethyl)benzenesulfonamide, allylamine, N-hydroxysuccinimide, 1,3-dicyclohexyl carbodiimide, cyclohexylisocyanate, N,N-dimethylacrylamide, and glybenzcyclamide were purchased from Aldrich Chemical (Milwaukee, WI). p-Aminophenyl a-D-glucopyranoside was obtained from Sigma Chem. Co. (St. Louis, MO). t-Butylperoctanoate (BPO) was obtained from Polysciences, Inc. (Warrington, PA). Poly(n-vinylpyrrolidone-co-acrylic acid) [poly(VP-co-AA), norminal MW=250,000, AA content = 50%] was obtained from International Specialty Products (Wayne, NJ). All other solvents were reagent grade.

Hanks' balanced salts, RPMI 1640 powder medium, sodium pyruvate, L-glutamine, penicillin-streptomycin solution (10000 IU/mL penicillin and 10000 μg/mL streptomycin), amphotericin B, collagenase type XI, Ficoll-DL400, N-2-hydroxyethylpiperazine-N'-2-ethanesulfonic acid (HEPES) were obtained from Sigma Chemical Co. (St. Louis, MO). Fetal bovine serum (FBS) was purchased from Hyclone Laboratories, Inc. (Logan, UT).

Islet culture medium was RPMI 1640 supplemented with 11.1 mM glucose (unless otherwise noted), 23.8 mM sodium bicarbonate, 2 mM sodium pyruvate, 2 mM L-glutamine, 6 mM HEPES, antibiotics (100 IU/mL penicillin, 100 mg/mL streptomycin, and 2.5 mg/mL amphotericin B), and 10 % FBS.

Methods

Sulfonylurea polymers by copolymerization

A sulfonylurea compound with polymerizable vinyl group was synthesized following Scheme 1. This compound has structure similar to glyburide which is one of the most effective second generation drug used for the treatment of non-insulin dependent diabetes mellitus (NIDDM). The structure of this monomer was confirmed by ^{1}H-NMR according to the assignment of glyburide (31). The sulfonylurea monomer (**VII**) was

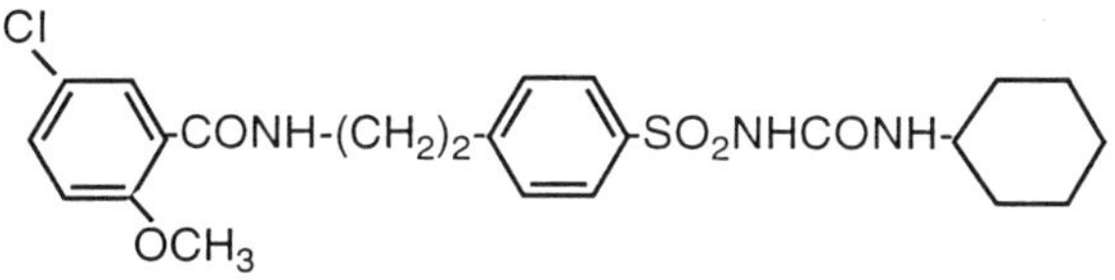

Figure 1. Structural formula of glyburide

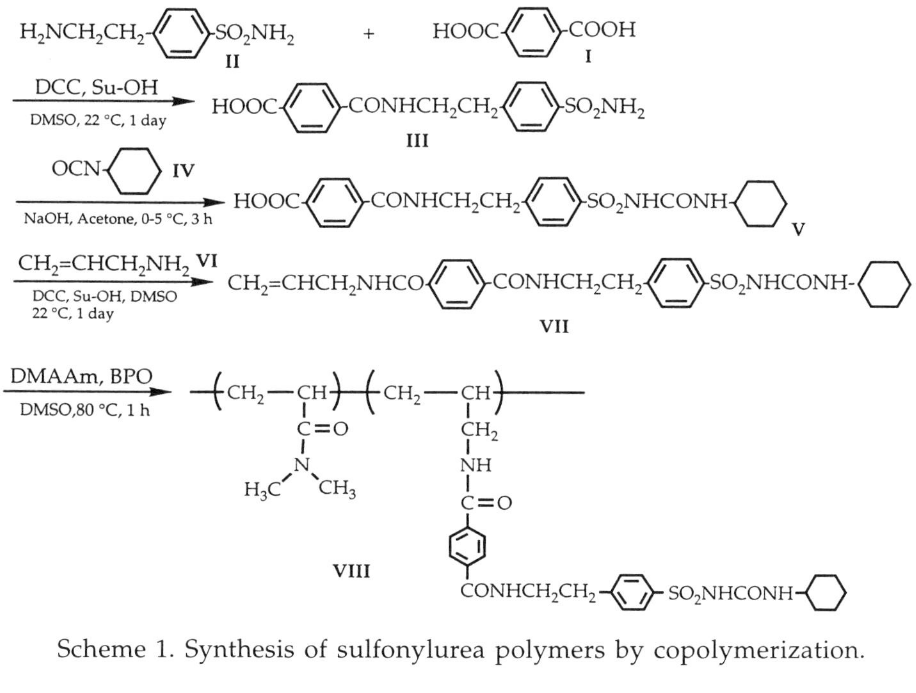

Scheme 1. Synthesis of sulfonylurea polymers by copolymerization.

then copolymerized with hydrophilic monomer, DMAAm in DMSO by using a radical initiator. Copolymer compositions were determined by H-NMR. In the following *in vitro* experiments, the sulfonylurea copolymer with sulfonylurea content of 19.6 wt% was used.

Synthesis of III (32-34). Twenty mmol of terephthalic acid (**I**), 20 mmol of N-hydroxysuccinimide (Su-OH) was dissolved in 40 mL dimethyl sulfoxide (DMSO) at room temperature. To this solution, 20 mmol of 1,3-dicyclohexylcarbodiimide (DCC) solution in DMSO (20 mL) was added and stirred to form active ester of **I** for 1 h at 20~23 °C. Then, 20 mmol of 4-(2-aminoethyl)benzenesulfonamide (**II**) in DMSO solution (20 mL) was added and stirred for overnight to prepare **III**. Formed dicyclohexylurea (DCU) was removed by filtration, and the filtrate was poured dropwise into an excess amount of distilled water. The precipitate was collected by filtration, then dissolved in 400 mL of dilute NaOH (25 mmol) aqueous solution. The insoluble part was removed by filtration, the filtrate was acidified by adding 4 M HCl dropwise while stirring. The precipitate was recovered by filtration and dried under vacuum. Yield 95 %.

Synthesis of V (31, 35). Ten mmol of **III** was mixed with 40 mL of acetone and 10 mL of 2 mole/dm^3 NaOH aqueous solution. Then, approximately 40 mL of water was added to this mixture while stirring. The solution was cooled on an ice-water bath to 0-5 °C, cyclohexylisocyanate (**IV**, 12 mmol) was added dropwise to this solution. After 3 h of stirring at 0-5 °C, the reaction mixture was diluted with 20 mL of methanol and 60 mL of distilled water followed by filtration. The filtrate was acidified with 4 M HCl. Precipitate (**V**) was collected by filtration and washed with distilled water followed by thorough drying under vacuum. Yield 71.6 %.

Synthesis of VII. To 7 mmol of **V** and 7 mmol of Su-OH in DMSO solution (25 mL), DCC (7 mmol) in DMSO solution (15 mL) was added and stirred for 1.5 h at 22 °C to form active ester of **V**. Seven mmol of allylamine was then added to this solution, which was allowed to stir for 1 day at 22 °C. Formed dicyclohexylurea (DCU) was removed by filtration, the filtrate was poured into an excess amount of distilled water to precipitate the product (**VII**). The precipitate was collected by filtration, followed by dissolution into NaOH (10 mmol) aqueous solution. After removal of insoluble part, the filtrate was acidified with HCl aqueous solution. The product was filtered and washed with distilled water followed by vacuum drying. Yield 70.8 %.

Poly(N,N-dimethylacrylamide-co-VII) (VIII). Predetermined amounts of N,N-dimethylacrylamide (DMAAm) and sulfonylurea monomer (**VII**) were dissolved in DMSO at a monomer concentration of 10 w/v%. Butylperoxyoctoate (BPO, Polysciences, Inc., Warrington, PA) was used as an initiator (10 mmol/dm^3). After 15 min of dried N_2 gas bubbling, polymerization was proceeded at 80 °C for 1 h with vigorous stirring. Polymers were precipitated in an excess amount of diethyl ether to separate from unreacted chemicals and dried under vacuum. Yield of the copolymers was ranged from 81 to 93 %. The amount of sulfonylurea unit in the copolymer was determined from H-NMR spectra using DMSO-d_6 as a solvent. The result of copolymerization was summarized in Table I.

Sulfonylurea polymers by graft modification of poly(VP-co-AA)

A more water soluble polymer having glucose and sulfonylurea units was synthesized by Scheme 2. Glucose unit was introduced to enhance water solubility.

Synthesis of IV. To 4.2 mmol of **V** and 4.2 mmol of Su-OH in DMF solution (15 mL), DCC (4.2 mmol) in DMSO solution (15 mL) was added and stirred for 1.5 h at 22 °C to form active ester of **V**. Ethylenediamine solution (42 mmol in 50 mL DMF) was then added to this solution, which was allowed to stir overnight at 22 °C. Formed DCU was removed by filtration, the filtrate was poured into an excess amount of distilled water to precipitate the product (**IV**). The precipitate was dissolved in DMSO and reprecipitated in distilled water. The precipitate was collected by filtration, followed by dissolution into NaOH (10 mmol) aqueous solution. After removal of insoluble part by filtration, the filtrate was acidified with HCl aqueous solution. The product was filtered and washed with distilled water followed by vacuum drying. Yield 65.6 %. The amino group content determined by amine titration was 92%.

Synthesis of X. Poly(VP-co-AA) (0.5 g) was dissolved in 10 mL DMF and 0.4 mmole isobutyl chloroformate and equimolar amount of tributylamine were added. This mixture was stirred under dried N_2 at ice temperature for 15 min and additional 0.48 mmole tributylamine was added. Separately, 0.38 mmole of p-aminophenyl a-D-glucopyranoside was dissolved in 10 mL water. A portion (0.5 mL) of this solution was mixed with the polymer solution while stirring at ice temperature and maintaining the pH at 8.7. After stirring this mixture at room temperature for 3 hr, the solution was extensively dialyzed against distilled water and lyophilized.

Synthesis of XI. **X** (0.3 g) was dissolved in DMSO and 0.25 mmole isobutyl chloroformate and equimolar amount of tributylamine were added. This mixture was stirred under dried N_2 at ice temperature for 15 min and additional 0.48 mmole tributylamine was added. Separately, 0.15 mmole of **IV** was dissolved in 10 mL DMSO. A portion (0.5 mL) of this solution was mixed with the polymer solution. After stirring this mixture at room temperature for 3 hr, the solution was extensively dialyzed against distilled water, filtered and lyophilized.

Table I. Copolymerization of Sulfonylurea Monomer (VII) with N,N-dimethylacrylamide (DMAAm)

	Feed Composition		VII Feed Ratio		Yield	Copolymer Composition	
Run	DMAAm (mmol)	VII (mmol)	(mol %)	(wt.-%)	(%)	(mol %)	(wt.-%)
1	23.97	0.24	1.0	5.0	81	1.2	5.8
2	23.38	0.36	1.5	7.3	93	1.2	5.9
3	22.27	0.57	2.5	11.7	93	2.1	10.0
4	19.82	1.04	5.0	21.4	88	4.5	19.6

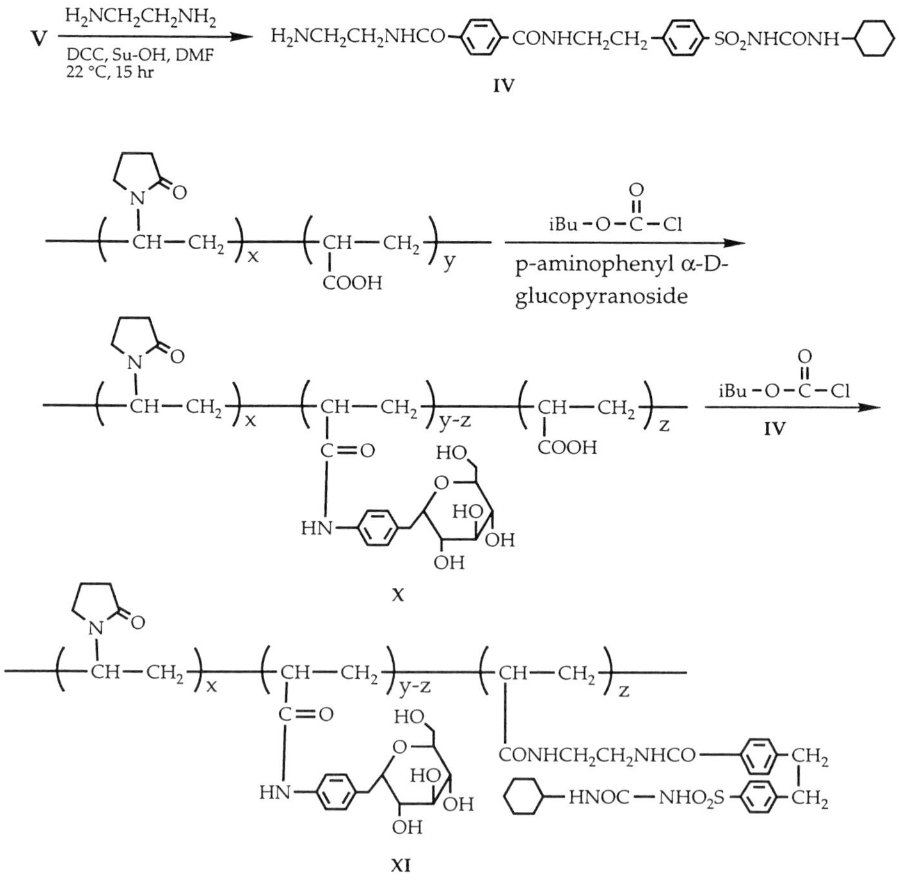

Scheme 2. Sulfonylurea polymers by graft modification of poly(VP-co-AA).

Isolation and Purification of Langerhans Islets of Rats. Langerhans islets were obtained from pancreases of male Sprague-Dawley rats (Sasco, Omaha, Nebraska, 250-300 g body weight) by collagenase digestion technique (36) followed by density gradient centrifugation (37). Brief procedures are as follows; under anesthesia, abdomen of rat was opened. Both proximal and distal ends of common bile duct were clamped with hemostatic forceps, then freshly prepared cold collagenase solution (Collagenase type XI, 13 mg/mL in HBSS with 10 mM HEPES, pH 7.4) was injected to pancreas through common bile duct using 25 G needle connected to syringe via polyethylene tubing (Cat.# PE-205, INTRAMEDIC, Clay Adams, Parsippany, NJ). Swollen pancreas was dissected and incubated in collagenase solution at 37 °C for 20 min. After incubation, cold HBSS was added to stop enzyme activity. Digested tissue was further broken by mechanical pipetting 50-100 times, then washed twice with cold HBSS. Tissue suspension was passed through a 860 mm stainless steel mesh followed by a 520 mm mesh to remove large tissue fragments of pancreas.

After centrifugation, crude islets were resuspended into 8 mL of 27 % Ficoll-HBSS, then 5 mL of 23 %, 20.5 %, and 11 % Ficoll-HBSS were layered on 27 % Ficoll-HBSS suspension, successively. This density gradient was centrifuged at 3500 rpm for 15 min to isolate islets from exocrine cells and cell debris. Islets were collected by Pasteur capillary pipette, then washed twice with cold HBSS. The islets were cultured in RPMI 1640 medium supplemented with 10 % FBS and antibiotics at 37 °C in a humidified atmosphere of 7 % CO_2. The following day after islet isolation, round-shaped islets were hand-picked under a microscope, and subcultured in RPMI 1640 supplemented with 10 % FBS. Two hundred to 400 islets per rat pancreas were collected for each isolation.

***In Vitro* Bioactivity of Sulfonylurea Compounds.** The subcultured islets were suspended in three different RPMI 1640 media with glucose concentrations of 50 mg/dL, 100 mg/dL, and 200 mg/dL, respectively, without serum, then placed into a 24-well tissue culture plate (Falcon 3047, Becton and Dickinson, Lincoln Park, NJ) at islet concentration of 40-50 islets/mL in each well. Sulfonylurea monomer (**VII**), sulfonylurea copolymer (**VIII**) (sulfonylurea content of 19.6 wt%), and glyburide, commercially available second generation compound, were dissolved in DMSO at a definite concentration of sulfonylurea. The copolymer of **XI** was dissolved in PBS. Ten mL of these solutions were added to each well and incubated in a humidified atmosphere at 37 °C for 2 h. Final DMSO concentration in the medium was less than 1 % (v/v) and at this concentration no apparent DMSO toxicity to islets was observed by islet viability. Before and after incubation, the number of islets was counted. The islet viability after 2 h incubation was above 90%. Islet suspension was spun down, then supernatant was collected and kept frozen until insulin radio-immunoassay was performed.

Insulin concentration in the supernatant was determined by INSULIN RIA kit (Cat.# D1804, ICN Biomedicals, Costa Mesa, CA). Data were expressed as mean with standard error of mean (S.E.M.) (μIU/islet/2 h).

Stimulation index (SI) is defined as the ratio of insulin secreted from rat islets in the presence of sulfonylurea compounds to that in the absence of sulfonylurea compounds at each experimental condition.

SI = (insulin secreted with sulfonylurea)/(insulin secreted without sulfonylurea)

Results and Discussion

Solubility of Sulfonylurea Monomer and Copolymer. Solubility of glyburide in aqueous milieu was limited because of its high hydrophobicity. Solubility of this drug was increased in alkaline aqueous solution due to the formation of salts (4 mg/mL at pH 4, 600 mg/mL at pH 9) (31, 38).

Due to the structural similarity, solubility of sulfonylurea monomer (**VII**) was close to that of glyburide. **VII** was soluble in DMSO, alkaline aqueous solution (pH>10), and partly soluble in DMF, and dioxane. Solubility of **VII** in phosphate buffered saline (isotonic PBS, pH 7.4) was 25 mg/mL. Copolymers of sulfonylurea monomer with DMAAm were also soluble in alkaline aqueous solution (pH>10), THF, DMF, and DMSO. The copolymer (**XI**) were readily soluble in PBS (isotonic pH 7.4) at 50 mg/mL based on the amount of sulfonylurea monomer unit.

Bioactivity of Sulfonylurea Compounds *In Vitro*.. To evaluate the bioactivity of the sulfonylurea monomer as an islet stimulant, sulfonylurea monomer solution in DMSO was added to the islet culture, and incubated for 2 h at 37°C. Glucose concentration of the culture medium was varied from 50 mg/dL to 200 mg/dL. Figure 2 shows the amount of insulin secreted from rat islets stimulated by **VII** plotted against glucose concentration. As expected, insulin secretion from Langerhans islets increased with increasing glucose concentration of the medium in the absence of sulfonylurea monomer. At low glucose concentrations, i.e. 50 mg/dL and 100 mg/dL, the addition of **VII** to islet culture resulted in the increase in insulin secretion from islets compared to the control. This demonstrates the same trend as the results when glyburide was added to islet culture (Table II). At these glucose concentrations, insulin secretion from sulfonylurea-stimulated islets was significant compared to insulin secretion from unstimulated islets. However, the difference in insulin secretion was not significant at a glucose concentration of 200 mg/dL regardless of sulfonylurea concentration. It has been reported that the difference between insulin secretion from normal islets in the absence and the presence of sulfonylureas was relatively smaller at high glucose

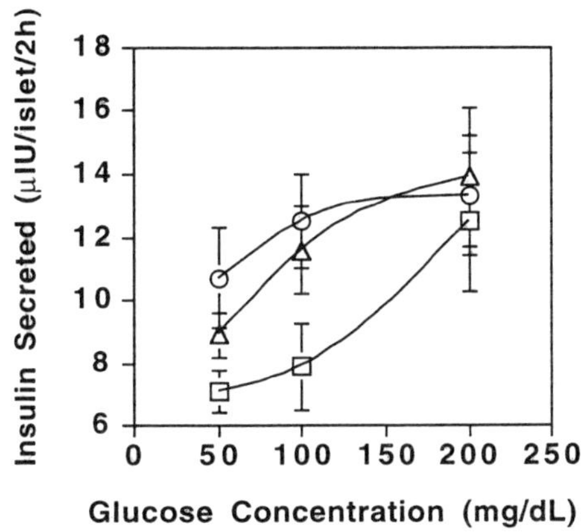

Figure 2. Insulin secretion from rat Langerhans islets stimulated by sulfonylurea monomer (**VII**). Rat islets were cultured in RPMI 1640 medium without serum for 2 h at 37°C in the absence and the presence of **VII**. Glucose concentration of the medium was changed from 50 mg/dl to 200 mg/dl. ❑; 0 μg/mL **VII**, Δ; 1 μg/mL **VII**, ○; 10 μg/mL **VII**. Data are expressed as a mean with standard error of mean (S.E.M.) of tripricate experiments.

Table II. Effects of Glyburide, Sulfonylurea Monomer (VII), and Sulfonylurea Polymer (VIII) on Insulin Secretion from Rat Islets

	SI[a)]		
Glucose Concentration	50 mg/dL	100 mg/dL	200 mg/dL
Control (without sulfonylurea)[b)] (n=2 ~ 5)	1.0	1.0	1.0
Glyburide (n=2) 1μg/mL	1.13 ± 0.07	1.43 ± 0.18	1.21 ± 0.02
10 μg/mL	1.30 ± 0.12	1.42 ± 0.05	1.13 ± 0.22
Sulfonylurea monomer (VII) (n=3) 1 μg/mL	1.35 ± 0.29	1.50 ± 0.19	1.14 ± 0.14
10 μg/mL	1.65 ± 0.48	1.61 ± 0.15	1.09 ± 0.10
Sulfonylurea in copolymer (Run #4) (n=5) 1 μg/mL[c)]	1.04 ± 0.05	0.99 ± 0.06	0.94 ± 0.04
10 μg/mL[c)]	1.44 ± 0.22	1.12 ± 0.03	0.97 ± 0.02
50 μg/mL[c)]	1.27 ± 0.11	1.07 ± 0.08	0.89 ± 0.05

a) Data are presented as mean ± S.E.M.
b) Refer Figure 2 for the data range of control experiment.
c) Sulfonylurea concentrations were calculated based on sulfonylurea unit (VII) in the copolymer.

Table III. Effects of IX on Insulin Secretion from Rat Islets as a Function of Incubation Period

Incubation Period	SI
30 min	1.67
2 hr	1.47
24 hr	1.39

Glucose Concentration: 50 mg/dL
SU unit in polymer: ~10 μg/mL

concentration than the difference at low glucose concentration (39 - 41). Glucose itself is the stimulant for Langerhans islets to secrete insulin, and thus, there was little effect of sulfonylurea at 200 mg/dL glucose. It is obvious from the results as summarized in Table II, that rat islets were stimulated to secrete insulin with **VII** at 50 mg/dL and 100 mg/dL glucose concentrations. These values are equivalent to or more than the glyburide effect. No significant difference in stimulation index is seen at a glucose concentration of 200 mg/dL regardless of sulfonylurea concentration. At glucose concentrations of 50 mg/dL and 100 mg/dL, the amount of insulin secreted from islets was increased by the addition of a small amount of sulfonylurea compound (1 μg/mL, ca. 2 μM). At a higher concentration (20 μM) of either glyburide and **VII**, the stimulation effect was not improved.

Consequently, the newly synthesized sulfonylurea monomer is concluded to be active in stimulating Langerhans islets insulin secretion *in vitro* and the result suggests that sulfonylurea monomer incorporated into a polymer may have stimulatory effect on Langerhans islets.

VII is then incorporated into polymers by radical copolymerization with DMAAm. Bioactivity of this sulfonylurea polymer was explored by adding the copolymer into islet culture. The sulfonylurea concentration was calculated by monomer unit in the copolymer. The glucose dependent insulin secretion activity of the islets in the presence of the copolymer was normal. Insulin secretion was increased by the addition of the sulfonylurea copolymer at a glucose concentration of 50 mg/dL. At a glucose concentration of 50 mg/dL and sulfonylurea concentration of 10 μg/mL, the sulfonylurea copolymer exhibited approximately 80 % bioactivity of **VII**. At glucose concentrations of 100 mg/dL and 200 mg/dL, however, no significant effects of sulfonylurea copolymer were observed. In other words, the effect of sulfonylurea in the polymer chain is considered to be lower when compared to **VII** at the same sulfonylurea concentration.

The preliminary in vitro result of the sulfonylurea polymer obtained from side chain modification of poly(VP-co-AA), which is more soluble in PBS, also demonstrated similar bioactivity and its stimulation index reduced as increasing incubation time in a static condition as summarized in Table III.

It is considered that the stimulatory effect of sulfonylurea on insulin secretion from Langerhans islets was concealed by the stimulatory effect of glucose at higher glucose concentrations. When sulfonylurea is added to islet cell culture, sulfonylurea is partitioned into the lipid phase of the cellular membrane or in the proteins because of their high hydrophobicity (10, 23, 35, 42). It is suggested that sulfonylurea receptors on the ATP-sensitive K^+ channels are located in the lipid phase of the cellular membrane (16, 24), and thus sulfonylureas are likely to access their receptors in the lipid phase of the β-cell membrane via lateral diffusion. Taking into account this consideration, the sulfonylurea moiety in the copolymer can be partitioned into the cell membrane lipid phase, it is difficult to access the sulfonylurea receptors because they are conjugated to a relatively hydrophilic polymer backbone. The polymer backbone may interfere with the lateral diffusion of sulfonylurea within the b-cell membrane to its receptor on the ATP-sensitive K^+ channels, which may lead to reduced binding of sulfonylurea to its receptor. In addition, it is plausible that the penetration of sulfonylurea polymer within islets is also limited due to its large size in a hydrated state, resulting in the reduced contact with b-cells inside of Langerhans islets. Therefore, the stimulatory effect of sulfonylurea copolymer can not be seen at glucose concentrations of 100 mg/dL and 200 mg/dL. This deficit may be improved by conjugating sulfonylurea moiety to polymer via a spacer.

Conclusion

Polymers having sulfonylurea unit in side chains were synthesized in two methods; copolymerization and side chain graft modification. *In vitro* bioactivity test demonstrated that the sulfonylurea compounds (glyburide analogue) as a polymerizable monomer has the stimulatory effect on rat islets, equivalent to glyburide. The sulfonylurea polymer synthesized by copolymerization stimulated Langerhans islets to secrete insulin at a low glucose concentration of 50 mg/dL with 80 % bioactivity of its monomeric compound and showed no significant stimulation at higher glucose concentrations (100 mg/dL and 200 mg/dL). The preliminary in vitro result of the sulfonylurea polymer obtained from side chain modification of poly(VP-co-AA) also showed bioactivity and its stimulation index reduced as increasing incubation time in a static condition.

Although further detailed investigation is necessary, it is proven that the synthesized monomer and polymers have bioactivity to islets. This approach may find a potential application to a bioartificial pancreas and as a tool for investigating drug-cell surface receptor interactions.

Literature Cited

1. Colton, C. K.; Avgoustiniatos, E. S. Bioengineering in development of the hybrid artificial pancreas. *J. Biomech. Eng.* **1991**, *113*, 152-170.

2. Reach, G. Bioartificial pancreas: status and bottlenecks. *Intern. J. Atrif. Organs.* **1990**, *13*, 329-336.

3. Reach, G. Artificial and bioartificial replacement of the endocrine pancreas. *Artif. Organs.* **1992**, *16*, 61-70.

4. Maki, T.; Ubhi, C. S.; Sanchez-Farpon, H.; Sullivan, S. J.; Borland, K.; Muller, T. E.; Solomon, B. A.; Chick, W. L.; Monaco, A. P. Successful treatment of diabetes with the biohybrid artificial pancreas in dogs. *Transplantation.* **1991**, *51*, 43-51.

5. Sullivan, S. J.; Maki, T.; Borland, K. M.; Mahoney, M. D.; Solomon, B. A.; Muller, T. E.; Monaco, A. P.; Chick, W. L. Biohybrid artificial pancreas: Long-term implantation studies in diabetic, pancreatectomized dogs. *Science.* **1991**, *252*, 718-721.

6. Lim, F.; Sun, A. M. Microencapsulated islets as bioartificial endocrine pancreas. *Science*. **1980**, *210*, 908-910.
7. O'Shea, G. M.; Goosen, M. F. A.; Sun, A. M. Prolonged survival of transplanted islets of Langerhans encapsulation in a biocompatible membrane. *Biochim. Biophys. Acta*. **1984**, *804*, 133-136.
8. Sun, A. M.; Lim, F.; Rooy, H. V.; O'Shea, G. Long-term studies of microencapsulated islets of Langerhans: A bioartificial endocrine pancreas. *Artif. Organs*. **1981**, *5(Suppl.)*, 784-786.
9. Warnock, G. L.; Rajotte, R. V. Critical mass of purified islets that induce normoglycemia after implantation into dogs. *Diabetes*. **1988**, *37*, 467-470.
10. Groop, L. C. Sulfonlureas in NIDDM. *Diabetes Care*. **1992**, *15*, 737-754.
11. Gorus, F. K.; Schuit, F. C.; In't Veld, P. A.; Gepts, W.; Pipeleers, D. G. Interaction of sulfonylureas with pancreatic b-cells -A study with glyburide. *Diabetes*. **1988**, *37*, 1090-1095.
12. Hellman, B.; Sehlin, J.; Taljedal, I.-B. The pancreatic b-cell recognition of insulin secretagogues. II. Site of action of tolbutamide. *Biochem. Biophys. Res. Commun*. **1971**, *45*, 1384-1388.
13. Schmid-Antomarchi, H.; Weille, J. D.; Fosset, M.; Lazdunski, M. The receptor for antidiabetic sulfonylureas controls the activity of the ATP-modulated K^+ channel in insulin-secreting cells. *J. Biol. Chem*. **1987**, *262*, 15840-15844.
14. Henquin, J.-C. Glucose-induced electrical activity in b-cells -Feedback control of ATP-sensitive K^+ channels by Ca^{2+}. *Diabetes*. **1990**, *39*, 1457-1460.
15. Sturgess, N. C.; Ashford, M. L. J.; Cook, D. L.; Hales, C. N. The sulfonylurea receptor may be an ATP-sensitive potassium channel. *Lancet*. **1985**, *2*, 474-475.
16. Schwanstecher, C.; Dickel, C.; Ebers, I.; Lins, S.; Zünkler, B. J.;Panten, U. Diazoxide-sensitivity of the adenosine 5'-triphosphate-dependent K^+ channel in mouse pancreatic b-cells. *Br. J.Pharmacol*. **1992**, *107*, 87-94.
17. Schwanstecher, M.; Brandt, Ch.; Behrends, S.; Schaupp, U.; Panten, U. Effect of MgATP on pinacidil-induced displacement of glibenclamide from the sulphonylurea receptor in a pancreatic b-cell line and rat cerebral cortex. *Br. J. Pharmacol*. **1992**, *106*, 295-301.
18. Schwanstecher, M.; Löser, S.; Brandt, Ch.; Scheffer, K.; Rosenberger F.; Panten, U. Adenine nucleotide-induced inhibition of binding of sulphonylureas to their receptor in pancreatic islets. *Br. J. Pharmacoly*. **1992**, *105*, 531-534.
19. Boyd, A. E. III. Sulfonylurea receptors, ion channels, and fruit flies. *Diabetes*. **1988**, *37*, 847-850.
20. Gaines, K. L.; Hamilton, S.; Boyd, A. E. III. Characterization of the sulfonylurea receptor on beta cell membranes. *J. Biol. Chem*. **1988**, *263*, 2589-2592.
21. Misler, S.; Gee, W. M.; Gillis, K. D.; Scharp, D. W.; Falke, L. C. Metabolite-regulated ATP-sensitive K^+ channel in human pancreatic islet cells. *Diabetes*. **1989**, *38*, 422-427.

22. Panten, U.; Burgfeld, J.; Goerke, F.; Rennicke, M.; Schwanstecher, M.; Wallasch, A.; Zünkler, B. J.; Lenzen, S. Control of insulin secretion by sulfonylureas, meglitinide and diazoxide in relation to their binding to the sulfonylirea receptor in pancreatic islets. *Biochem. Pharmacol.* **1989**, *38*, 1217-1229.
23. Panten, U.; Schwanstecher, M.; Schwanstecher, C. Pancreatic and extrapancreatic sulfonylurea receptors. *Horm. Metab. Res.* **1992**, *24*, 549-554.
24. Zünkler, B. J.; Trube, G.; Panten, U. How do sulfonylureas approach their receptor in the B-cell plasma membrane? *Naunyn-Schmiedeberg's Arch.Pharmacol.* **1989**, *340*, 328-332.
25. Schwanstecher, C.; Dickel, C.; Panten, U. Cytosolic nucleotides enhance the tolbutamide sensitivity of the ATP-dependent K^+ channel in mouse pancreatic B cells by their combined actions at inhibitory and stimulatory receptors. *Mol. Pharmacol.* **1992**, *41*, 480-486.
26. Schwanstecher, M.; Behrends, S.; Brandt, C.; Panten, U. The binding properties of the solubilized sulfonylurea receptor from a pancreatic b-cell line are modulated by the Mg^{++}-complex of ATP. *J.Pharmacol. Exp. Therapeut.* **1992**, *262*, 495-502.
27. Cook, D. L.; Ikeuchi, M.; Fujimoto, W. Y. Lowering of pH_i inhibits Ca^{2+}-activated K^+ channels in pancreatic B-cells. *Nature.* **1984**, *311*, 269-271.
28. Nelson, T. Y.; Gaines, K. L.; Rajan, A. S.; Berg, M.; Boyd, A. E. III. Increased cytosolic calcium -A signal for sulfonylurea-stimulated insulin release from beta cells. *J. Biol. Chem.* **1987**, *262*, 2608-2612.
29. Turk, J.; Gross, R. W.; Ramanadham, S. Amplification of insulin secretion by lipid messengers. *Diabetes.* **1993**, *42*, 367-374.
30. Obereigner, B.; Buresová, M.; Verána, A.; Kopecek, J. Preparation of polymerizable derivatives of N-(4-aminobenzenesulfonyl)-N'-butylurea. *J. Polym. Sci.: Polym. Symp.* **1979**, *66*, 41-52.
31. Takla, P. G. in *Glibenclamide. Analytical Profiles of Drug Substances Vol. 10.*; Florey, K., Ed.; Academic Press; New York, 1981; 337-355.
32. Atherton, E.; Sheppard, R. C. in *Solid phase peptide synthesis -a practical approach*-The Practical Approach Series; Rickwood, D.; Hames, B. D. Eds.; IRL Press: Oxford, 1989;pp.1-12.
33. Bodanszky, M.; Bodanszky, A. in *The practice of peptide synthesis.* Springer-Verlag: Berlin Heidelberg, 1984
34. Jones, J. in *Amino Acid and Peptide Synthesis.* Davies, S. G., Ed.; Oxford University Press: Oxford, 1992.
35. Neth. Pat. Appl. 6,610,580 Jan 30, 1967; Ger. Appl. July 27, 1965. *Chemical Abstracts.* **1968**, *68*, 12733h.
36. Lacy, P. E.; Kostianovsky, M. Method for the isolation of intact islets of Langerhans from the rat pancreas. *Diabetes.* **1967**, *16*, 35-39.
37. Lindall, A.; Steffes, M.; Sorenson, R. Immunoassayable insulin content of subcellular functions of rat islets. *Endocrinology.* **1969**, *85*, 218-223.

38. AHFS. *Glyburide. American Hospital Formulary Service Drug Information.* The American Society of Hospital Pharmacists, Inc.; Bethesda, MD, 1993; 1973-1978.

39. Kiekens, R.; In't Veld, P.; Mahler, T.; Schuit, F.; Winkel, M. V. D.; Pipeleers, D. Differences in glucose recognition by individual rat pancreatic B cells are associated with intercellular differences in glucose-induced biosynthetic activity. *J. Clin. Invest.* **1992**, *89*, 117-125.

40. Pipeleers, D. G. Heterogeneity in pancreatic b-cell population. *Diabetes.* **1992**, *41*, 777-781.

41. Schuit, F. C.; In't Veld, P. A.; Pipeleers, D. G. Glucose stimulates proinsulin biosynthesis by a dose-dependent recruitment of pancreatic beta cells. *Proc. Natl. Acad. Sci. USA.* **1988**, *85*, 3865-3869.

42. Pearson, J. G. Pharmacokinetics of glyburide. *Am. J. med.* **1985**, *79*, 67-71.

RECEIVED January 3, 1996

Chapter 6

Administration of Ovalbumin Encapsulated in Alginate Microspheres to Mice

T. L. Bowersock[1], H. HogenEsch[1], M. Suckow[2], John Turek[1], E. Davis-Snyder[1], D. Borie[1], R. Jackson[3], Haesun Park[3], and Kinam Park[3]

[1]School of Veterinary Medicine, [2]Laboratory Animal Program, and [3]School of Pharmacy, Purdue University, West Lafayette, IN 47907

Sodium alginate microparticles were tested as a delivery system for oral vaccines. Ovalbumin was incorporated into alginate microparticles. The release of ovalbumin from the microparticles was determined. Electron microscopy was performed to evaluate surface characteristics of the microparticles. Plain microspheres or ovalbumin incorporated within alginate microspheres were administered orally to groups of mice; another 2 groups were given either ovalbumin in microspheres or in Freund's adjuvant by subcutaneous injection. Antibodies specific to ovalbumin were measured in serum and antibody-secreting cells specific for ovalbumin were enumerated in spleen cells. Microparticles released ovalbumin within 5 days in vitro. Alginate microparticles were uniform in density and spherical in shape by electron microscopy. Mice had increased numbers of antibody secreting cells and serum antibody titers specific for ovalbumin regardless of the method of inoculation. These studies suggest that alginate microparticles could be used for the delivery of antigens. Further studies of this delivery system are warranted.

Many infectious diseases begin at mucosal surfaces. Pathogens attach to mucosal surfaces, release toxins that damage host epithelial cells, colonize the mucosa, and may then invade into deeper tissues. The prevention of infectious diseases has relied on the administration of parenteral vaccines. This strategy has several drawbacks. First, several injections are required to stimulate protective immunity. Poor patient compliance may result in inadequate immunity. Second, humoral immunity (antibodies in general circulation) is primarily induced. These antibodies do not necessarily cross to mucosal surfaces. Humoral antibodies mainly attack infectious agents only after they have invaded to deeper tissues. Prevention of the initial infection is not affected. Immunity at mucosal sites can prevent

0097–6156/96/0627–0058$15.00/0

infections from starting where they begin. There is a need to improve vaccine strategies to provide immunity at mucosal sites, and to stimulate long standing immunity with fewer inoculations.

The importance of mucosal immunity in preventing infectious diseases has been recognized. Unfortunately, induction of immunity at mucosal sites is not always easily accomplished. Stimulation of mucosal immunity can be accomplished by delivering the vaccine directly to the site where protection is needed. However, the administration of antigens to mucosal sites is challenging. Administration of vaccines to sites such as the lower lung, reproductive tract, mammary gland, and even the intestine is difficult. Antigens in vaccines can be inactivated by enzymes, microbial degradation, or by the low pH enroute to, or present at, mucosal sites. Moreover, mucosal immunity is of short duration and therefore requires multiple inoculations to induce long-term protection.

An easy way to stimulate mucosal immunity is by oral administration of antigens in the feed or water. Orally administered antigens are taken up in the intestines by the Peyer's patches, the largest accumulation of lymphoid tissue in the body. Antigen specific lymphocytes are induced which leave the Peyer's patches, enter the general circulation, and migrate to all other mucosal sites in the body. In this way oral vaccines can stimulate local (mucosal) immunity at any site in the body. However, delivery of antigens to the Peyer's patch is difficult. Antigen must be protected from the low pH and enzymes of the intestinal tract as well as dilution in the ingesta that could decrease presentation to the Peyer's patches. Microencapsulation is a unique way to protect antigens as well as to stimulate their uptake by Peyer's patches (1). The most common material used to encapsulate antigens has been poly(DL-lactide-co-glycolide) (DL-PLG) (2). Although DL-PLG particles are readily taken up by Peyer's patches, and can readily deliver antigens to the Peyer's patches, there are limitations. Production of these microspheres requires the use of high temperatures and organic solvents, conditions that could damage fragile antigens, and make incorporation of live organisms impossible. Therefore, there is a need to investigate other polymers for the incorporation and delivery of antigens. We propose that sodium alginate is one material that could be useful. Sodium alginate is a natural polysaccharide formed by sea weeds that polymerizes into a solid material when mixed with divalent cations. Alginate has been used for incorporation of beta-islet cells for the experimental treatment of diabetes (3) as well as for use in production of antibodies by hybridomas (4). We propose that sodium alginate can be used to incorporate any antigen for oral delivery to any animal host. To test this hypothesis we prepared alginate microspheres containing a simple antigen ovalbumin and administered them to mice.

Materials and methods

Microparticle production - Microspheres were produced as previously described (5). A 2.0% (w/v) solution of sodium alginate (Kelco, Chicago, IL., USA) was dissolved in distilled water and clarified by filtration using a 0.22 μm filter. The clarified sodium alginate was mixed to a final concentration of 1 mg/ml of ovalbumin (Sigma,

St. Louis, MO, USA). The ovalbumin/alginate was then infused by syringe pump (Harvard Instruments, South Natick, MA, USA) into an atomizer (Turbotak Inc, Waterloo, Ont, Canada) and sprayed under pressure (40 PSI of bottled nitrogen) into a 0.5% $CaCl_2$ solution. Resultant microparticles were coated with poly-*l*-lysine (mean molecular weight 45,000, Sigma, St. Louis, MO, USA) to enhance stability and surface hydrophobicity. Microparticles were stored in water at 4^oC until used.

Particle size analysis - Samples of alginate microspheres were washed thoroughly in water, sonicated at low frequency to disrupt clumps, and degassed by placing in a vacuum. The samples were then inserted into a particle size analyzer (Microtrac, Leeds and Northrup Instruments, North Wales, PA, USA).

Electron microscopy of alginate microspheres - Microspheres were prepared using techniques previously described (6). Briefly, 100 microliters of microspheres were allowed to settle to the bottom of a 1.5 ml microfuge tube. The supernatant $CaCl_2$ was removed and replaced with 1% of OsO_4 in distilled water. The microspheres were gently dispersed in the OsO_4 by shaking every 10 minutes and held at 20^oC for 1 hour. The microspheres were then allowed to settle by gravity followed by centrifugation at 200 times gravity for 1 minute. Supernatant was removed and spheres washed 2 times with distilled water. For preparation for examination by transmission electron microscopy, the microspheres were then layered onto 3% agar at 55^oC and centrifuged at 200 times gravity for 1.5 minutes to disperse them through the agar. The agar was allowed to solidify and then cut into 1 mm^3 pieces. These pieces of agar were washed in increasing concentrations (30-100%) of alcohol, embedded in epoxy resin (Polybed, Polysciences, Warrington, PA, USA) and cut in 70 nm sections for viewing by a transmission electron microscope (JEOL-100CX, Tokyo, Japan). For preparation for scanning electron microscopy, spheres were dehydrated through a 30-100% ethanol gradient, washed twice in freon-113, and then pipetted onto a nucleopore filter for mounting on a specimen stub. The spheres were sputter coated with gold and viewed on an scanning electron microscope (model ISI-100A, Topcon Technologies, Inc., Paramus, NJ, USA).

Elution of ovalbumin from microspheres - A volume of microspheres with a total ovalbumin concentration of OVA of 100 μg was placed in 1 mL of sterile saline in a screw cap tube and held at 37^oC. At 24 hour intervals, the saline was removed and replaced with an equal volume of saline. The eluents were stored at 4^oC until assayed for total protein. An equal volume of microspheres without any antigen were prepared and tested at the same time. Eluents were tested for protein content using the biochinic acid microassay (Pierce Chemical Company, Rockford, IL, USA) using BSA as the standard as previously described (7).

Animals - Ten-twelve week old female BALB/c mice from the Purdue University Department of Biological Sciences research animal facility were used in this study. Mice were housed in the biological science building and provided with food and water ad libidum. Each experimental group was housed in a separate cage. Use of the animals was approved by the institutional animal care and use committee.

Inoculation of mice - Four groups of 3 mice each were used: The groups received one of the following vaccine regimens: 1) ovalbumin (OVA) in microspheres orally, 2) OVA in microspheres subcutaneously (SC), 3) microspheres (no antigen) orally, and 4) OVA in complete Freund's adjuvant (CFA) SC. Mice were inoculated at 0 and 3 weeks with the antigen and route indicated. Each dose administered consisted of a total volume of 100 µl. Microspheres were suspended in sterile water and administered by oral feeding needle directly into the stomach of each mouse. Subcutaneous inoculations were administered in the nape of the neck.

Collection of samples for immunoassay. Serum was separated from blood obtained prior to inoculation by aspiration from the periorbital plexus from each mouse. At week 4, each mouse was euthanized and blood obtained by exsanguination from the brachial plexus. The spleen from each mouse was removed and placed immediately into chilled RPMI-1640 tissue culture medium (Sigma Laboratories, St. Louis, MO, USA). Each spleen was macerated in a petri dish to release lymphocytes. The macerated tissue was placed in a conical centrifuge tube on ice and large cellular debris was allowed to settle for 10 minutes. Mononuclear cells (primarily lymphocytes) were recovered from the suspension above the cellular debris. These cells were used in ELISPOT assays.

Immunological assays - *ELISA*. Serum was assayed for IgG, IgM, and IgA by ELISA. ELISA was performed using methods previously described (8). Briefly, OVA (Grade V, Sigma, St. Louis, MO, USA) was dissolved in phosphate buffered saline at 10 µg/mL. Fifty microliters of this solution was placed in each well of a 96 well polystyrene microtiter plate (Immulon 2, Dynatech Laboratories, Chantilly, NY, USA) and placed at $4^{o}C$ overnight. The plate was washed 3 times with PBS containing 0.5% Tween (PBS-tween), and 50 µl of a 0.1% bovine serum albumin (BSA Fraction V, Sigma, St. Louis, MO, USA) placed in each well and incubated at $37^{o}C$ for 1 hour. Fifty microliters of each serum sample were placed in wells in triplicate at a 1:25 dilution and incubated overnight at $4^{o}C$. The plate was washed as described above and goat anti-mouse antibody conjugated with horse radish peroxidase (Bethyl Laboratories, Montgomery, TX, USA) was added to each well and the plate was incubated at room temperature for 3 hours. The plate was washed and orthophenyldiamine substrate (Sigma, St. Louis, MO, USA) added to each well. The plate was incubated at room temperature for 30 minutes and concentrated H_2SO_4 added to each well to stop color development. The optical density of each well was determined using a microtiter plate spectrophotometer (Titertek, Multiscan, ICN/Flow Laboratories, Costa Mesa, CA, USA) at a wavelength of 490 nm.

ELISPOT Assay. Spleen cells were tested for antibody secreting cells (ASC) specific for OVA using techniques described previously (9). Briefly, cells were washed 2 times in RPMI-1640 containing 20 mM hepes buffer, 10 % fetal calf serum, and penicillin and streptomycin antibiotics (complete RPMI). The cells were adjusted to 10^6 cells per mL and 100 µl added per well of a microtiter plate (Immulon 2, Dynatech Laboratories, Chantilly, NY, USA). One hundred µl of OVA at 100 μg/ml in sterile PBS had been placed in each well and the plate incubated overnight at 4°C. A 0.1% BSA solution was used to block unbound areas on the plate as described above for the ELISA assay. The cells were incubated at 37°C and 5% CO_2 for 4 hours. Cells were washed off the plate and 75 μl of goat anti-mouse immunoglobulin was then placed in each well. The plate was then incubated overnight at 4°C. The plate was then washed and rabbit anti-goat IgG conjugated to alkaline phosphatase was place in each well and the plate was incubated at 20°C for 1 hour. The plate was washed and 100 μl of the substrate 5-bromo-4-chloro-3-indolyl phosphate mixed in agarose warmed to 50°C placed in each well. The plate was incubated at 20°C in the dark for 15 minutes to allow color to develop. The plate was then refrigerated for 2 hours and the resultant spots were counted using an inverted microscope .

Results and Discussion

Electron micrographs of alginate microspheres. Microspheres varied in size from 1-75 μm in diameter with 70% of microspheres less than 10 μm. Most microspheres containing OVA had a uniform grainy appearance when viewed by transmission electron microscopy as shown in Figure 1. Some had one to several vacuoles in within the center on cut section. It is not clear whether the vacuoles were air bubbles injected at the time of formation or were the result of fixation. Most microspheres were spherical in shape with a smooth surface as shown by scanning electron microscopy as shown in Figure 2. These results show that alginate containing a protein antigen can form solid microspheres. The microspheres used in this study were stored in $CaCl_2$ for 3 months. Stability for longer periods of time must be determined for practical vaccine application. The long term stability of microspheres in $CaCl_2$ or alternative liquids must be examined. One alternative for storage may be lyophilization. We have lyophilized alginate microspheres and recovered viable bacteria (unpublished results). However, the integrity of such microspheres has not been determined.

Elution of ovalbumin from microspheres. The total load of OVA in the microspheres was eluted over 5 days as shown in Figure 3. No OVA was determined until day 4 when 44% was eluted. By day five 99% of the OVA was eluted from the microspheres. These microspheres were tested in this manner three times with nearly identical results. These microspheres were 2 weeks old at the time of the study. Release of materials from microspheres at various time intervals will be needed to determine long term stability for vaccine applications. The release profile of different batches of microspheres must also be performed to determine the variability in the release profile.

Immunoassays - The OVA specific IgG and IgA antibodies were quantified by determining the increase in optical density from pre to post immunization (Table 1). There was a statistically significant increase ($p \leq .001$) in OVA specific serum IgG

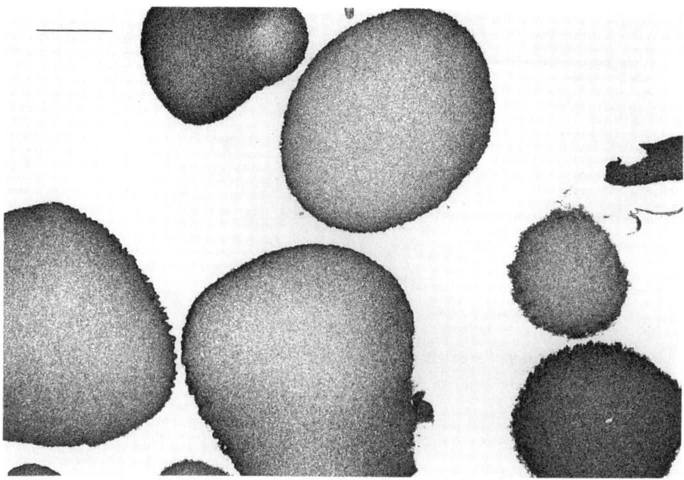

Figure 1. Transmission electron micrograph of alginate microparticles. Particles are mostly spherical in shape with intact surface and uniform consistency on cut section suggesting more or less uniform polymerization through the sphere. Microparticles produced as described in text. Magnification 4000 X. Reference bar equals 5 μm.

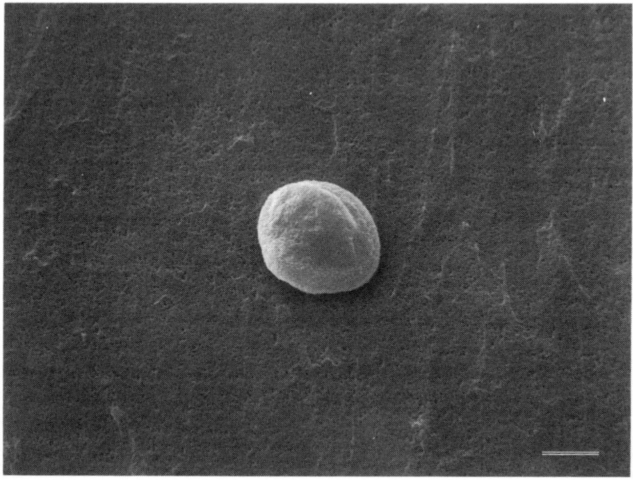

Figure 2. Scanning electron micrograph of alginate microparticle. Particle is spherical in shape with an intact surface. Particles produced as described in text. Magnification 3000 X. Reference bar equals 5 μm.

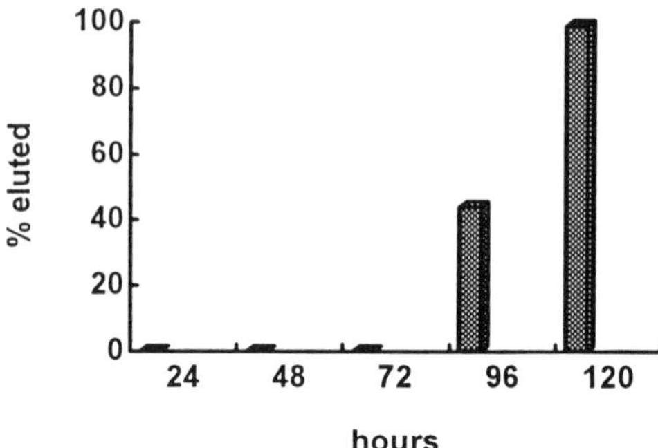

Figure 3. Daily total of ovalbumin eluted from sodium alginate microspheres. Results represent 3 aliquots run in triplicate on 3 different samples of microspheres.

for each group of mice inoculated with OVA regardless of the route of administration compared to the sham vaccinated group that received an oral inoculation of microspheres without antigen. Serum IgA specific for OVA was only detected in mice inoculated SC with OVA in microspheres or in CFA. The number of OVA specific ASC per 10^6 spleen cells is reported also in Table 1. Each OVA inoculated group had a statistically significant number of ASC compared to the sham vaccinated group ($p \leq .001$). Similar numbers of ASC were detected in the spleens of mice inoculated SC or orally with OVA in alginate microspheres with mice inoculated with OVA in CFA having the greatest mean number of ASC.

Table 1. Increase in anti-ovalbumin serum IgG and IgA activity, and antibody secreting cells specific for ovalbumin in mice

Group (Inoculation)	**Change in (IgG) Optical Density**	**Change in (IgA) Optical Density**	**ASC/10^6 cells**
OVA in microspheres/orally	340 ± 315	0	38 ± 10
OVA in microspheres/SC	230 ± 42	152 ± 47	52 ± 51
OVA in CFA/SC	380 ± 28	214 ± 112	95 ± 103
Microspheres/orally	0	0	0

OVA = Ovalbumin SC = subcutaneously CFA = complete Freund's adjuvant
ASC = antibody secreting cells
Mice inoculated with OVA had significantly increased antibodies compared to sham (microspheres/orally) mice as determined by analysis of variance $p < .001$.

In the present study, the oral or subcutaneous administration of ovalbumin encapsulated in alginate microspheres were equally effective in stimulating a systemic immune response in mice. The immune response of mice inoculated with OVA administered in microspheres suggests that alginate microspheres may possess adjuvant qualities since the immune response was comparable to that seen in mice inoculated with OVA in CFA. The lack of IgA in serum of mice inoculated orally with OVA in microspheres is not clear. It may be that the immune response was below our detection level or that memory cells were primarily induced that were detected in the ELISPOT assay. The ELISPOT was performed to detect all antibody specific cells so it is not known what proportion of these cells were specific for IgA. In future studies it would be desirable to determine the isotypes of lymphocytes elicited by this method of immunization.

It is interesting that both orally and subcutaneously administered alginate microparticles induced similar immune responses in mice. The oral administration of antigens would eliminate the handling of individual animals required with injectable vaccines as well as injection site reactions that can reduce the value of carcasses and hides in feedlot animals. For human vaccines, it would also increase compliance by patients who have a fear of needles. This is especially important for vaccines that must be given several times for optimal immunity. Lyophilized or temperature stable oral vaccines could be most practical in situations where many people must be vaccinated with limited personnel and limited access to refrigeration as in developing countries.

Ovalbumin is notoriously poorly immunogenic when administered orally. Incorporation of more potent antigens in alginate microspheres may stimulate an even greater immune response. An alternative explanation for the lack of detection of IgA in serum may be the relatively short time period of the experiment. For some antigens, it may take a longer for antibodies to be produced in sufficient quantity to be detected. In some studies in rodents, no detectable antibodies were detectable for up to 6 months after oral inoculation of microspheres (10).

Conclusions

Currently there is growing interest in producing microparticle vaccines for use in single dose parenteral as well as oral vaccines. Orally administered antigens must be able to retain immunogenicity following ingestion. Microspheres offer a unique way to both protect antigens from the adverse conditions of the gastrointestinal tract while also providing an easily administered delivery system. In addition to these advantages, microspheres provide a means of efficiently delivering antigen to the Peyer's patches, the lymphoid tissue of the gastrointestinal tract. It now appears that microspheres less than 10 μm in size are taken up most efficiently. Alginate microparticles can easily and economically be produced as spheres in this size range. Although alginate particles have been investigated for incorporation of cells to produce biologically active particles in vivo, no one has investigated their use for the delivery of oral vaccines. Our results are among the first to demonstrate that alginate microparticles release antigen in vivo in a manner to stimulate an immune response. These results support a recent study that showed that alginate microspheres containing antigens of a respiratory bacterium administered orally through the water can induce protective immunity in rabbits challenged later by an intranasal dose of viable bacteria(11).

Alginate microspheres show promise as a means of encapsulating and delivering antigens orally to animals. Further studies will be needed to determine usefulness of this polymer for use in humans as well. The oral or subcutaneous administration of ovalbumin encapsulated within sodium alginate microspheres was as effective in stimulating an immune response as ovalbumin combined with conventional adjuvants and administered by parenteral injection. This study indicates the need for further investigations on the efficacy of alginate microspheres in the oral administration of antigens to stimulate immunity at mucosal sites.

Acknowledgments

The authors wish to thank Ms. Deborah Van Horn technical assistance with preparation of electron microscopic samples.

Literature Cited

1. O'Hagan, D.T.; Rahman, D.; McGee, J.P.; Jeffery, J.; Davies, M.C.; Williams, P.; Davis, S.S.; and Challacombe, S.J. *Immunol.* **1991**; 73, 239-242.
2. Eldridge, J.H.; Hammond, C.J.; Meulbroek, J.A.; Staas, J.K.; Gilley, R.M.; and Tice, T.R. *J Control Rel.* **1990**, 11, 205-211.
3. Lim, F.; and Sun, A.M. *Science* **1980,** 210, 908-910.
4. Bano, M.C.; Cohen, S.; Visscher, K.B.; Allcock, H.R.; and Langer, R. *Biotechnol.* **1991**, 9, 468-471.
5. Kwok, K.K.; Groves, M.J.; and Burgess, D.J. *Pharm Res.* **1991**, 8, 341-344.
6. Bullock, G.R. *J. Micros.* **1984**, 133, 1-15.
7. Nugent, D.W.; Doyle, C.; and Fottrell, P.F. *Clin Chem.* **1987**, 33, 1671.
8. Voller, A.; Bidwell, D.E.; and Bartlett, A. In: *The enzyme linked immunosorbent assay;* Nuffield Laboratories of Comparative Medicine, The Zoological Society of London: Regent's Park, London, **1979**, pp. 23-39.
9. Czerkinsky, C.; Nilsson, L.A.; Nygren, J.; Ouchterlony, O.; and Tarkowski, A. *J Immunol Meth.* **1983**, 65, 109-121.
10. Richardson, J.L.; McGee, J.P.; Gumaer, D.; Potts, B.; Wang, C.Y.; Koff, W.; and O'Hagan, D.T. *Proc Int Sympos Control Rel.* **1994**, 21, 69-870.
11. Suckow, M.; Bowersock, T.L.; and Park, K. *Proc Int Sympos Control Rel.* **1994**, 21, 843-844.

RECEIVED January 3, 1996

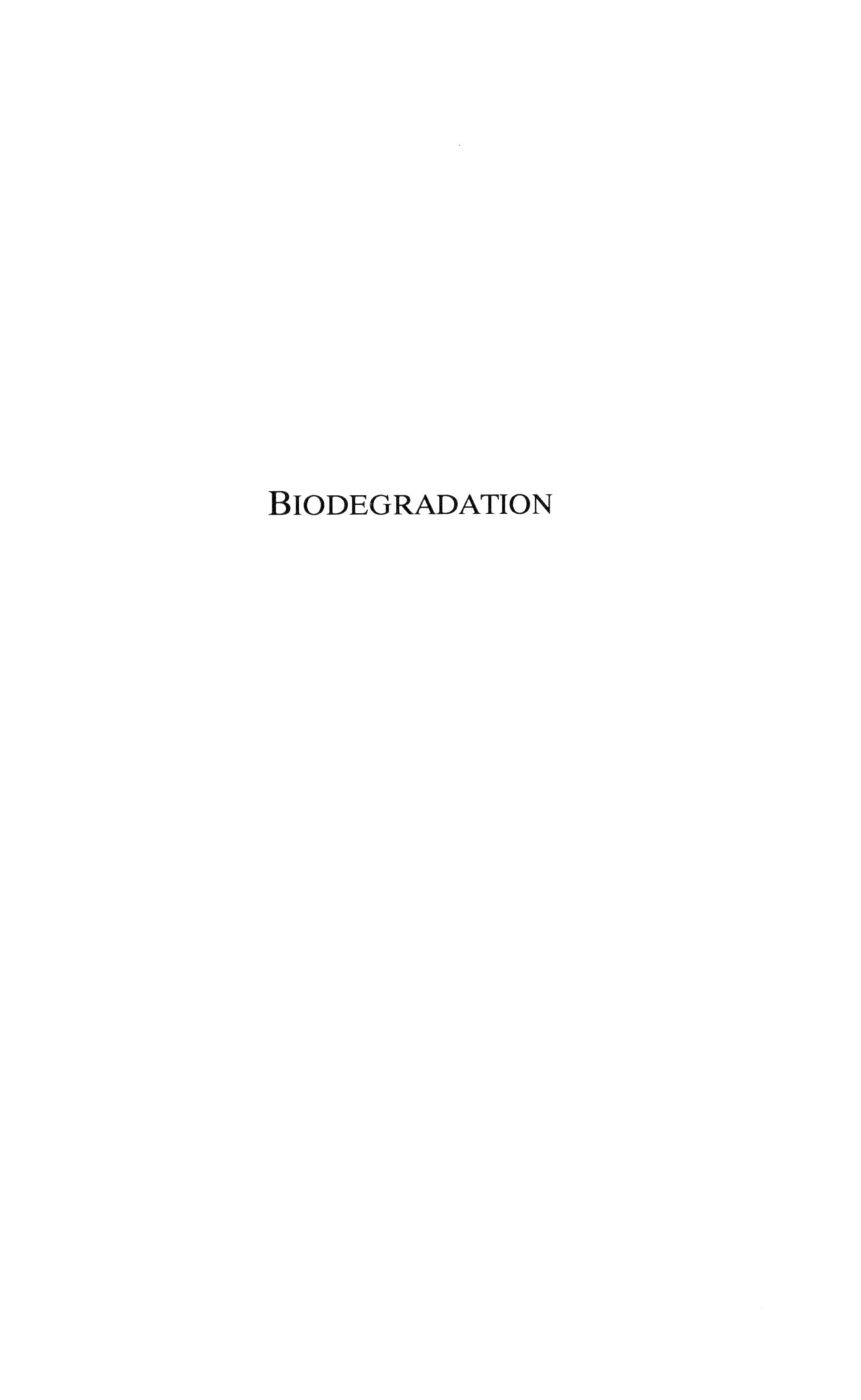

BIODEGRADATION

Chapter 7

Activities of Extracellular and Intracellular Depolymerases of Polyhydroxyalkanoates

L. J. R. Foster[1,2,3], R. C. Fuller[1], and R. W. Lenz[2]

[1]Department of Biochemistry and Molecular Biology and [2]Department of Polymer Science and Engineering, University of Massachusetts, Amherst, MA 01003

A diverse family of microbially produced polyesters, polyhydroxyalkanoates, PHAs, are receiving considerable attention as biodegradable, biocompatable thermoplastics. The rates and mechanisms of biodegradation of these biopolymers are, therefore, of great interest. Polymer degradation is catalyzed by depolymerase enzymes. In this chapter we review the activities of both extracellular PHA depolymerases secreted by a variety of microorganisms, and intracellular depolymerases associated with polymer inclusion bodies isolated from the PHA producing microbes. A number of examples are used to compare and contrast different enzymes and emphasize the considerably faster biodegradation of these novel biopolymers compared to more conventional, commercially available synthetic polymers.

In 1925, Maurice Lemoigne, a microbiologist working at the Lille branch of the Pasteur Institute, began a series of reports detailing his investigations into the refractile granule-like inclusions he had observed within the cytoplasm of the bacterium *Bacillus megaterium* (*1-4*). Lemoigne demonstrated that these inclusions were composed of a polyester, poly-3-hydroxybutyrate, (PHB), with the empirical formula of $(C_4H_6O_2)_n$. His observations regarding this biopolymer and its importance can only be considered as 'ahead of their time'. Indeed, over the next three decades PHB was studied primarily as an academic curiosity.

In the 1950's interest in this novel material increased, so that by the turn of the decade W.R. Grace & Co. began production of PHB for commercial evaluation and subsequently developed several articles such as sutures and prosthetic devices (*5*). In the 1970-80's ICI and their division, Marlborough Biopolymers (now Zeneca), in an effort to find naturally occurring substitutes for synthetic plastics, began production of PHB and its copolymers of PHB and poly-3-hydroxyvalerate, (PHBV, trade name Biopol), using *Alcaligenes eutrophus* (*6-9*). In 1990, Wella, in conjunction with Marlborough Biopolymers, released on trial in Germany, what was marketed as the world's first totally biodegradable product: a biodegradable shampoo in a Biopol container, with edible dye for labelling (*10*) (ICI personal communication 1989). In the past five years PHB and Biopol based plastics have

[3]Corresponding address: 42 Kingston Lane, West Drayton, Middlesex UB7 9EB, England

0097–6156/96/0627–0068$16.25/0

been used in the production of a number of commercial devices; interestingly, these are predominantly from Japan. At the time of this book going to press, the ubiquitous McDonalds is developing a method of coating paper with Biopol, thus preventing the oils from its various culinary delights despoiling the consumer, whilst maintaining the biodegradability of its packaging (Langert R. Personal communication 1994). On this, the 70th. anniversary of Lemoigne's discovery, interest in PHB and bacterial polyesters can be said to have truly reached global proportions!

Polyhydroxyalkanoates...PHAs

PHB and PHBV are part of a large family of bacterial polyesters called polyhydroxyalkanoates (PHAs). PHAs are natural polymers produced by a wide variety of microorganisms, and function as intracellular reserves of carbon and energy as well as ion sinks (*11-13*). These polymers are receiving increased attention for possible applications as biodegradable, melt processable polymers. These thermoplastic biopolymers can vary from rigid brittle plastics to more flexible materials with good impact properties to strong elastomers, depending on the size of the pendant group R in the polymer composition shown below: (*14-16*).

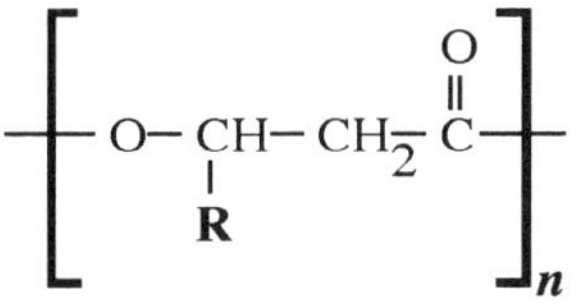

$$\left[-O-\underset{\mathbf{R}}{\underset{|}{CH}}-CH_2-\overset{O}{\overset{\|}{C}}- \right]_n$$

When bacterial cells are grown under conditions of excess carbon substrate and a limiting essential nutrient, the PHA formed can account for up to 90% of the dry cell weight (*17-20*).

Recently the principles of microbial PHB synthesis have been utilized and adapted to produce other bacterial polyesters with side chains possessing various lengths and a variety of functional groups (*21-30*). To date, there are approximately 70 different types of PHAs. This number continues to increase as we expand our knowledge about PHAs and the microorganisms continue to amaze us in their versatility to utilize the most unlikely of carbon based substrates. PHAs have been categorized into three groups as shown below and in table 1:

(a) PHA(SCL) - those possessing side chains of relatively few carbon numbers such as PHB (C4) and PHBV (C4-C5).

(b) PHA(LCL) - PHAs possessing relatively long, or as is sometimes referred to, medium, side chain lengths, eg. polyhydroxyoctanoate (PHO - C8) and polyhydroxynonanoate (PHN - C9).

(c) PHA(FSC) - This group contains those LCL-PHAs with functional groups in the side chain. eg. poly-3-hydroxyundecanoate (PHU), having a carbon-carbon double bond in the side chain and poly(-3-hydroxy-5-phenylvalerate) (PHPV), possessing a phenyl group in the side chain.

The exact composition of PHAs produced during fermentation procedures are dependent upon numerous conditions, primarily the bacterial species and carbon feedstock; for example, Brandl et al. report on the production of PHO with a composition of C6 - 14%, C8 - 78% and C10 - 8% using *Pseudomonas oleovorans* and octanoic acid as a feedstock (*21*). However, Gagnon and coworkers using similar conditions report the PHO produced as having a composition of 11, 84 and 5% respectively (*31*). More recently, Foster et al. show a change in the composition to C6 - 7.5% and C8 - 92.5% as a result of sequential feeding (Foster et al in progress).

Table 1: Classification and Examples of Microbially produced PHAs

Name	Composition **R** Groups	Polymer	PHA Type
PHB	$—CH_3$	C4 Homopolymer	PHA (SCL)
PHO	$—(CH_2)_2—CH_3$ $—(CH_2)_4—CH_3$ $—(CH_2)_6—CH_3$	C6, C8 & C10 Terpolymer	PHA (LCL)
PHOU	$—(CH_2)_x—CH_3$ $—(CH_2)_x—CH=CH_2$	A Copolymer of PHO & Polyhydroxyundecanoic acid, 17% Unsaturated.	PHA (FSC)
PHPV	$—CH_2—C_6H_5$	Homopolymer	PHA (FSC)

Depolymerases and Biodegradation.

In this chapter we provide a concise review of PHA depolymerase activity. One of the primary factors stimulating interest into PHAs is their potential biodegradability. There is considerable confusion in the literature as to the exact definition of the term 'biodegradable'. Taylor has defined biodegradation as 'the breakdown of polymeric materials by living organisms or their secretions' (*32*) and this view is similar to those expressed by Potts (*33*), Williams (*34*), Griffin (*35*) and Kopet et al. (*36*). In contrast Pitt et al. (*37*) and Gilbert et al. (*38*) define biodegradation as 'the simple hydrolytic breakdown of polymeric materials'. Doi (*14*) states it as 'the simple hydrolytic and enzymatically induced hydrolytic degradation processes occurring in polymers', whilst Lenz defines it as 'chemical degradation caused by biochemical reactions, especially those catalyzed by enzymes produced by microorganisms under either aerobic or anaerobic conditions' (*39*). Further definitions by Gilding (*40*), Zaikov (*41*) and others all vary according to the particular work concerned. To complicate matters further, the terms 'bioerosion' and 'bioerodible' are also widely use when dealing with drug release devices, as in the papers by Langer et al. (*42*) and Heller (*43*). One shudders to think of the variations that exist in the political and legal definitions.

Biodegradation can be of two kinds: enzymatic catalyzed hydrolysis or simple chemical hydrolysis. Enzymatic degradation can be further divided into two categories: intracellular and extracellular; both break the relatively large polymer chain into smaller, more manageable sub-units. Endo-depolymerases do this by randomly attacking the polymer chain backbone at any ester linkage; this results in random chain cleavage and a concomitant decrease in polymer molecular weight. Alternatively, exo-depolymerases specifically attack the polymer chain ends. In this instance the immediate effects on the molecular weight are less noticeable, as it is characterized by the release of terminal units such as monomers, dimers and trimers, as shown in Figure 1.

Extracellular PHA Depolymerases

Environmental Biodegradation. The hydrolytic degradation of PHB and PHBV samples is influenced by a number of factors, including temperature, pH, hydroxyvalerate content and sample structure and production. The surface area to volume ratio of samples has been shown to play a primary role in influencing their hydrolytic degradation (*44-46*). Similarly, blending with various additives can significantly alter sample degradation, although this also affects sample properties (*47-51*). By these means it has been suggested that it may be eventually possible to tailor the degradation profiles of samples to suit individual needs. This is of particular interest for PHB and PHBV based devices used as biomedical and surgical implants.

The factors affecting PHB and PHBV hydrolytic degradation also influence their biodegradation in the macroenvironment of the ecosphere. Environmental biodegradation of PHAs can occur by a combination of enzymatic and chemical hydrolysis. Similarly, other environmental factors such as mechanical stresses, weathering and non-biological erosion also facilitate degradation.

The biodegradation of PHAs has been examined in a variety of environments including their burial in soils, composts etc. and immersion in fresh and salt water. These studies are important to demonstrate the suitability of these novel thermoplastics in comparison to the more conventional synthetic plastics such as polyethylene (PE) and poly(vinyl chloride) (PVC). In most instances degradation was monitored by gravimetric weight changes; however, the practical problems in sample removal can have considerable repercussions in experimental accuracy.

PHA Degradation in Soil and Compost. Studies have shown that after twelve months burial in soil, solvent cast films of polycaprolactone (PCL), bacterial cellulose (BC) and PHBV were biodegraded, whilst poly(vinyl chloride) (PVC), poly(methyl-L-glutamate) (PMLG) and high density polyethylene (HDPE) were not. In each case, examination of the PVC, PMLG and HDPE films revealed clear, clean samples similar to new, undegraded films. The PHBV specimen exhibited numerous holes, indicating a mainly bacterial biodegradation. In contrast, the PCL samples revealed the presence of many hyphae, suggesting a predominantly fungal degradation (*52*).

The environmental biodegradation of injection moulded PHB and PHBV samples in five different soil types: hardwood, pinewood, sandy, clay and loam, have also been investigated. Weight loss of the plastics occurred linearly but varied considerably from 0.03 to 0.64% per day depending upon soil conditions, incubation temperature and plastic composition (Figure 2). From these samples Mergaert et al. reported the isolation of two hundred and ninety-five dominant microbial strains capable of degrading PHB and PHBV, including *Bacillus*, *Streptomyces*, *Aspergillus* and *Penicillium* (*53*).

In a leaf compost, compression moulded samples of PHB and PHBV showed a significantly faster degradation rate compared to samples of PCL blended with low density polyethylene (PCL/LDPE) and low density polyethylene and polypropylene blended with corn starch, (St/LDPE and St/PP, respectively). Undegraded films of PHB/V were flat and featureless when viewed by SEM. However, samples exposed to decaying leaves for 1 month exhibited visible signs of fungal colonization; after 2 months exposure, washed samples revealed deep grooves and pits. One hundred and ninety-three bacterial and twenty-eight fungal species were isolated from the PHBV samples, of which only twenty-two of the

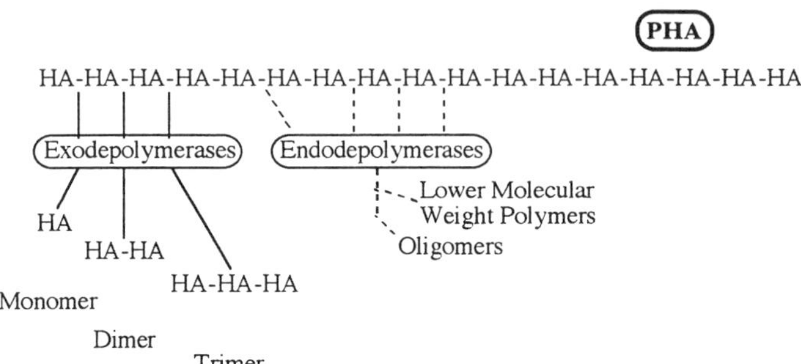

Figure 1. Endo and exo PHA depolymerase activity.

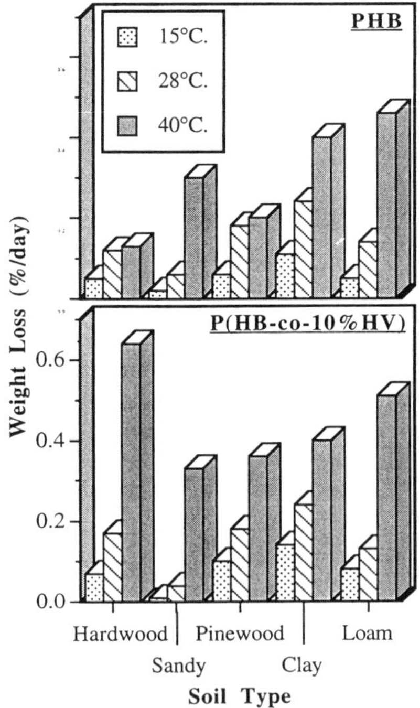

Figure 2. Degradation of PHB and PHBV samples in different types of soils and incubation temperatures. (Modified from ref. 53)

fungal isolates demonstrated depolymerase activity. Polymer degradation was considerably greater when a 'mixture' of these microorganisms was used in preference to singlc isolate studies (*54*). In contrast, PHBV samples in household composts appear to degrade by the action of both bacterial and fungal depolymerases (*55*).

The predomination of bacteria or fungi as the main culprits for polymer degradation appears to be subject to a variety of influencing factors, most noticeably temperature. In soil at 15°C, the microflora consisted mainly of Gram-negative bacteria and strptomycetes; as the soil temperture was increased to 28°C, Gram-positives and some moulds were also isolated. At 40°C, fungi and streptomycetes predominated (*56*). Since temperatures in active composts frequently exceed 60°C, it is unsurprising that in this environment fungi have been found to be the primary PHBV degraders.

Surface rugosity and the 'gloss factor' have been used to measure the biodegradation of solvent cast films of various PHBV copolymers blended with poly(ethylene adipate), poly(ethylene succinate) and PCL, buried in different soils around the world (*57*). This technique provides a semiquantitative assessment of film erosion within a relatively short period of time (7-21 days). Similarly, the optical density of PHBV based films has also been utilized to measure the initial stages of biodegradation where gravimetry is less accurate. The biodegradation in activated sewage sludge of these PHBV samples blended with cellulose acetate butyrate and cellulose acetate propionate has been investigated by Gilmore et al. (*58*). They reported rapid degradation rates, with unblended PHBV copolymers showing complete degradation (t_{100}) after approximately 65 days. Extrusion blow-moulded samples were reported to take significantly longer (60 weeks), similar samples in soil took 75 weeks. Complete degradation was achieved in only 6 weeks when placed in anaerobic sewage (*59*). 50μm thick packaging films are reported to totally degrade within a 2 week period in anaerobic sludge, 7 weeks in aerobic sewage, 10 weeks in soil (25°C) and 15 weeks in seawater (15°C) (*60*).

The molecular weight changes of PHB/V samples during environmental biodegradation in soils and composts match those of similar samples subjected to simple chemical hydrolysis (*54,58*). This, together with gravimetric data and qualitative observations, supports the conclusion that environmental degradation of these biopolymers proceeds via the surface by extracellular depolymerase action and throughout the polymer sample by simple chemical hydrolysis.

PHA Degradation in Fresh and Saline Waters. Plastics and plastic waste tend to find their way into a wide variety of environmental niches. Hundreds of thousands of tons of plastics are annually discarded into marine environments and are responsible for the death of vast numbers of large marine animals (*61-63*).

Injection moulded PHB homopolymer and PHBV copolymer samples with 10 and 20% hydroxyvalerate (HV) content showed less than 10% degradation after approximately 26 weeks immersion in a fresh water pond and canal. In comparison, similar homopolymer and copolymer samples submersed in a marine environment exhibited around 30 and 40% degradation, respectively (*56*).

The degradation in seawater of solvent cast films, melt-extruded plates and melt-spun monofilament fibres of PHB, PHBV and P(3HB-4HB) have been investigated by Doi et al. (*64*). After 3 weeks exposure in the marine environment, the PHBV polymer films, originally 50-100μm in thickness, lost 13-16μm. Similar P(3HB-4HB) samples lost 31-33μm in 8 weeks during the months of January to March, when the water temperature was 14±2°C, and 55-60μm in August to October (24±3°C). After 8 weeks submersion, the monofilament fibres, having an original diameter of 260μm, showed 65% degradation and a complete loss of mechanical integrity. Electron microscopy showed no noticeable changes within the films; also, there was little change in their molecular weights. These results are consistent with a primarily enzymatic degradation. Control samples in pretreated seawater showed little change, indicating a lack of simple chemical hydrolysis. Liddel reported that extrusion blow-moulded samples of PHB in seawater took 350 weeks for complete degradation. However, in the comparatively specialized, and mainly anaerobic environment of estuarine sediments, similar samples took only 40 weeks (*59*).

The commercial trial of the biodegradable Wella shampoo bottle in 1990, led to the frequently quoted research of Brandl and Püchner; who investigated the biodegradation of the bottles in a freshwater lake (*65*). They found that at 85m depth, the bottles degraded at a rate of approximately 0.2% per week. From these studies Brandl and Püchner predicted a life span for the bottles of around 10 years, depending upon depth, temperature, oxygen content and hydrostatic pressure (Figure 3). The Biopol bottles required only 15 weeks in a compost pile for complete biodegradation. However, significant degradation was only observed after 40 weeks in a commercial landfill (*66*).

Depolymerase Isolation. It becomes clear from the environmental biodegradation studies that PHB and PHBV based plastics readily degrade under a wide variety of conditions. The results show that even under environmental extremes, Biopol items maintain their biodegradability. A lack of standardization techniques for the evaluation of biodegradation makes it difficult to compare results. It is hoped that this situation may soon be remedied, as it has been suggested that Biopol may be used as a standard reference for the correlation of environmental and laboratory testing of potentially biodegradable materials (*67-68*). However, it is, as yet, quite impossible to predict, with any degree of accuracy, the environmental biodegradation of PHA based plastics.

Microorganisms that secrete extracellular PHA depolymerases are usually detected and isolated by the use of clear-zone techniques. In this method, a fine colloidal suspension of the PHA in distilled water is mixed with agar, poured into a sterile petri dish and permitted to solidify. Environmental samples such as soil or water are inoculated onto the plates and organisms capable of utilizing the polymer carbon produce colonies with distinct clear halos in the whitish polyester layer (Figure 4). PHA degraders have been isolated by this method and used to produce liquid cultures, the supernatant of which may provide a concentrated source of the depolymerase(s) (*53,55,56,64,69-71*).

Despite the increasing number of different types of PHAs, environmental testing appears limited to those with short side chains, namely PHB/V. This is primarily due to the relatively small yields of the long/functionalized side-chained PHAs compared to the more readily available PHB/V; which can also be purchased commercially. The use of clear zones provides a means to direct initial investigations into the potential biodegradability of these more unusual PHAs. Ramsay et al. used the technique to isolate a number of LSC-PHA degraders from a compost (*70*); similarly, Schirmer et al. have also used this method for the initial isolation of the PHO degrader *P. fluorescens* from activated sludge (*72*).

A modification of the clear zone technique provides quantitative information as to the factors affecting depolymerase secretion. In this adaption, an agar base layer containing growth media with essential nutrients and various carbon sources was utilized to ensure colony growth irrespective of the organism's depolymerase ability. *P. maculicola*, isolated from soil, was grown in a liquid preculture with 20mM of a desired carbon source; prior to the onset of the stationary growth phase, samples were removed and diluted to the same optical density, thus ensuring similar live cell quantities. Fixed volumes were then inoculated onto the clear zone plate.

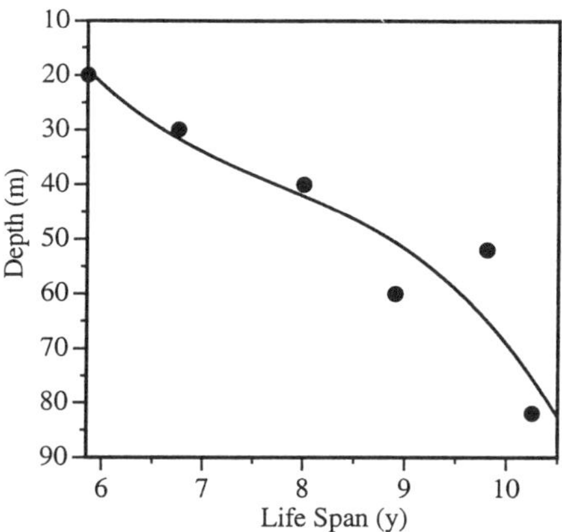

Figure 3. Estimated life spans of Biopol biodegradable shampoo bottles in a fresh water lake, as a function of depth. (Modified from ref. 66)

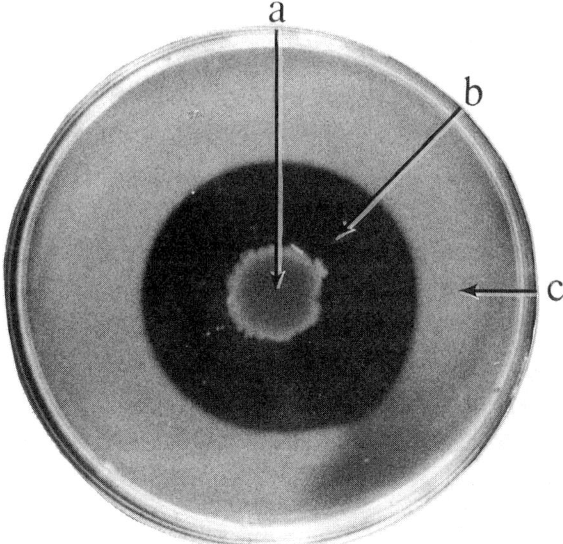

Figure 4. Photograph of clear zone plate, illustrating: **a:** *P. maculicola* colony growth, **b:** Clear halo of depolymerase activity and **c:** Undegraded PHA overlay.

The area of clearing in the polymer overlay can be used to calculate the degree of polymer degradation (*73*).

The extracellular depolymerase system of *P. maculicola* was found to degrade a number of LSC-PHAs including PHO, PHN and copolymers thereof, and PHOU. Degradation was also affected by the source and concentration of secondary carbon in the base layer, with a simultaneous utilization of both the polymer and base layer carbon (Figure 5). There was also evidence to suggest that the bacteria initially secrete a basal amount of enzyme to 'detect' degradable polymers. The *P. maculicola* enzyme system was unable to degrade SCL-PHAs, (PHB and PHBV copolymers), nor was it capable of degrading the functionalized polymer PHPV. The colony growth for *P. maculicola* grown on the PHPV overlay was substantially reduced compared to that grown on the PHB. It has been suggested that the phenyl group was either toxic or that it physically interfered in the organisms biochemical pathways as a result of its bulky size (*73*).

P. fluorescens is also reported as being unable to degrade SCL-PHAs (*72*), whilst the extracellular depolymerase from *Comamonas testosteroni*, *P. lemoignei* and other SCL-PHA degraders displayed no depolymerase activity towards the LCL-PHA, PHO (*74-76*). This division between organisms capable of degrading short and long side chain PHAs, is also reflected in those microorganisms that produce PHAs. There are no reports of microorganisms which can synthesize PHAs and produce extracellular depolymerases. Similarly, it was considered that those microbes which secrete SCL-PHA depolymerases were unable to degrade LSC-PHAs, and vice-versa. This apparently distinct division between PHAs with long and short side chains has been challenged by Brandl et al. who report the synthesis by *Rhodospirillum rubrum* of a terpolymer of PHBV and poly-3-hydroxyhexanoate (PHHex - C6) (*77*). A copolymer of PHB and PHHex has recently been produced by Shiotani and Kobayashi (*78*). Shimamura et al have recently reported that films of this copolymer were apparently degraded by the extracellular depolymerase from *Alcaligenes faecalis* (*79*).

Depolymerase Activity and Kinetics. The depolymerase from *P. lemoignei* has been isolated and chromatographically purified into four isoenzymes; A1, A2, B1 and B2; each possessing molecular weights of 54 and 58K for the A and B fractions respectively. All the isoenzymes hydrolyzed PHB to monomer and dimer units, whilst both the A isoenzymes produced small quantities of trimers. In contrast, timeric units were the principle product of PHB degradation by the B isoenzymes. Both the A and B enzymes showed no activity towards the dimeric ester (*80-81*). Genetical studies by Briese et al. report *P. lemoignei* as having at least five depolymerases (Briese et al. personal communication ISBP Montreal 1994). This identification of small oligomers as the primary degradation products from PHB tends to suggest an exo-attack.

The extracellular depolymerase PHB depolymerase from *A. faecalis* was found to act as an endo-type hydrolase against the pentamer and higher oligomers of 3-hydroxybutyric acid. Shirakura et al. investigated the frequency of bond cleavage using radiolabelled oligomers, from trimer to octomer. These oligomers were treated with the *A. faecalis* depolymerase and the radioactivity of each product formed during the initial stages of degradation was determined. In the case of the trimer and tetramer, only the second ester linkage from the hydroxy terminus was cleaved. In the case of the pentamer, the third ester bond was also hydrolyzed, but at a lower rate than the frequency of second bond cleavage. In the higher oligomers, all the ester bonds, except for the final one at the hydroxyl terminus, were cleaved with various frequencies (*82*). These results tend to suggest that the *A. faecalis* depolymerase has endo-type capabiliites. More information on the endo and exo-depolymerase capability of *P. lemoignei* and other microbial systems has been derived from their activity against synthetic PHB with different tacticities (*83-86, 90*) .

Synthetic PHB differs from its microbial counterpart in the presence of both R and S optical units; this affects the material properties and hence, the biodegradability of the polymer. Microbially produced PHAs consist exclusively of R units. Figure 6a illustrates the degradation of PHB solvent cast films of varying tacticities by the *P. lemoignei* extracellular depolymerase system. The synthetic samples of racemic PHB showed weight losses of up to 80% of their initial mass. Exo-attack by the enzymes accounts for less than 50%, since the enzymes would be unable to penetrate the S stereoblocks. Such relatively high degradation rates therefore, have led the suggestion that the depolymerase system has some endo capability (*83*). Predominantly isotactic and atactic PHB polymers were hydrolyzed by the depolymerase, but syndiotactic polymers were hardly degraded.

Abe et al. using *P. pickettii* and *A. faecalis* provided similar evidence that degradation rates for PHB films with 68-92% isotacticites were greater than microbially produced PHB (Figures 6b and 6c). The highest rate was observed with PHB films possessing 76% isotactic diads and was approximately 7 times faster than that of bacterial PHB (*85*). Kemnitzer et al. investigated the enzymatic degradation of PHB stereoisomers by *P. funiculosum* and found that the degradation rates of PHB with 81-95% R content were lower than that of bacterial PHB; but with 67-77% R units the degradation rates were dramatically higher (*87*). Marchessault and coworkers reports that synthetic racemic PHB degraded by *P. lemoignei*, *A. faecalis* and *Aspergillus fumigatus* (Figure 6d) depolymerases were an order of magnitude less than microbial PHB (*83,84*). Experiments have shown that decreasing the crystallinity by the addition of S units into the polymer chain increases the degradability (*86,88,89*), but decreasing the length of R sequences decreases degradability (*86,87,90*). These opposing factors reach an optimal balance in isotactic racemic PHB; at higher isotacticities, the comparatively higher crystallinity has a greater effect than isotacticity and thus a negative influence is exerted on the polymer degradability. In syndiotactic racemic PHB the decreasing isotacticity has a corresponding increase in crystallinty, since both these factors have a negative effect on degradability, little degradation is observed.

These experiments emphasize the differences in activity between different extracellular depolymerases. However, the evidence suggests that all those investigated to date are capable of both endo and exo cleavage but express a relative preference for the latter.

Figure 7 illustrates the degradation of a solvent cast bacterial PHBV film by the *P. lemoignei* depolymerase system. The erosion observed is similar to that seen in samples from environmental testing. The A and B fractions of *P. lemoignei* depolymerases, exhibit a maximum degradation rate at pH 8 (*91*). Similarly, *A. faecalis* depolymerase works best at pH 7.5 (*92*), but *P. stutzeri* and a *Comamonas* species isolated from fresh water lakes, exhibited pH optimums of 9.5-10 and 9.4, respectively (*93,74*).

Mukai et al. have tested the degradation of solvent cast PHB films by a variety of isolated extracellular depolymerases (Figure 8). The differences in degradation rates shown by the various enzymes in Mukai's study, may be a result of different hydrophobicities. *P. lemoignei*, *P. pickettii*, *A. faecalis* and *C. testosteroni* were isolated from soil, air, activated sludge and seawater, respectively and revealed considerably lower hydrophobic coefficients than those of the *Comamonas* species and *P. stutzeri* (*93*). The high hydrophobicities of these enzymes may play an essential role in their efficient adherence to PHA surfaces.

Muller and Jendrossek have recently isolated a PHV depolymerase from *P. lemoignei* (*76*); this depolymerase and the A and B isoenzyme fractions of the PHB depolymerase system have similar molecular weights to the *A. faecalis* and *P.*

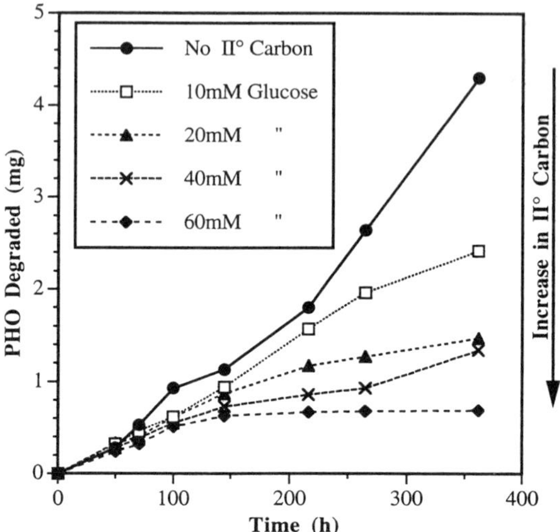

Figure 5. Variation in *P. maculicola* LCL-PHA depolymerase activity with time, as a function of changes in concentration of glucose as a secondary carbon source in the base layer. (Modified from ref.73)

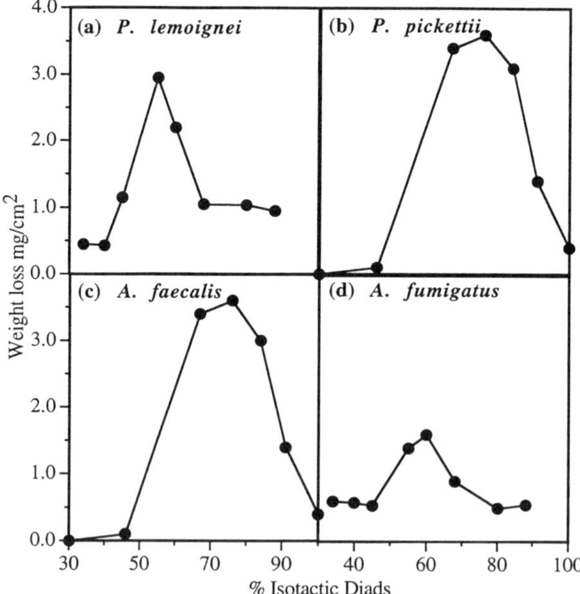

Figure 6. Degradation of synthetic PHB with changes in isotacticity: **b & c;** films incubated at 37°C for 3 hours with 2µg/ml of depolymerase. (Modified from Abe et al. *Can. J. Microbiol.* in press). **a & d;** films incubated at 30°C for b and 45°C for c, with 3.6 units of enzyme activity. (Modified from ref. 83)

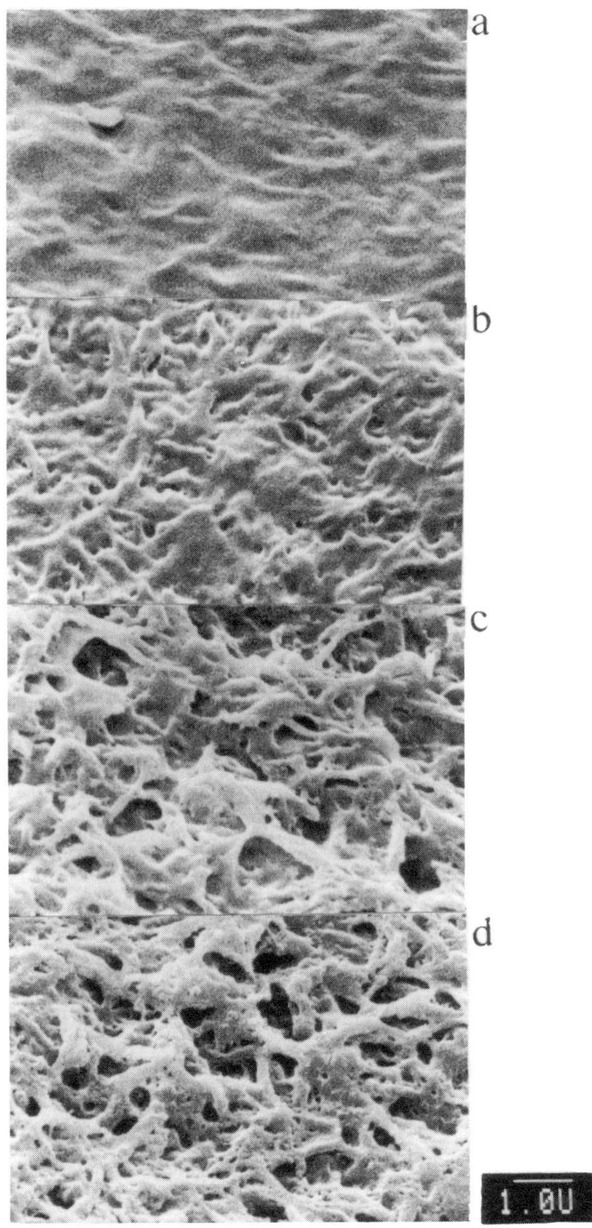

Figure 7. SEMs of solvent cast PHB films during degradation by *P. lemoignei* depolymerase: **a:** Undegraded, **b:** 15 minutes, **c:** 85 min, **d:** 150 min. (incubated at 30°C & pH 8).

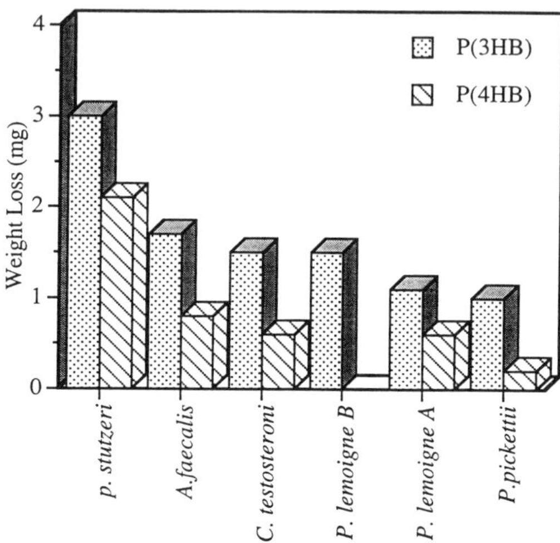

Figure 8. Degradation of PHB films by various extracellular depolymerases, films were incubated at 30°C & pH 7.4 with optimal enzyme concentrations. (Modified from ref. 93)

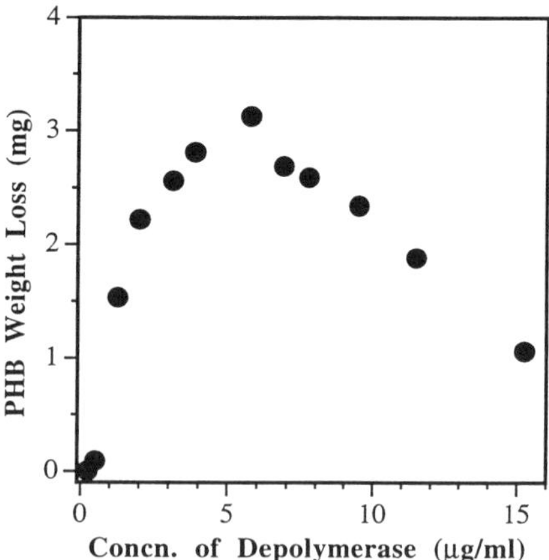

Figure 9. Variation in PHB degradation rates as a function of changes in concentration of *P. stutzeri* depolymerase, films were incubated at 37°C & pH 7.4 for 5 hours. (Modified from ref. 93)

stutzeri enzymes. Similarly, their activities are characterized by the inhibitory effects on degradation of non-ionic surfactants, such as Tween-20 and Triton X-100, and serine base inhibitors, such as PMSF (phenylmethylsulphonyl fluoride) (*75,76,92,93*). These results tend to indicate that the enzymes possessed hydrophobic binding sites and serine residues at their active sites. In contrast, Jendrossek et al report that PMSF had no effect on the depolymerase from the *Comamonas* species. This led the authors to conclude that this enzyme represents a novel type of PHB depolymerase (*74*). The same research group have also isolated a depolymerase from *P. fluorescens*, a LCL-PHA degrader; results from degradation experiments have shown that this depolymerase was also unaffected by PMSF (*72*). However, this was later contradicted by research into the primary structure of the depolymerase which showed the presence of serine groups at the active sites (*94*) (Jendrossek personal communication ISBP Montreal 1994). The relative rates of the forward and reverse hydrolysis reactions are controlled by the water content of the reaction mixture. To optimize enzyme activity lipases possess a 'flap' which protects the enzyme, when the enzyme detects the substrate the flap is 'opened' to reveal the active site (*95*) (Reyes personal communication, 1993). It can be suggested, therefore, that a similar structural mechanism exists in some of the PHA depolymerases. This would help to explain the serine dilemma that studies regarding *P. fluorescens* have raised. Indeed, further studies suggest that PHAs may also be degraded by various bacterial lipases. This may provide an explaination as to why environmental biodegradation proceeds relatively more quickly in multi micro-organism systems compared to single isolate studies.

Other factors which appear to vary between depolymerases is the effects of calcium and/or magnesium on degradation. PHB degradation by the *P. lemoignei* depolymerase system is increased by the presence of the calcium or magnesium salts (*75*); this is not the case with the *A. faecalis* enzyme (*92*). Such differences are also seen in the intracellular depolymerase studies.

Tanio et al. have investigated the degradation of a colloidal suspension of PHB by the extracellular depolymerase from *A. faecalis* and report a degradation rate constant of 13.3μg/ml (0.78μM) for PHB with a number average molecular weight of 17,000. This rate was reduced to 4.5μM for the enzyme activity against the trimeric ester, indicating a higher affinity towards the polymer (*92*). Other reports quote rate constants of 413μg/ml (*75*) for PHB and 65μg/ml and 77μg/ml for PHB and PHV respectively (*76*). The PHB depolymerases A and B from *P. lemoignei* are reported to have rate constants of 70 and 92μg/ml respectively (*91*). These results emphasize that standard Michaelis-Menton kinetics cannot be applied to heterogeneous enzymatic reactions such as PHA degradation. A kinetic model for PHA degradation has been proposed by Mukai et al.; evidence for this model is based on enzyme concentration changes. Figure 9 shows the change in the PHB degradation rate as the concentration of depolymerase from *P. stutzeri* was varied. The rate of degradation increased with an increase in depolymerase concentration; however, as the enzyme concentration continued to increase the degradation rate decreased (*96*) Similar trends have been reported for the depolymerases of *A. faecalis*, *P. pickettii* and *P. lemoignei*. (*96,97*). Initial results from intracellular depolymerase studies suggest similar trends (Foster et al. in progress). The presence of binding and catalytic domains in the depolymerases tends to indicate that enzymatic hydrolysis of PHAs is a two stage process: adsorption and hydrolysis. At relatively low enzyme concentrations the rate of hydrolysis increases proportionately with the concentration of enzyme; the rate reaches a maximum when all the available sites for binding and hydrolysis on the polymer are filled by enzyme. As the enzyme concentration continues to increase, the available sites for binding are full and no available sites for hydrolysis remain (*96*). The hydrophobic properties of the binding domains in the PHA depolymerases are dependent upon the environmental niches of the PHA degraders.

There are relatively few reports available concerning the depolymerase activity and biodegradation of LCL-PHAs. One to note however, is the recent work of de Koning et al. (*98*). PHO is a thermoplastic elastomer but its low melting point and cystallinity seriously impair its processability and commercial viability (*99-102*). These limitations can be somewhat overcome, by the use of irradiation to enhance cross-linking and modify the material properties of the polymer. Crosslinking was found to seriously alter these properties while maintaing its biodegradability to the isolated depolymerase from *P. fluorescens*.

Intracellular PHA Depolymerases.

PHA Inclusion Bodies. In contrast to the comparatively large amount of work focusing on the extracellular depolymerases, little research into the activity of their intracellular counterparts has been reported.

Early studies on the structure and properties of isolated inclusion bodies of PHB were reported by Eller et al. and Dunlop and Robards (*103,104*). The original study by Dunlop and Robards proposed a model of a granule where a core was surrounded by a membrane which contained the biosynthetic complex and other non-PHB material (*103*). Work by Williamson and Wilkinson (*105*) and from the laboratories of Marchessault, Merrick and Lundgren (*106-108*) suggest that *in vivo* inclusion bodies possess a lipid material that is lost upon isolation or purified polymer extraction. More recently, Bernard and Sanders have investigated the structure of *in vivo* PHB inclusion bodies of methylobacterium using radiolabelled NMR spectroscopy. They concluded that PHB in its native form is neither a solid, mobile liquid nor a solution; it exists in cells as an amorphous elastomer (*109-110*). This has been confirmed by the reports of Marchessault and others (*100-107*). *In vivo* and isolated *P. oleovorans* inclusion bodies containing LCL-PHAs have also been examined using ^{13}C NMR spectroscopy; this confirms the amorphous nature of these PHAs (Figure 10) (Browne personal communication 1993).

The ultrastructure of isolated *P. oleovorans* granules of PHO have been investigated by Fuller et al. (*113*) and Stuart et al. (*Can. J. Microbiol* in press). Normarski computer-enhanced imaging freeze fracture electron microscopy of isolated PHO granules has revealed a highly organized para-crystalline network located on the surface of the granules (*113*). This proteinaceous lattice displayed a pattern with repeating units 135Å in width, which resembled the regular arranged outer membrane proteins (rOmp) observed in *P. acidovorans* (*114*). Ultrastructural studies by the same group clearly demonstrate that the lattice network surrounding a granule consists of two layers. Curiously, the inner layer apparently does not completely enclose the polymer core. SDS-PAGE (sodium dodecyl-sulfatte polyacrylamide gel electrophoresis) of these coats reveals the presence of a number of proteins with approximate molecular weights of 59, 55, 43 32 and 18KDa. Huisman and Witholt (*115,116*) have used a recombinant strain of *Pseudomonas* to express the genes responsible for two polymerases and discovered protein bands at 58 and 56KDa, suggesting that the 59 and 55KDa protein bands observed in the gels constitute part of the polymerase system. The DNA sequence of esterase also suggests the possibility that the 32KDa polypeptide is the intracellular depolymerase. The functionality of this 32KDa protein has recently been confirmed as the depolymerase (Foster et al. in progress). The function of the 43 and 18KDa proteins remains unknown, but may have a structural function responsible for the containment of the polymer core.

Immunogold labelling using rabbit antibodies generated against the 59, 55 and 43KDa proteins, has shown a definite ordered distribution of the proteins in the

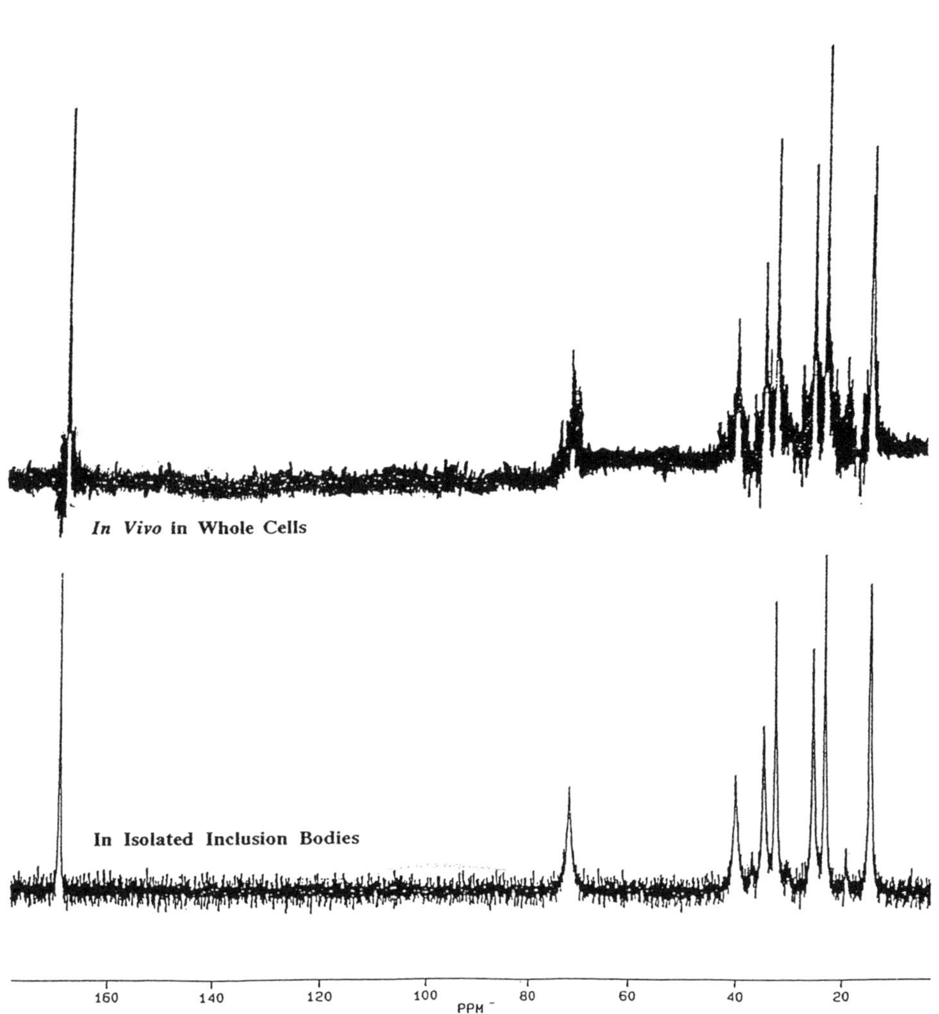

Figure 10. NMR spectrographs of *in vivo* and isolated *P. oleovorans* inclusion bodies containing LCL-PHAs, illustrating their amorphous nature.

two lattice layers. These results have led to the recent proposal of a model for the macromolecualr architecture of the polymer granule assembly (Stuart et al. *Can. J. Microbiol* in press). Therefore, it can be concluded that PHA inclusion bodies and isolated granules represent a complex organization of amorphous polyester contained in a double protein layer, which also holds the polymerase and depolymerase systems.

Depolymerase Activity. Fairly extensive studies on the biosynthesis of PHAs in a variety of microorganisms have been reported (*117-128*), but relatively little is yet known about the mechanisms and activity of intracellular depolymerases, owing to the granule complexity. The depolymerase systems break down the polymer chains into more manageable subunits which are eventually taken into the cell and utilized for various biochemical purposes. The fate of these subunits has been investigated in a variety of microorganisms (*117-129*).

In 1964, Merrick and Doudoroff investigated the intracellular depolymerase system in isolated PHB granules from *B. megaterium*.. Isolated granules were added to a reaction mixture containing a crude extract of *R. rubrum* cells which had been permitted to expend all their accumulation of PHB, the 'digestion' of the granules was then followed by the titration of the acid products, hydroxybutyric acid formation and suspension turbidity. This crude extract from *R. rubrum* was purified and found to consist of two seperate compounds: a thermolabile depolymerase and a thermostable 'activator'. Neither of these components alone was found to affect depolymerization; however, when both were added to the reaction mixture simultaneously, or when the activator was added after the depolymerase fraction, an initial lag period in degradation was observed. When the granules were first incubated with the activator prior to the depolymerase, this lag period was eliminated. Merrick and Doudoroff also reported increases in polymer hydrolysis by the use of 'activation elements': low concentrations of magnesium, sodium and calcium (*129*). This research was later complemented by Hippe and Schlegel (1967), who described the intracellular degradation of PHB in *A. eutrophus* (*131*). In *A. eutrophus*, the sole product of intracellular depolymerase activity was the monomer, but *B. megaterium* was found to produce a mixture of monomers and dimers (*132*). Intracellular dimer hydrolases have been subsequently isolated from *B. megaterium* (*132*), *Z. ramigera* (*133*) and *R. rubrum* (*134*).

More recently, Saito and coworkers have investigated the the activities of PHB intracellular depolymerases of *Zoogloea ramigera* and *A. eutrophus* (*134*) (Saito et al. personal communication ISBP Montreal 1994). Isolated granules from *Z. ramigera* showed negligible degradation when incubated in buffers of pH 7.4 to 8.5; further increases in pH were followed by increases in PHB hydrolysis, as detected by the release of monomer (Figure 11a). Triton X-100 and PMSF inhibited granule degradation. In contrast to the work of Merrick and Doudoroff, the presence of sodium and magnesium, whilst acting as activation elements, were also essential for degradation activity. A supernatant fraction of centrifuged PHB granules isolated from *Z. ramigera* was added to protease treated PHB granules isolated from *A. eutrophus*, these subsequently showed an increase in polymer degradation (*134*).

The degradation of native PHB granules from *A. eutrophus* was also highly dependent upon pH and showed two peaks of degradation activity: one around pH 6-7 and the second at a more alkaline pH. Similar to the *R. rubrum* and *B. megaterium* intracellular depolymerase studies, the presence of magnesium and sodium salts resulted in an increase in degradation; however, this was limited to the neutral pH region (Figure 11b). In contrast to the *Z. ramigera* studies by the same group, these activation elements were not essential for degradation activity. Similarly, both Triton X-100 and PMSF inhibited polymer degradation. The

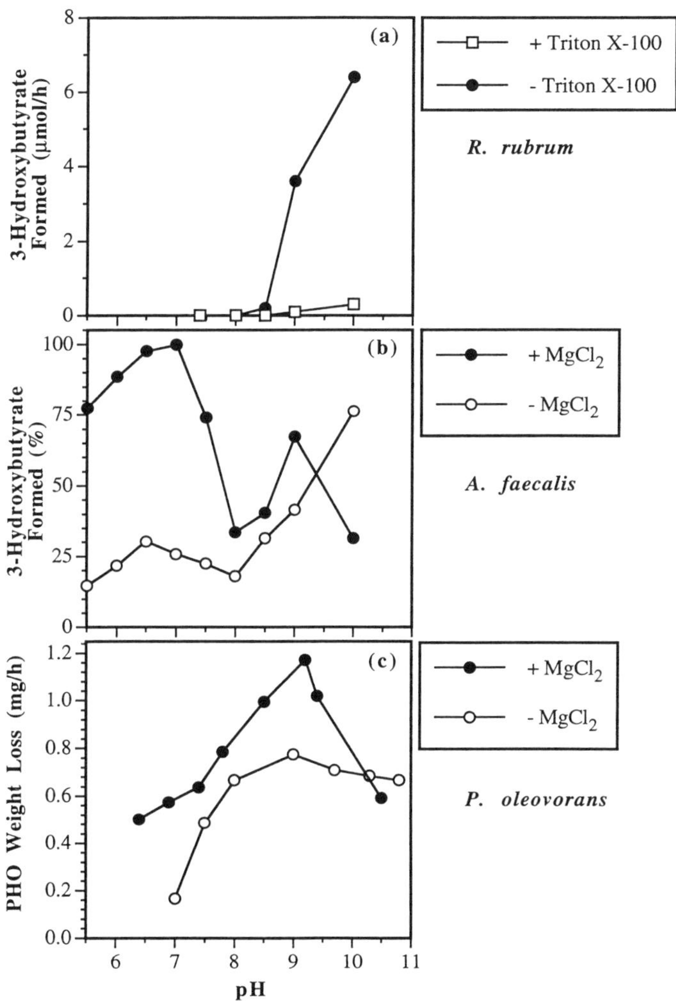

Figure 11. Optimum pH for intracellular depolymerase activity in isolated inclusion bodies, **a:** PHB granules (Modified from ref. 134) **b:** PHB granules (Modified from Saito et al. *Can. J. Microbiol.* in press) **c:** PHO granules (Foster et al. in progress)

authors concluded from this that the intracellular depolymerase system in *A. eutrophus* consisted of two enzymes (Saito et al. personal communication ISBP Montreal 1994).

Research into the intracellular depolymerase activity has been restricted, in part, by the absence of an appropriate quantitative degradation monitoring procedure. In the case of many novel LCL-PHAs, polymer yields are comparatively low, the use of gravimetry to monitor degradation is therefore inaccurate and unsuitable. Other monitoring techniques include optical density changes, titrimetric assays, radiolabelling etc., and in the case of PHB, a spectrophotometric assay for the evolution of NADH (reduced nicotinamide dinucleotide) (*46,92*) (Foster and Tighe *Biomaterials* in press). However, these methods are somewhat restrictive and provide only an indirect measurement of enzyme activity. Foster et al have recently reported an accurate, quantitative technique to monitor small weight changes in PHAs. This technique is readily adaptable and has been used in our laboratory to investigate the intracellular depolymerase activity in a number of LCL-PHAs produced by *P. oleovorans*, including PHO (*135*), PHN and PHPV. *P. oleovorans* has been demonstrated to be a versatile micro-organism capable of synthesizing many different types of LCL-PHAs (*26-29*).

Isolated *P. oleovorans* inclusion bodies of PHO exhibited a linear degradation rate of 0.41mg/h when incubated at 30°C at pH 8.6 (Figure 12) The depolymerase showed noticeable hydrolytic activty throughout the pH range studied, with a single optimum peak around pH 9 (Figure 11c). The presence of magnesium increased polymer degradation but was not essential for activity. Similar to the intracellular PHB depolymerases of *Z. ramigera* and *A. eutrophus*, degradation was inhibited by PMSF and Triton X-100, indicating the presence of hydrophobic binding sites and serine residues at its active site (*136*). This contrasts with the the reports on the extracellular PHO depolymerase of *P. fluorescens* where degradation was unaffected by PMSF. Obviously the 'flap' observed in many lipases and speculated to be present in the *P. fluorescens* depolymerase, is not a neccessary requirement for the *P. oleovorans* intracellular enzyme, which is organized in the protein coats encompasing the polymer granule. Similarly, the molecular weight of this intracellular enzyme is considerably less than its extracellular counterpart.

It is interesting to note that whilst the extracellular depolymerases from *P. lemoignei* and *A. faecalis* hydrolyzed purified PHB with a highly crystalline structure, they were unable to degrade the amorphous polymer in native granules (*75,92*).

Concluding Remarks.

This review has given some indication at the complexity and diversity of properties that exist in PHA depolymerases. Research into these bacterial polyesters continues to focus on the more readily, and commercially, available PHB copolymers. We are only just beginning to investigate other more novel PHAs.

A wide variety of microorganisms are capable of producing a number of different PHAs. These thermoplastics have great commercial potential. Faced with increasing legislation against plastic waste and the requirements of biodegradable, biocompatable characteristics for numerous plastic devices, research has increased into the production and utilization of these novel biopolymers. It is vital, therefore, to investigate and understand the degradation of these polyesters. Results from environmental testing and the activities of extracellular depolymerases play an important part in determining, and manipulating, their suitability. Investigations into the activities of the intracellular depolymerases will help us to control the synthesis

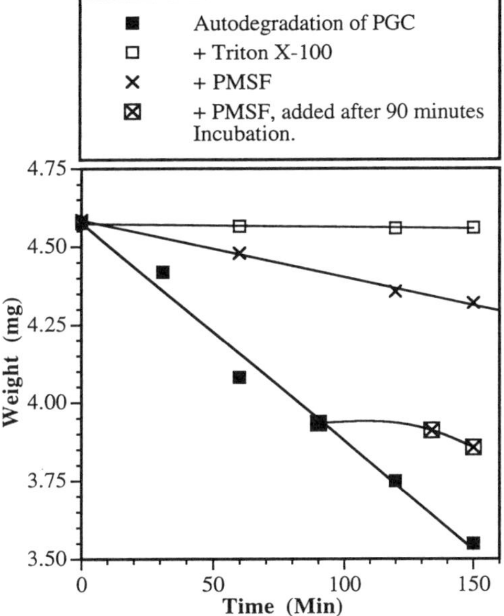

Figure 12. Degradation of *P. oleovorans* inclusion bodies containing LCL-PHAs by an associated intracellular depolymerase system, incubated at 30°C & pH 8.6. (Modified from ref. 135)

of novel PHAs whilst optimizing their production, as well as influencing the degradation of the final devices.

Seventy years after Lemoigne's discovery, we have only just begun to realize the great potential of these novel bacterial polyesters.

Acknowledgments.

We are indebted to Dr. M. Timmins for the SEM photographs (Figure 7) and Dr. S. Browne of Mt. Holyoke College for the NMR data (Figure 11).

Literature Cited.

(1) Lemoigne, M. *C.R. Acad. Sci.* **1925,** *180*, 5139
(2) Lemoigne, M. *Ann. Inst. Pasteur* **1925**, *39*, 144-173
(3) Lemoigne, M. *Bull. Soc. Chim. Biol.* **1926**, *8*, 770-782
(4) Lemoigne, M. *Ann. Inst. Pasteur* **1927**, *41*, 148-165
(5) Baptist, J.N. & Ziegler, J.B. US patent appl. **1965**, 3-225-766
(6) Howells, E.R. *Chem. Ind.* **1982**, *15*, 508-511
(7) Holmes, P.A.; Wright, L.F. & Collins, S.H. European patent appl. **1982**, 0052-459
(8) Holmes, P.A. *Phys. Technol.* **1985**, *16*, 32-36
(9) Anon. *Chemistry in Britain* **1987**, *23(12)*, 1157
(10) Anon. *New Scientist* **1990**, *126*, 36
(11) Dawes, E.A. *Bacteriol. Rev.* **1964**, *28*, 126-149
(12) Doudoroff, M. In: Current Aspects of Biochemical Energetics; Kaplan, N.O. & Kennedy, E.P. Eds., Academic Press, New York, USA, **1966**, 385-400
(13) Dawes, E.A. Microbial Energetics, Blackie press, Glasgow, UK, **1986**, 145-165
(14) Doi, Y. Microbial Polyesters, VCH Publishers, New York, USA, **1990**, 99-134
(15) Anderson, A.J. & Dawes, E.A. *Microbial Rev.* **1990**, *34*, 450
(16) Brandl, H.; Gross, R.A.; Lenz, R.W. & Fuller, R.C. *Adv. Biochem. Eng. Biotech.* **1990**, *41*, 77
(17) Baptist, J.N. US patent appl. **1962**, 3-036-959
(18) Dawes, E.A. & Senior, P.J. *Adv. Microbiol. Physiol.* **1973**, *10*, 203-266
(19) Ward, A.C.; Rawley, B. & Dawes, E.A. *J. Gen. Microbiol.* **1977**, *102*, 61
(20) Müller, H.M. & Seebach, D. *Angew. Chem. Internat. Ed. Eng.* **1993**, *32*, 477-502
(21) Brandl, H.; Gross, R.A.; Lenz, R.W. & Fuller, R.C. *Appl. Environ. Microbiol.* **1989**, *Aug*, 1977-1982
(22) Doi, Y.; Kunioka, M.; Nakamura, Y. & Soga, K. *Macromolecules.* **1988**, *21*, 2722-2727
(23) Kunioka, M.; Nakamura, Y. & Doi, Y. *Polymer Comm.* **1988**, *29*, 174-176
(24) Gross, R.A.; DeMello, C.; Lenz, R.W.; Brandl, H. & Fuller, R.C. *Macromolecules.* **1989**, *22*, 1106-1115
(25) Ballisteri, A.; Montaudo, G.; Impallomeni, G.; Lenz, R.W.; Kim, Y.B. & Fuller, R.C. *Macromolecules.* **1990**, *23*, 5059-5064
(26) Fritzsche, K.; Lenz, R.W. & Fuller, R.C. *Makromol. Chem.* **1990**, 191, 1957-1965
(27) Lenz, R.W.; Kim, Y.B. & Fuller, R.C. *J. Bioact. Compat. Polym.* **1991**, *6*, 382-392

(28) Lenz, R.W.; Kim, Y.B. & Fuller, R.C. *FEMS Microbiol. Rev.* **1992**, *103*, 207-214
(29) Kim, Y.B.; Lenz, R.W. & Fuller, R.C. *Macromolecules.* **1992**, *25*, 1852-1857
(30) Hori, K.; Soga, K. & Doi, Y. *Biotechnol. Lett.* **1994**, *16(5)*, 501-506
(31) Gagnon, K.D. PhD Thesis, University of Massachusetts, Amherst-MA, USA, **1993**
(32) Taylor, L. *Chem. Tech.* **1979**, *Sept*, 542-548
(33) Potts, J.E. In; Aspects of Degradation and Stabalization of Polymers, Jellinek, H.H.G. Ed. **1978**, Elselevier Scientific Publishing Co., Amsterdam, Holland, pp. 617-657
(34) Williams, D.F. *J. Mater. Sci.* **1982**, *17*, 1233-1246
(35) Griffin, G.J.L. *Pure Appl. Chem.* **1980**, *52*, 399-407
(36) Kopet, J. & Ulbrich, K. *Pro. Polym. Sci.* **1983**, *9*, 1-58
(37) Pitt, C.G.; Marks, T.A. & Schindler, A. In; Biodegradable Drug Delivery Systems Based On Aliphatic Polyesters: Applications to Contraceptives and Narcotic Antagonists, Willette, E. & Barnett, G. Eds. Naltrexane Research Monograph 28, National Institute of Drug Abuse, UK., **1980**, pp. 232-253
(38) Gilbert, R.D.; Stannett, V.; Pitt, C.G. & Schindler, A. In; Developments in Polymer Degradation: 4, Applied Science Publishers, London, UK, **1982**, pp. 259-293
(39) Lenz, R.W. In: Advances in Polymer Science, Springer-verlag, Berlin, Germany, **1992**, pp. 107
(40) Gilding, D.K. *Biocompat. Clin. Implant. Mater.* **1981**, *2*, 209-232
(41) Zaikov, G.E. *J. Macromol. Sci. Rev. Macromol. Chem. Phys.* **1985**, C25(4), 551-597
(42) Langer, R.S. & Peppas, N. *Biomaterials.* **1981**, *2*, 201-214
(43) Heller, J. *ACS Symp. Ser.* **1983**, *212*, 373-392
(44) Yasin, M.; Holland, S.J. & Tighe, B.J. *Biomaterials.* **1990**, *11*, 431-454
(45) Foster, L.J.R. PhD Thesis, Aston University, Birmingham, UK, **1992** vol.2
(46) Foster, L.J.R. & Tighe, B.J. *J. Environ. polym. Deg.* **1994**, *2(3)*, 185-194
(47) Yasin, M.; Holland, S.J.; Jolly, A.M. & Tighe, B.J. *Biomaterials.* **1989**, *10*, 400-412
(48) Holland, S.J.; Yasin, M. & Tighe, B.J. *Biomaterials.* **1990**, *11*, 205-215
(49) Yasin, M. & Tighe, B.J. *Biomaterials.* **1992**, *13*, 9-16
(50) Yasin, M. PhD Thesis, Aston University, Birmingham, UK, **1988**
(51) Yasin, M. & Tighe, B.J. *Plastics, Rubb. Comp. Process. Appl.* **1993**, *19*, 15-27
(52) Kimura, M.; Toyata, K.; Iwatsuki, M. & Sawada, H. Report for International Biodegradable Polymer Society, proceedings, **1990**
(53) Mergaert, J.; Webb, A.; Anderson, C.; Wouten, A. & Swings, J. *App. Environ. Microbiol.* **1993**, *59(10)*, 3233-3238
(54) Gilmore, D.F.; Antoun, S.; Lenz, R.W.; Goodwin, S.; Austin, R. & Fuller, R.C. *J. Ind. Microbiol.* **1992**, *10*, pp. 199-206
(55) Mergaert, J.; Anderson, C.; Wouters, A. & Swings, J. *J. Environ. Polym. Deg.* **1994**, *2(3)*, 177-183
(56) Mergaert, J.; Anderson, C.; Wouters, A.; Swings, J. & Kersters, K. *FEMS Microbiol. Rev.* **1992**, 317-322
(57) Yasin, M. & Foster, L.J.R. Polymat'94, Conference Proceedings, London UK, **1994**
(58) Gilmore, D.F.; Letti, N.; Lenz, R.W.; Fuller, R.C. & Scandola, M. In: Biodegradable Polymers and Plastics, **1990**, pp. 251-254

(59) Liddel, J.M. In: The Chemical Industry - Friend to the Environment? **1991**, pp.10-25
(60) Smith, D.K.; Belson, N. & Kilp, T. A Study in Biodegradable Plastics, Queens Printers, Ontario, Canada, **1990**
(61) Pruter, A. *Mar. pollut. Bull.* **1987**, *18(6B)*, 305
(62) Laist, D. *Mar. pollut. Bull.* **1987**, *18(6B)*, 319
(63) Manning, A. USA Today, **1992**, 22nd. May
(64) Doi, Y.; Kanesawa, Y.; Tanahashi, N. & Kumagai, Y. *Polym. Deg. Stab.* **1992**, *36*, 173
(65) Brandl, H. & Püchner, P. In: Novel Biodegradable Microbial Polymers, Dawes, E.A., Ed., Kluwer Academic Publishers, Dordrecht, Netherlands, **1990**, pp. 421-422
(66) Püchner, P. & Müller, W.R. *FEMS Microbiol. Rev.* **1992**, *103*, 469-470
(67) August, J.; Müller, R.J. & Widdecke, H. *FEMS Microbiol. Rev.* **1992**, *103*, 477-479
(68) Moore, J.W. *Modern Plastics.* **1992**, *Dec.*, 58-63
(69) Mukai, K.; Yamada, K. & Doi, Y. *Polym. Deg. Stab.* **1993**, *41*, 85-91
(70) Ramsay, B.A.; Sarcovan, I.; Ramsay, J.A. & Marchessault, R.H. *J. Environ. Polym. Deg.* **1994**, *2(1)*, 1-7
(71) Kanesawa, Y.; Tanahashi, N.; Doi, Y. & Saito, T. *Polym. Deg. Stab.* **1994**, *45*, 179-185
(72) Schirmer, A.; Jendrossek, D. & Schlegel, H.G. *Appl. Environ. Microbiol.* **1993**, *59(4)*, 1220-1227
(73) Foster, L.J.R.; Zervas, S.J.; Lenz, R.W. & Fuller, R.C. *Biodegradation.* **1995**, *6*, 67-73
(74) Jendrossek, D.; Knoke, I.; Habribian, R.B.; Steinbüchel, A. & Schlegel, H.G. *J. Environ. Polym. Deg.* **1993**, *1(1)*, 53-63
(75) Mukai, K.; Yamada, K. & Doi, Y. *Int. J. Biol. Macromol.* **1992**, *14*, pp. 235-239
(76) Müller, B. & Jendrossek, D., *Appl. Microbiol. Biotechnol.* **1993**, 38, 487-492
(77) Brandl, H.; Knee, E.J.; Fuller, R.C.; Gross, R.C. & Lenz, R.W. *Int. J. Biol. Macromol.* **1989**, *11*, 49
(78) Shiotani, T. & Kobayashi, G. Japanese patent appl. **1993**, 93049
(79) Shiamura, E.; Kasaya, K.; Kobayashi, G.; Shiotani, T.; Shima, Y. & Doi, Y. *Macromolecules.* **1994**, *27*, 878-880
(80) Lusty, C.J. & Doudoroff, M. *Natl. Acad. Sci. USA.* **1966**, *56*, 960-965
(81) Nakayama, K.; Saito, T.; Fukui, T.; Shirakura, Y. & Tomita, T. *Biochim. Biophys. Acta.* **1985**, *827*, 63-72
(82) Shirakura, Y.; Fukui, T.; Saito, T.; Okamoto, Y.; Narikawa, T.; Koide, K.; Tomita, T.; Takemasa, T. & Masamura, S . *Biochem. Biophys. Acta.* **1986**, *880*, 46-53
(83) Hocking, P.J.; Marchessault, R.H.; Timmins, M.R.; Scherer, T.M.; Lenz, R.W. & Fuller, R.C. *Macromol. Rapid Commun.* **1994**, *15*, 447-452
(84) Marchessault, R.H.; Monasterios, C.O.; Jesudason, J.J.; Ramsay, B.; Sarcovan, I.; Ramay, J.A. & Saito, T. *Polym. Deg. Stab.* **1994**, *45*, 187-196
(85) Abe, H.; Matsubara, I.; Doi, Y.; Hori, Y. & Yamaguchi, A. *Macromolecules.* **1994**, *27*, 6018-6025
(86) Doi, Y.; Kumagai, Y.; Tanahashi, N. & Mukai, K. In: Biodegradable Polymers and Plastics, Vert, M.; Feilen, J.; Albertsson, A.; Scott, G. & Chiellini, E. Royal Society of Chemistry, Redwood Press, Cambridge, UK, **1992**

(87) Kemnitzer, J.E.; McCarthy, S.P. & Gross, R.A. *Macromolecules.* **1992**, *25*, 5927-5934
(88) Kumagai, Y.; Kanesawa, Y. & Doi, Y. *Makromol. Chem.* **1992**, *193*, 53
(89) Parikh, M.; Gross, R.A. & McCarthy, S.P. *Polym. Mater. Sci. Eng.* **1992**, *66*, 408
(90) Jesudason, J.J.; Marchessault, R.H. & Saito, T. *J. Environmen. Polym. Deg.* **1993**, *1*, 89
(91) Delafield, F.D.; Doudoroff, M.; Pallenoni, N.J.; Lusty, C.J. & Contopaulos, R. *J. Bacteriol.* **1965**, *90*, 1455
(92) Tanio, T.; Fukui, T.; Shirakura, Y; Saito, T.; Tomita, K.; Kaiho, T. & Masamune S. *Eur. J. Biochem.* **1982**, *124*, 71-77
(93) Mukai, K.; Yamada, K. & Doi, Y. *Polym. Deg. Stab.* **1994**, *43*, 319-327
(94) Jaeger, K.E.; Ransac, S.; Koch, H.B.; Ferrato, F. & Dijkstra, B.W. *FEBS Lett.* **1993**, *332*, 143-149
(95) Malcata, F.X.; Reyes, H.R.; Garcia, H.S.; Hill, C.G. & Amundson, C.H. *Enzyme Microb. Technol.* **1992**, *14*, 426-446
(96) Mukai, K.; Yamada, K. & Doi, Y. *Int. J. Biol. Macromol.* **1993**, *15*, 361-367
(97) Timmins, M.R. PhD Thesis, University of Massachusetts, Amherst-MA, USA, **1994**
(98) de Koning, G.J.M.; van Bilsen, H.M.M.; Lemstra, P.J.; Hazenberg, W.; Witholt, B.; Preusting, H.; van der Galiën, J.G.; Schirmer, A. & Jendrossek, D. *Polymer.* **1994**, *35*, 2090-2097
(99) Richtering, H.W.; Gagnon, K.D.; Lenz, R.W.; Fuller, R.C. & Winter, H.H. *Macromolecules.* **1992**, *25*, 2429-2433
(100) Curley J. PhD. Thesis, Dept. of Polymer Science & Engineering, University of Massachusetts, Amherst, USA. **1994**
(101) Gagnon, K.D.; Lenz, R.W. & Farris, R.J. *Rubber Chemistry and Technology.* **1993**, *65*, 761-777
(102) de Koning, G.J.M. PhD Thesis, Technische Universiteit, Eindhoven, Netherlands, **1993**.
(103) Eller, D.; Lundgren, D.G.; Okamura, K. & Marchessault, R.H. *J. Mol. Biol.* **1968**, *35*, 489
(104) Dunlop, W.F. & Robards, A.W. *J. Bacteriol.* **1973**, *114(3)*, 1271-1280
(105) Williamson, D.H. & Wilkinson, J.F. *J. Gen. Microbiol.* **1958**, *19*, 198-209
(106) Alper, R.; Lundgren, D.G.; Marchessault, R.H. & Coté, W.A. *Biopolymers.* **1963**, *1*, 545-556
(107) Merrick, J.M.; Lundgren, D.G. & Pfister, R.M. *J. Bacteriol.* **1964**, *89*, 234-239
(108) Lundgren, D.G.; Alper, R.; Schnaitman, C. & Marchessault, R.H. *J. Bacteriol.* **1965**, *89*, 245-251
(109) Bernard, G.N. & Sanders, J.K.M. *FEBS Lett.* **1988**, *231*, 16-18
(110) Bernard, G.N. & Sanders, J.K.M. *J. Biol. Chem.* **1989**, *24*, 3286-3291
(111) Revol, J.F.; Chamy, H.D.; Deslandes, Y. & Marchessault, R.H. *Polymer.* **1989**, *30*, 1973-1976
(112) Ramsay, B.A.; Saracovan, I.; Ramsay, J.A. & Marchessault, R.H.; *Appl. Environ. Microbiol.* **1992**, *58*, 744-746
(113) Fuller, R.C.; O'Donnell, J.P.; Saulnier, J.; Redlinger, T.E.; Foster, J. & Lenz, R.W. *FEMS Microbiol. Rev.* **1992**, *103*, 279-288
(114) Sleytr, U. & Messner, P. *Annu. Rev. Microbiol.* **1983**, *37*, 311-339

(115) Huisman, G.W.; Meime, R.; Wonink, E. & Witholt, B. In: Novel Biodegradable Microbial Polymers, Dawes, E.A. Ed; Kluwer Academic Publishers, Dordrecht, Netherlands, **1990**, pp.451
(116) Huisman, G.W.; Wonink, E.; Meima, R.; Kaxemier, B.; Terpstra, P. & Witholt, B. *J. Biol. Chem.* **1991**, *266*, 2191-2198
(117) Oeding, V. & Schelgel, H.G. *Biochem. J.*, **1973**, *134*, 239-248
(118) Schubert, P.; Steinbüchel, A. & Schlegel, H.G. *J. Bacteriol.* **1988**, *170*, 5837-5847
(119) Peoples, O.P. & Sinskey, A.J. *J. Biol. Chem.* **1989**, *264*, 15293-15303
(120) Haywood, G.W.; Anderson, A.J. & Dawes, E.A. *FEMS Microbiol. Lett.* **1989**, *57*, 1-6
(121) Senior, P.J. & Dawes, E.A. *Biochem. J.* **1973**, *134*, 225-238
(122) Ritchie, B.A.F.; Senior, P.J. & Dawes, E.A. *Biochem. J.* **1971**, *121*. 309-316
(123) Saito, T.; Fukui, T.; Ikeda, F.; Tanaka, Y. & Tomita, K. *Arch. Microbiol.* **1977**, *114*, 211-217
(124) Shuto, H.; Fukui, T.; Saito, T.; Shirakura, Y. & Tomita, K. *Eur. J. Biochem.* **1981**, *118*, 53-59
(125) Doi, Y.; Kawaguchi, Y.; Koyama, N.; Nakamura, S.; Hiramitsu, M.; Yoshida, Y. & Kimura H. *FEMS Microbiol. Rev.* **1992**, *103*, 103-108
(126) Kawaguchi, Y. & Doi, Y. *Macromolecules.* **1992**, *25*, 2324-2329
(127) Lageveen, R.G.; Huisman, G.W.; Eggink, G. & Witholt, B. *Appl. Environ. Microbiol.* **1988**, *54*, 2924-2932
(128) Doi, Y.; Kunioke, M.; Nakamura, Y. & Soga, K. *Macromolecules.* **1987**, *20*, 2988-2991
(129) Merrick, J.M. & Doudoroff, M. *J. Bacteriol.* **1964**, *38(1)*, 64-71
(130) Hippe, H. & Schlegel, G. *Arch. Microbiol.* **1967**, *56*, 278-299
(131) Gervaid, R.; Dahinger, A.; Hartecoeur, B. & Reynaud, C. *C. R. Hebd. Seances Acad. Sci. Paris.* **1966**, *263*, 1273-1275
(132) Tanaka, Y.; Saito, T.; Fukui, T.; Tanio, T. & Tomita, K. *Eur. J. Biochem.* **1982**, *118*, 177-182
(133) Merrick, J.M. & Yu, C.I. *Biochemistry.* **1966**, *5*, 3563-3568
(134) Saito, T.; Saegusa, H.; Miyata, Y. & Fukui, T. *FEMS Microbiol. Lett.* **1992**, *118*, 279-282
(135) Foster, L.J.R.; Lenz, R.W. & Fuller, R.C. *FEMS Microbiol. Lett.* **1994**, *118*, 279-282
(136) Foster, L.J.R.; Timmins, M.; Scherer, T.; Fuller, R.C. & Lenz, R.W. *Polym. Prepr.* **1994**, *35(2)*, 425-426

RECEIVED December 22, 1995

Chapter 8

Poly(L-lactic acid-*co*-amino acid) Graft Copolymers: A Class of Functional Biodegradable Biomaterials

Jeffrey S. Hrkach[1], Jean Ou[2], Noah Lotan[3], and Robert Langer[1]

[1]Department of Chemical Engineering, Massachusetts Institute of Technology, E25–342, Cambridge, MA 02139
[2]Department of Chemistry, Harvard University, Cambridge, MA 02138
[3]Department of Biomedical Engineering, Technion Israel Institute of Technology, Haifa 32000, Israel

The synthesis of poly(L-lactic acid-*co*-L-lysine), poly(L-lactic acid-*co*-β-benzyl-L-aspartate), and poly(L-lactic acid-*co*-D,L-alanine) comb-like graft copolymers and their characterization by ^{1}H NMR and elemental analysis are presented. The lysine ε-amino groups in linear poly(L-lactic acid-*co*-L-lysine) copolymers with approximately 1% lysine content were utilized to initiate the ring opening polymerization of the amino acid N-carboxyanhydrides to yield the graft copolymers.

The family of degradable polyesters: poly(glycolic acid), poly(lactic acid), and their copolymers, has received considerable attention as resorbable sutures,[1] in drug delivery,[2] tissue engineering,[3] and other biomedical applications, and FDA approval has been granted for some uses. The main advantages of these materials are their degradability in the physiological environment to yield naturally occurring metabolic products, and the ability to control their rate of degradation by varying the ratio of lactic acid to glycolic acid repeat units in the copolymers.

Tissue engineering is a multi-disciplinary field that focuses on restoring, maintaining, or improving tissue function for patients suffering from tissue loss or organ failure. One approach in tissue engineering involves the use of polymers as matrices to provide the necessary physical support for cell growth and implantation. Unfortunately, control over the cellular response to these degradable polyesters is minimal. This is due to the fact that these polymers do not possess functional groups (other than end groups) that will allow their chemical modification to change their characteristics. As a result, interactions between the polymers and cells are dictated solely by the cell's response to these polymers, and in many instances, this leads to unsatisfactory results. As such, much progress in the field of tissue engineering could be achieved if it becomes possible to retain the beneficial properties of the degradable polyesters while incorporating functionalities that promote specific, desirable interactions between the polymers and cells.

An additional application of these degradable polyesters is for drug delivery. Their hydrophobic properties allow for convenient incorporation and delivery of

0097–6156/96/0627–0093$15.00/0

lipophilic agents, but make it difficult to do so with hydrophilic or ionic agents. This limitation may be circumvented by attaching functional groups to the polymers that can make them more hydrophilic, and thus, expand their use as delivery devices for hydrophilic drugs as well. In addition, the degradation of these polymers into lactic or glycolic acid can lead to a highly acidic local environment which can have a detrimental effect on the agents being delivered.[4] Therefore, incorporation of basic functionalities that can neutralize the acids and control the pH is a possible route to overcome these problems.

The considerations presented above lead to the conclusion that there is a need in the field of biomaterials for the development of polymers that are specifically tailored to have the appropriate physical and chemical properties so as to fit the requirements of a particular biomedical application.

In this report we present the development of a new family of polyester-poly(amino acid) materials, particularly, the synthesis of poly(L-lactic acid-*co*-L-lysine), poly(L-lactic acid-*co*-β-benzyl-L-aspartate), and poly(L-lactic acid-*co*-D,L-alanine) comb-like graft copolymers. The incorporation of the amino acid chains into the poly(lactic acid) structure imparts new properties to poly(lactic acid) based materials, without losing the beneficial properties that make poly(lactic acid) so successful. The changes depend upon the particular amino acid used as well as the poly(amino acid) chain length and overall composition within the copolymer. An important advantage of this new class of materials stems from the fact that they possess functional groups which can be used as sites for further chemical modification, thus providing the opportunity for the attachment of biologically active molecules that can actively promote favorable cell-polymer interactions, a feature very much sought after in tissue engineering. Also, when used in drug delivery systems, these polymers can potentially be modified to increase the level of agent incorporated or provide greater stability for its delivery. In addition, they may be functionalized in an effort to target the agents for delivery to a particular site within the body.

Experimental Section

Materials THF and dioxane were distilled from sodium and stored over sodium-potassium alloy in the dry box. CH_2Cl_2 and $CHCl_3$ were refluxed over CaH_2, and then distilled onto and stored over CaH_2. All other solvents were used as received. N_ε-Carbobenzoxy-L-lysine (ZLYS), β-benzyl-L-aspartate (BASP), and D,L-alanine (ALA) (all from Sigma) and triphosgene (Aldrich) were stored in the freezer in the dry box and used as received. HBr (Aldrich, 30 wt% in acetic acid) was used as received. N,N-Diisopropylethyl amine (Aldrich) was stored over molecular sieves.

Equipment Reactions were set up or run in a Vacuum Atmospheres dry box (model HE-43-2). All copolymers were analyzed after precipitation, isolation, and drying under vacuum at room temperature. NMR spectra were collected on a Bruker AC250 FTNMR in $CDCl_3$, d_7-DMF, or d_6-DMSO. GPC analysis was conducted in DMF (Phenogel linear column; Phenomenex) or $CHCl_3$ (Phenogel guard, linear, and 1000Å columns in series; Phenomenex) using a Perkin Elmer model 250 pump and model LC30 differential refractive index detector. Molecular weights of pLAZL and pLAL were calculated based on polystyrene standards. Elemental analyses were performed by Quantitative Technologies, Inc., Whitehouse, NJ.

Synthesis of amino acid N-carboxyanhydrides (NCA's) Following a procedure by Daly,[5] three equivalents of triphosgene in THF were added to the stirred suspension of amino acid in THF at 60 °C. In all cases, a clear solution is obtained within 1h. After 3h, excess hexane was slowly added to precipitate the NCA. The mixture was stored overnight in a freezer (-30 °C) in the dry box, and vacuum filtered to yield approximately 85% NCA in the cases of ZLYS and BASP. The crystallization of ALA-NCA was much less efficient and lead to lower yields.

Synthesis of poly(L-lactic acid-*co*-amino acid) graft copolymers Poly(L-lactic acid-*co*-Z-L-lysine) (pLAZL), a linear copolymer containing approximately 1 mol% lysine in the backbone, was synthesized as previously described by Barrera et al.[6] PLAZL was deprotected to yield poly(L-lactic acid-*co*-L-lysine) (pLAL) by first dissolving it in $CHCl_3$, then adding HBr/HOAc (10-15 fold excess of polymer) and stirring under argon for 30-60 min. The polymer was precipitated with ether, washed with methanol and collected by vacuum filtration. Conversion from the hydrobromide salt to the free amine was carried out in $CHCl_3$ by reaction with excess diisopropylethyl amine. Precipitation from methanol and drying yielded pLAL with free ε-amine groups. Analysis of the polymer by ^1H NMR shows the disappearance of the Z-phenyl peak at 7.35 ppm to indicate deprotection is complete.

As an example of the grafting procedure, the polymerization using ZLYS-NCA will be described. In the dry box, pLAL (1 g, 0.14 mmol lysine) was dissolved in 10 mL dioxane, and then a dioxane solution of ZLYS-NCA (1.02 g in 5 mL dioxane, 3.33 mmol) was added. The graft copolymerization mixture was brought out of the dry box, purged with argon throughout the reaction or equipped with a drierite drying tube, and stirred at room temperature for 2-4 days. The graft copolymer was precipitated by slowly adding a large excess of methanol, vacuum filtered, and dried under vacuum on a lyophilizer, obtaining a yield of 1.5 g.

Results and Discussion

Synthesis of amino acid N-carboxyanhydrides The preparation and polymerization of amino acid N-carboxyanhydrides (NCA's) can be carried out in a variety of ways, and these areas are well documented by several reviews.[7-9] In the present work, NCA's were synthesized by reacting the appropriate amino acid with triphosgene (Scheme 1).[5] The advantage of using triphosgene over phosgene gas is that it is a solid material, and therefore, safer and easier to handle. It is soluble in THF and hexane, the solvents used in this procedure, so any excess is efficiently separated from the NCA's. The procedure is the same for all three amino acids used, with the isolated yield of ALA-NCA being lower due to poor crystallization of the racemic mixture. Although they are sensitive materials, the NCA's can be stored in a freezer (-30 °C) in the dry box and used effectively after several days.

Synthesis of poly(L-lactic acid-*co*-amino acid) graft copolymers The ring opening polymerization of NCA's can be initiated by nucleophilic initiators such as amines, alcohols, and water. In most cases, the primary amine initiated ring opening polymerization of NCA's, although not a truly "living" process, allows good control over the degree of polymerization when the M/I ratio is less than 150.

Barrera et al. have synthesized poly(L-lactic acid-*co*-Z-L-lysine), (pLAZL), a linear copolymer composed of L-lactic acid units and approximately 1% Z-L-lysine units.[6] Random incorporation of lysine into the polymer backbone provides sites along the polymer chain that can be chemically modified through reaction at the ε-amine group of the lysine units (after deprotection) to introduce a variety of functionalities.

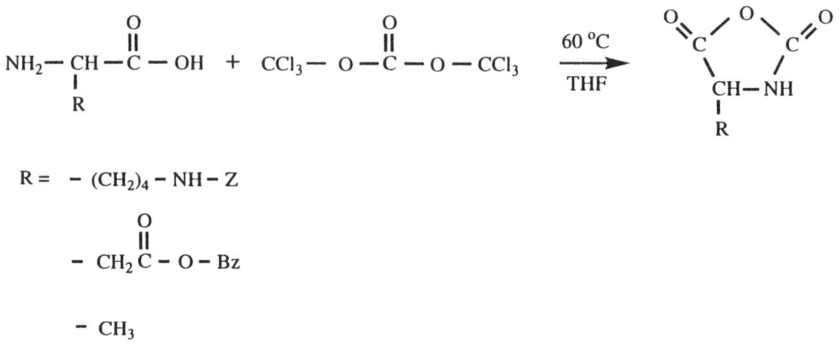

$$NH_2-CH(R)-C(=O)-OH \;+\; CCl_3-O-C(=O)-O-CCl_3 \xrightarrow[THF]{60\,^{o}C} \text{cyclic anhydride: } O=C\langle O \rangle C=O,\; CH(R)-NH$$

R = $-(CH_2)_4-NH-Z$

$-CH_2C(=O)-O-Bz$

$-CH_3$

Scheme 1.

In this study, the free ε-amine groups of the lysine repeat units in the backbone of pLAL have been utilized to initiate the ring opening polymerization of NCA's to yield graft copolymers composed of a poly(L-lactic acid) backbone and poly(amino acid) side chains. The concept of using lysine ε-amine groups as polymeric initiators for NCA polymerizations was demonstrated in the 1950's for the preparation of multichain poly(amino acids).[10] As shown in Scheme 2, the nucleophilic primary ε-amine of the lysine side chain attacks C-5 of the NCA leading to ring opening and formation of the amino acid amide, along with the evolution of CO_2. Propagation takes place via further attack of the amine group of the amino acid amides on subsequent NCA molecules. The degree of polymerization of the poly(amino acid) side chains and the corresponding amino acid content in the graft copolymers can be controlled by changing the M/I ratio for the NCA polymerization - that is, changing the ratio of NCA to lysine ε-amine groups of pLAL. In this work, grafted poly(amino acid) chains with degrees of polymerization ranging from approximately 10 to 100 were obtained.

The grafting of poly(amino acids) from the pLAL backbone was carried out in dioxane at room temperature for 2-4 days. With ZLYS-NCA and ALA-NCA, the polymerization solutions remained clear throughout, but with BASP-NCA the solutions became cloudy after 2 days. It was assumed that the cloudiness arises from the eventual insolubility of the pLAL-BASP copolymers as the BASP chain length increases. This was confirmed by the observation that the homopolymerization of BASP initiated by *n*-butyl amine under the same conditions also lead to precipitation from the polymerization solution. In all systems, the viscosity of the polymerization solutions increased with time as the polymerizations progressed. The highest viscosity was observed in the ZLYS systems, and those producing copolymers with high lysine content became almost gel-like within 1 day. For all systems, GPC analysis of the precipitated graft copolymers showed only one peak, indicating that no unreacted pLAL remains and that no linear homopoly(amino acids) are found.

Poly(L-lactic acid-*co*-L-lysine) Poly(L-lactic acid-*co*-Z-L-lysine) graft copolymers (pLAL-ZLYS) have been synthesized with lysine contents in the range of 7-72 mol%.[11] Table 1 lists all of the poly(L-lactic acid-*co*-amino acid) graft copolymers prepared, along with their molecular weights and amino acid content. The molecular weights of the graft copolymers can vary greatly, depending upon the initial molecular weight of pLAL and the amount of amino acid incorporated. Removal of the Z groups from pLAL-ZLYS has been accomplished using HBr/HOAc in a manner similar to that used to deprotect the pLAZL linear copolymers. In this case, too, ^{1}H NMR of the resulting copolymer shows no Z phenyl peak indicating that complete deprotection was achieved. At this stage, the ε-amines of the lysine units of the copolymer are in the form of the hydrobromide salts. Further reaction with diisopropylethyl amine yields the free amines, which can be used as reactive sites for the attachment of biologically active molecules.

Of particular interest for tissue engineering are the peptides possessing an RGD (arginine-glycine-aspartic acid) amino acid sequence. Found in proteins such as fibronectin, the RGD sequence has been shown to be active in promoting cell adhesion and growth.[12] As such, incorporation of RGD sequences as part of the copolymer structure is expected to enhance cell growth on the copolymers.

In neutral aqueous solution the ε-amine groups are protonated. We intend to take advantage of this characteristic of the copolymers in order to bind negatively charged molecules such as DNA, polysaccharides, and heparin to them.

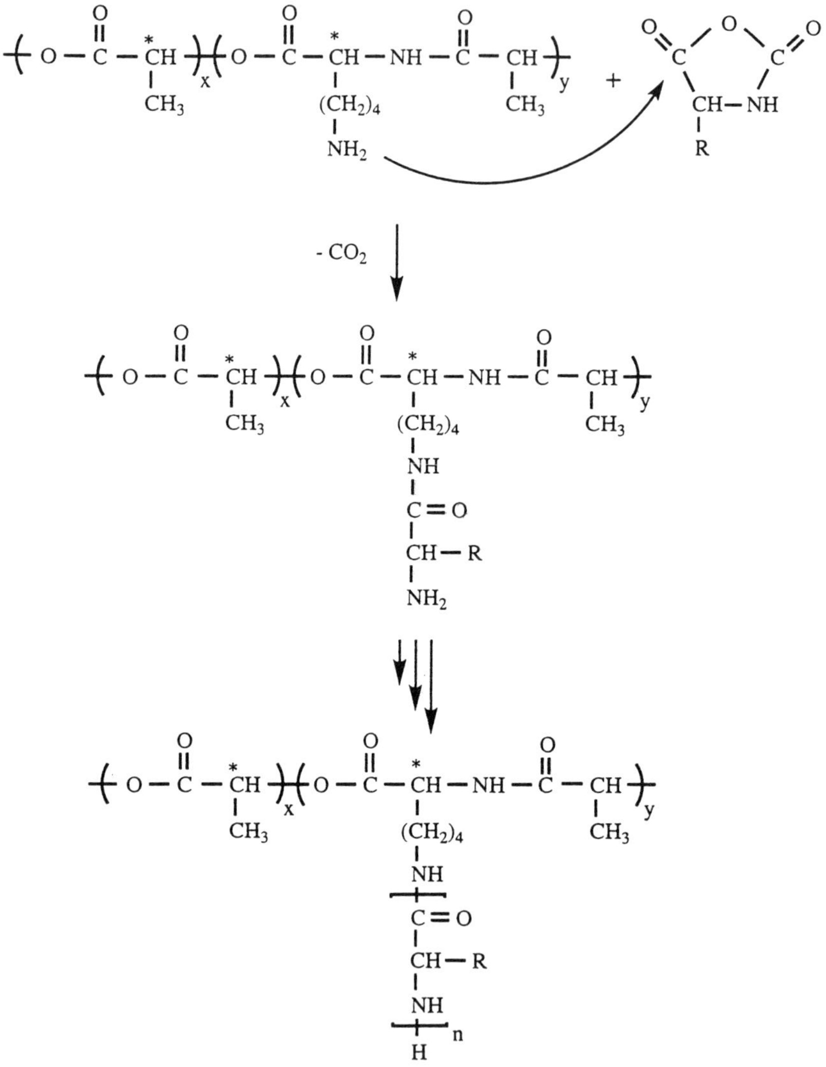

$$\left[O-\overset{O}{\overset{\|}{C}}-\overset{*}{\underset{CH_3}{CH}} \right]_x \left[O-\overset{O}{\overset{\|}{C}}-\overset{*}{\underset{\underset{NH_2}{|}}{\underset{(CH_2)_4}{CH}}}-NH-\overset{O}{\overset{\|}{C}}-\underset{CH_3}{CH} \right]_y + \text{(cyclic anhydride: } O=C-O-C=O,\ CH(R)-NH\text{)}$$

$-CO_2$

$$\left[O-\overset{O}{\overset{\|}{C}}-\overset{*}{\underset{CH_3}{CH}} \right]_x \left[O-\overset{O}{\overset{\|}{C}}-\overset{*}{CH}((CH_2)_4-NH-C(=O)-CH(R)-NH_2)-NH-\overset{O}{\overset{\|}{C}}-\underset{CH_3}{CH} \right]_y$$

$$\left[O-\overset{O}{\overset{\|}{C}}-\overset{*}{\underset{CH_3}{CH}} \right]_x \left[O-\overset{O}{\overset{\|}{C}}-\overset{*}{CH}((CH_2)_4-NH-[C(=O)-CH(R)-NH]_n-H)-NH-\overset{O}{\overset{\|}{C}}-\underset{CH_3}{CH} \right]_y$$

Scheme 2.

Poly(L-lactic acid-*co*-β-benzyl-L-aspartate) The precipitation observed in the synthesis of poly(L-lactic acid-*co*-β-benzyl-L-aspartate) graft copolymers results in a lower BASP content in the copolymers than that calculated from the ratio of BASP-NCA to lysine ε-amine groups of pLAL (the M/I ratio for the NCA polymerization). As seen in Table 1, elemental analysis shows that the BASP contents are about 55-70% of the amount calculated from the M/I ratio. Despite this difficulty, pLAL-BASP graft copolymers with BASP contents in the range of 6-35 mol% have been synthesized. Further studies will involve finding a more suitable solvent for the preparation of these copolymers.

The poly(L-lactic acid-*co*-β-benzyl-L-aspartate) graft copolymers were analyzed by ^1H NMR. A representative portion of the NMR spectrum of the copolymer with 35% BASP content is shown in Figure 1a. The peaks for the benzyl protecting groups of the BASP repeat units are seen at 7.4 ppm and 5.1 ppm. The small, broad peaks at 8.3 ppm and 4.7 ppm correspond to the NH and α-CH, respectively, in the BASP backbone. The quartet of the methine peak of the lactic acid repeat units is seen at 5.3 ppm, whereas the methyl peak at 1.5 ppm is not shown. Likewise, the BASP CH_2 at 2.8 ppm is not shown. The peak at 8 ppm corresponds to DMF.

Removal of the benzyl protecting groups will lead to free carboxylic acid groups, thus converting the BASP to aspartic acid residues. The carboxylic groups are quite versatile and can be utilized for the attachment of biologically active molecules, such as the RGD peptides mentioned previously. In addition, in a neutral aqueous environment, the poly(aspartic acid) side chains will be deprotonated, offering a negatively charged environment, which can allow the incorporation of positively charged molecules.

Poly(L-lactic acid-*co*-D,L-alanine) Figure 1b shows a representative region of the ^1H NMR spectrum of a poly(L-lactic acid-*co*-D,L-alanine) graft copolymer composed of 21% ALA content. PLAL-ALA graft copolymers with 35% ALA have also been prepared (see Table 1). The broad NH and CH peaks of the ALA backbone are seen at 7.95 ppm and 4.2 ppm, respectively. The quartet of the methine peak of the lactic acid repeat units is seen at 5.2 ppm. The methyl peaks for the lactic acid repeat units and ALA at 1.45 ppm and 1.2 ppm, respectively, are not shown.

Although it is not charged, poly (D,L-alanine) is a water soluble poly(amino acid), since the racemic mixture does not allow the formation of an internally hydrogen-bonded secondary structure that will prevent its solubility. As a result, the pLAL-ALA copolymers will be more hydrophilic than poly(lactic acid). This can be advantageous for applications where a hydrophilic material is required, yet a strongly charged environment cannot be tolerated. In addition, the ALA chains do possess a terminal amine group that can be utilized for the attachment of biologically active molecules.

As mentioned previously, the solutions for the grafting of ALA from pLAL remained clear throughout the polymerization. In contrast, poly(ALA) formed in the homopolymerization of ALA-NCA initiated by *n*-butyl amine under the same conditions precipitated out of the polymerization solution. Therefore, the pLAL keeps the ALA chains soluble in dioxane. This result demonstrates an important point. We are not only modifying the properties of poly(lactic acid) in the synthesis of the pLAL-amino acid graft copolymers. The connectivity of the amino acid polymer chains through the pLAL backbone has an influence on the properties of the poly(amino acids) themselves. Therefore, these copolymers can potentially also be applied toward improving some of the properties of poly(amino acids) and expanding the areas in which they may be used.

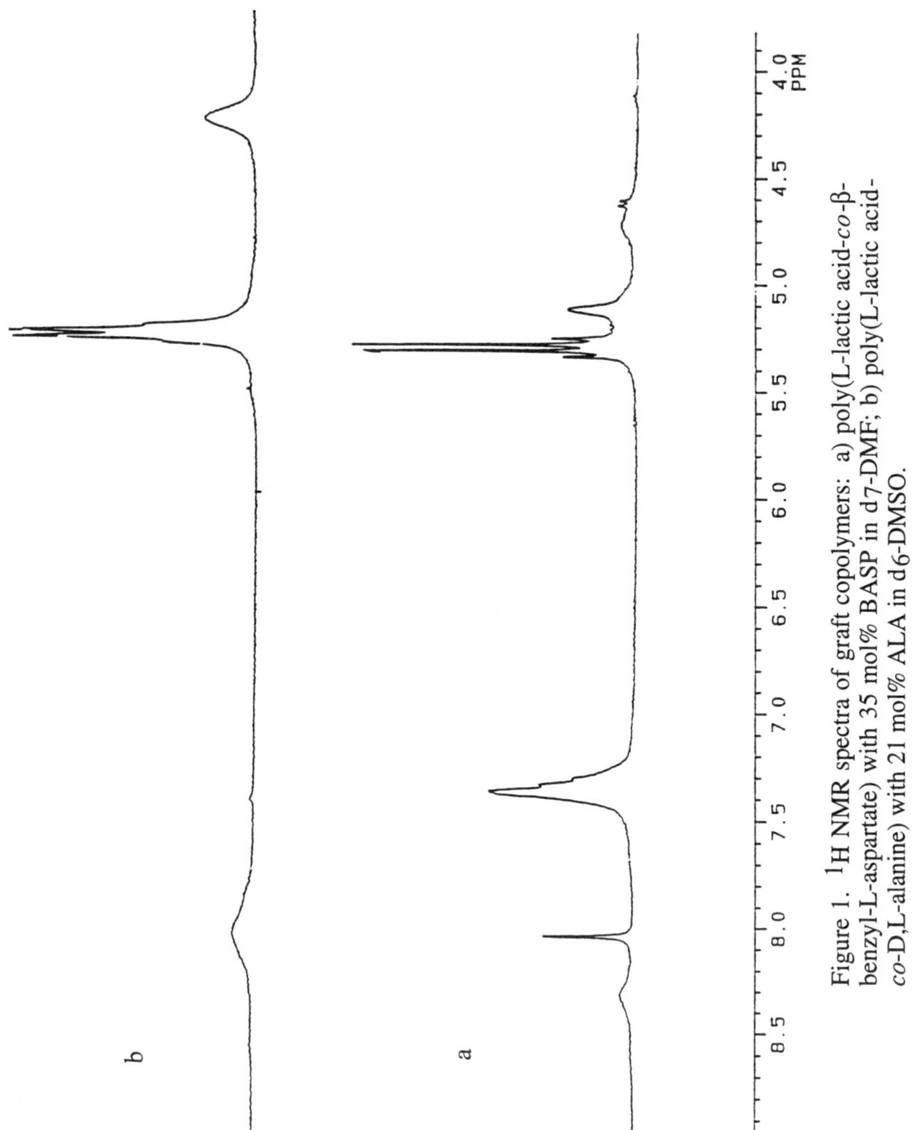

Figure 1. ^{1}H NMR spectra of graft copolymers: a) poly(L-lactic acid-*co*-β-benzyl-L-aspartate) with 35 mol% BASP in d_7-DMF; b) poly(L-lactic acid-*co*-D,L-alanine) with 21 mol% ALA in d_6-DMSO.

Table 1. Amino acid content and molecular weight of copolymers

Copolymer	%N[a]	Amino Acid Content (mol%)		MW (M_n x 10^{-3})	
		actual[b]	theor.[c]	pLAL	copolymer[d]
pLAL	0.40	1	---	50	---
pLAL-ZLYS	2.40	7	8	6	7.7
	5.39	22	17	50	100.8
	9.31	65	66	50	386.9
	9.64	72	81	6	61.2
pLAL-BASP	0.98	6	10	21.2	24.8
	2.02	13	20	21.2	30.1
	3.97	35	50	50	119.4
pLAL-D,L-ALA	4.03	21	15	21.2	26.7
	6.91	35	38	50	77.0

[a]Weight % N from elemental analysis. [b]Mol% amino acid calculated from weight %N. [c]Expected amino acid content based on pLAL $[LYS]_o$ and $[NCA]_o$ assuming all NCA is polymerized and incorporated in copolymer. [d]Molecular weight of copolymer calculated from weight %N and M_n pLAL.

Conclusions

We are developing a new class of biomaterials, in which poly(amino acid) chains are grafted from an appropriately modified polyester backbone. As the first stage in this program, we have synthesized poly(L-lactic acid-*co*-Z-L-lysine), poly(L-lactic acid-*co*-β-benzyl-L-aspartate), and poly(L-lactic acid-*co*-D,L-alanine) comb-like graft copolymers.

The amino acid content of the copolymers, and their resulting chemical and physical characteristics, can be varied in a controlled fashion by changing the molar ratio of the amino acid-NCA concentration to lysine units in the linear pLAL backbone. This new class of copolymers possess a variety of functional groups which provide the opportunity to further modify them by chemically attaching a variety of biologically active molecules. These modifications can be carried out so as to tailor the resulting biomaterials in order to meet the specific needs of particular biomedical applications.

Acknowledgment

We gratefully acknowledge Advance Tissue Sciences (La Jolla, CA) and NSF (NSF Grant No. BCS-9202311) for their financial support of this research.

References

1. Gilding, D. K.; Reed, A. M. *Polymer*, **1979**, *20*, 1459.
2. Lewis, D. H. in *Biodegradable Polymers as Drug Delivery Systems*, Chasin, M.; Langer, R., Eds.; Marcel Dekker: New York, 1990; p 1.
3. Mooney, D. J.; Organ, G.; Vacanti, J. P.; Langer, R. *Cell Transpl.*, **1994**, *2*, 203.
4. Li, S.M.; Garreau, H.; Vert, M. *J. Mater. Sci. Mater. Med.*, **1990**, *1*, 131.
5. Daly, W. H.; Poche', D. *Tetrahedron Lett.*, **1988**, *29*, 5859.
6. Barrera, D. A.; Zylstra, E.; Lansbury, P. T., Jr.; Langer, R. *J. Am. Chem. Soc.*, **1993**, *115*, 11010.

7. Kricheldorf, H. R. In *Models of Biopolymers by Ring-Opening Polymerization*; Penczek, S., Ed.; CRC Press: Boca Raton, 1990; Chapter 1.
8. Kricheldorf, H. R. α-Aminoacid-N-Carboxy-Anhydrides and Related Heterocycles; Springer-Verlag: Berlin, 1987.
9. Imanishi, Y. In *Ring-Opening Polymerization*; Ivin, K. J.; Saegusa, T., Eds.; Elsevier: London, 1984; Volume 2; Chapter 8.
10. Sela, M.; Katchalski, E.; Gehatia, M. *J. Am. Chem. Soc.*, **1956**, *78*, 746.
11. Hrkach, J.S.; Ou, J.; Lotan, N.; Langer, R. *Macromolecules*, in press.
12. Massia, S. P.; Hubbell, J. A. *J. Cell. Biol.* **1991**, *114*, 1089.

RECEIVED January 3, 1996

Chapter 9

Chemical and Morphological Changes in Peroxide-Stabilized Poly(L-lactide)

Anders Södergård

Department of Polymer Technology, University of Åbo Akademi, Porthansgatan 3–5, 20500 Turku, Finland

The melt degradation of poly(L-lactide) was studied at 180 °C using a melt rheometer for monitoring changes in the melt viscosity. Further studies of the degraded PLLA were made using SEC, ^{1}H-NMR and DSC . The thermal degradation was found to proceed in a random main-chain scission procedure. The mechanisms include depolymerization reactions and oxidative degradation in the presence of oxygen. However, hydrolysis as well as inter- and intramolecular reactions could not be excluded. Peroxide addition was found to stabilize the polymer against thermal degradation. The modified material was analyzed using SEC, XPS, DSC, TGA and by microscopic studies. Hydroperoxides, peroxy acid, peroxy ester, dialkanoyl peroxide, diaroyl peroxide and dialkyl peroxydicarbonate stabilized the melt during mixing. Non-stabilizing peroxides were dialkyl and diaralkyl peroxide. The stabilization mechanism was deactivation of the species that catalyze the reversible esterification reaction. The melt temperature of the peroxide modified polymer and the crystallinity decreased.

All polymers are susceptible to degradation, which often precludes the use of polymers in many fields. However, during the last decades the growing interest for polymers with a poor stability has resulted in several new applications.

The degradation processes of a polymer can be divided into two types: the first type are those involving absorption of energy, which then causes the propagation of molecular degradation by secondary reactions. This process can be initiated by heat, electromagnetic radiation, mechanical stress or ultrasonic vibration. Secondly, there are hydrolytic mechanisms that may result in molecular fragmentation, usually in heterochain polymers where the process can be seen as the reverse of polycondensation (*1*).

0097–6156/96/0627–0103$15.00/0

Poly(L-lactide) is an aliphatic polyester, which degrade easily according to the above mentioned mechanisms. The poly(L-lactide) has during the last decades been used in such applications where a poor stability is an advantage, mainly in various medical devices (*2*). Ring-opening polymerization of L-lactide, prepared by cyclic dimerisation of lactic acid, is the preferred polymerization route for high molar mass poly(L-lactide). The polymerization of lactide has successfully been carried out using melt-polymerization, bulk-polymerization, solution-polymerization and suspension-polymerization techniques (*3*). Each of these methods has its advantages and disadvantages, but generally the melt-polymerization is considered as the most simple and most reproductive method, though the melt-polymerization causes degradation during the process (*3*). A large number of metal compounds have been used as initiators for the ring-opening polymerization, particularly organo-metallic compounds containing tin compounds have been proved to be the most effective initiators (*4*). Stannous-2-ethylhexanoate (tin-octoate) is accepted by the FDA (Food and Drug Administration, Washington DC), which have led to a more common use of this specific catalyst (*4*). The polymerization of L-lactide in presence of tin-octoate starts with a complexation of tin-octoate with the exocyclic oxygen atom, followed by insertation of the L-lactide into the Sn-O bond accompanied by cleavage of the monomer acyl-oxygen bond. The polymerization proceeds by successive insertations of coordinated L-lactide dimers into the Sn-O bond (Figure 1) (*4*).

Polyesters are intrinsically unstable, which is utilized in the applications of poly(L-lactide). However, a tendency to degradation also appears during the manufacturing of the polymers. The melt processing, i.e. molding, extrusion and fiber drawing of polyesters is, in general, accompanied by an undesired melt degradation of the polymer (*5, 6*). The thermal degradation of polylactide has been suggested to be a random main chain-scission reaction of first order (*7*). According to earlier reported results the reactions involved in the melt degradation process of polylactide can be hydrolysis (*8*), depolymerisation (*9, 10*), thermooxidative degradation (*11, 12*) as well as intra- and intermolecular transesterifications (*9, 11*). Acidic end-groups, residual catalyst, moisture, residual monomers and other impurities have been reported to enhance the thermal degradation (*8, 9*). The melt degradation may cause problems with the control of the processing and lead to a decrease in the mechanical properties of the end products.

The present work deals with the melt stability of poly(L-lactide). One objective was to study the degradation process in the polymer during melt mixing using an internal mixer equipped with a rheometer, and by using standard analysis techniques for polymers. The second part of the work deals with the development of a stabilization process for poly(L-lactide) in the melt, and studies of the changes in the stabilized polymer.

Materials and Methods

The poly(L-lactide) used in this study was a pilot scale sample prepared by Neste Oy, Finland. Commercial grade L-lactide (Purac) was polymerized in the melt using tin-octoate (Sigma-Aldrich) as catalyst. The catalyst added in all experiments was 0.11 wt-%. The weight average molar mass of the base polymer was about 190 000.

The moisture content of the polymer after the polymerization was reported to be 0.02 wt-%. The polymer was permanently stored in a desiccator over P_2O_5 and used as received without further drying except for one experiment when the PLLA was dried in vacuum at 80 °C for 72 hours in order to examine the influence of polymer pre-drying. The drying conditions were chosen according to recommendations given by Boeringer Ingelheim, Germany. The dibenzoyl peroxide used in the study was from Fluka (moisture content about 25 wt-%), and used without further treatments. Sn-octoate from Sigma-Aldrich Corp. was used in the experiments where extra catalyst was added. The Sn-octoate contained water, which was separated by distillation before the use of the catalyst.

The sample preparation and melt viscosity measurements of PLLA during mixing was performed using a Brabender internal mixer (Plasti-corder PL 2000) equipped with a pneumatic loading chute. The mixer was supplied with a fluid inlet in the front, that also allowed gas to be flushed through the melt-zone. All experiments were conducted under constant gas flow, and nitrogen atmosphere was used in all experiments except for the one where the influence of oxygen was studied. The melt temperature and torque data were collected at 60 rpm. An amount of 40 g of PLLA was used in every experiment as this amount polymer should, according to Brabender AG, give the optimal mixing. In the experiments of peroxide addition the peroxide was introduced to the mixer together with the polymer.

The molar mass measurements were obtained by Size Exclusion Chromatography (SEC). The system was based on a Spectroflow 400 pump from Kratos, and a pre-column (AN Gel Guard Column) and a linear column (AN Gel Linear) from American Polymer Standards Corporation. The mobile phase in the system was tetrahydrofuran (chromatography grade) and the operating conditions in the system were: flow rate 0.3 ml/min at ambient temperature; sample concentration 3 mg/ml. For the calibrations polystyrene standards were used (Perkin Elmer). The injected sample volume was 75 μl and the sample solution was filtered through a 5 μm filter before the injection. A refractive index detector from Knauer, Germany, was used. The ^{1}H-NMR spectra were collected with a 60 MHz spectrometer (10 wt-% PLLA in $CDCl_3$). The formation of gel was determined according to ASTM D 2765, with the exception of using chloroform as solvent instead of xylene. As a reference method for further confirmations of the gel formation a particle size analyzer Malvern 2600c was used to detect network agglomerates sized in the region between 1.9 and 188 μm.

Thermal analysis of the samples was carried out using a Perkin Elmer Differential Scanning Calorimeter (DSC) System 4. The samples were scanned at the rate of 10 °C/min and after the first run the samples were cooled down rapidly (320 °C/min) a second run was performed. The annealing studies of the unstabilized and stabilized PLLA were performed using the Perkin Elmer DSC under isothermal heating of the samples at 100 °C for times between 5 minutes and 18 hours. The values reported from the DSC measurements are based on the values recorded at the second run, except in the annealing studies. The reported values of the degree of crystallinity were calculated using 93.6 J/g as value for 100 % crystallinity (*13*). Lamellar thickness distribution was calculated for dibenzoyl peroxide-PLLA mixtures from DSC data according to a method described elsewhere (*14*).

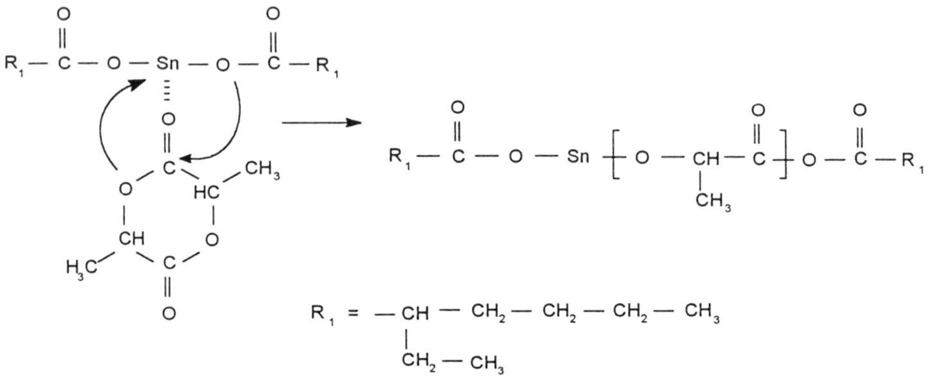

$$R_1-\overset{O}{\overset{\|}{C}}-O-Sn-O-\overset{O}{\overset{\|}{C}}-R_1 \;\; + \;\; \text{lactide} \longrightarrow R_1-\overset{O}{\overset{\|}{C}}-O-Sn\left[O-\underset{CH_3}{\underset{|}{CH}}-\overset{O}{\overset{\|}{C}}\right]O-\overset{O}{\overset{\|}{C}}-R_1$$

$$R_1 = -\underset{CH_2-CH_3}{\underset{|}{CH}}-CH_2-CH_2-CH_2-CH_3$$

Figure 1. Mechanisms for the tin-octoate initiated polymerization of lactide.

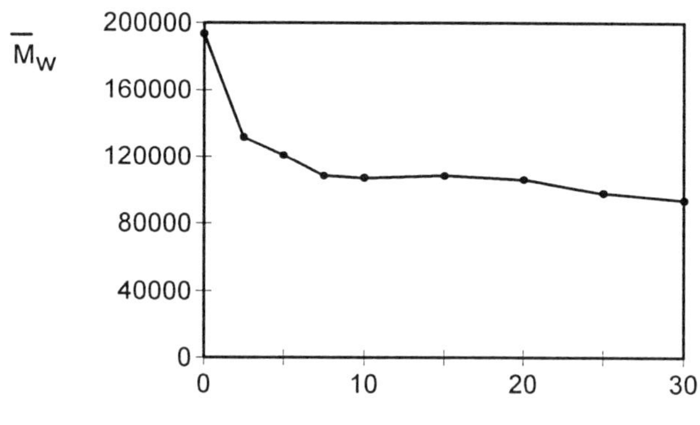

Figure 2. Molar mass reduction during melt mixing PLLA at 180 °C. (Reprinted with permission from reference 16. Copyright 1994, Elsevier Science Limited).

For the microscopic studies thin films were cast from a chloroform - PLLA solution (1 wt-%) onto an objective glass and allowed to dry slowly over chloroform and annealed for one hour at 130 °C. The change in the chemical state of tin was examined using a Perkin Elmer 5400 ESCA spectrometer. In order to receive a stronger signal from the tin element the samples for XPS measurements were in one case prepared by adding 1 wt-% stannous octoate to the polymer and in an other case by adding 1 wt-% stannous octoate and 1 wt-% dibenzoyl peroxide.
The experimental methods involved in the study of the influence of different peroxide types are described elsewhere (*15*).

Melt Degradation of Poly(L-lactide)

Poly(L-lactide) has been reported to degrade rapidly above its melting point (*16*). The reduction of the molar mass during thirty minutes of melt mixing at 180 °C is presented as a function of the time in Figure 2. The decrease in molar mass was rapid during the first five minutes, but after about eight minutes the change in molar mass leveled out.
No apparent differences could be noticed in the shape of the molar mass distribution curves of the basepolymer and the polymer kneaded for 30 minutes at 180 °C. This verifies that random main chain scission degradation occurs in the melt. During the mixing, volatile compounds were noticed to condense into rod-shaped crystals outside the mixer. ^{1}H-NMR measurements of the condensation product showed that the peaks were shifted compared to the peaks of the basepolymer (Table I).

Table I. Chemical shifts in ^{1}H-NMR of mono- and poly(L-lactide)

	Chemical shifts, δ	
Ester compound	CH	CH_3
L-lactide[a]	1.66	5.09
poly(L-lactide)[a]	1.56	5.17
PLLA basepolymer	1.60	5.22
PLLA condensation product	1.69	5.10

a) Values reported by Kricheldorf et al. *(17)*.

The shifts of the condensation product were close to those reported for L-lactide (*17*), which proved that the polymer has depolymerized into lactide during the melt mixing. DSC measurements of the condensation product gave a melting point of 96.3 °C, which further confirmed the depolymerization.

The changes in molar mass of a linear polymer can be monitored by measuring the melt viscosity of the polymer. Different melt viscosity curves of PLLA at 180 °C are shown in Figure 3, where the upper curve is the melt viscosity curve when no extra Sn-octoate was added, and the two other curves are the melt viscosity of PLLA when 0.5 and 1.0 wt-% Sn-octoate was added, respectively.

The initial rapid decrease in the melt viscosity could, based on the corresponding molar mass measurements (Figure 2), be related to the melting process of the granules as well as to the melt degradation. From these curves it is apparent that the more Sn-octoate present in the polymer, the more rapid is the depolymerization during the thermal processing of PLLA.

Polymer predrying of the PLLA has been reported to on the one hand give a more stable polymer, and on the other hand not to influence the thermal degradation. From these contradictory results one can assume that the influence of predrying depends on the history of the PLLA (polymerization routine, post-polymerization treatments). In order to study the influence of predrying on the poly(L-lactide), the polymer was dried at 80 °C for 72 hour. The melt viscosity curve of the predried PLLA was not different from the one of the undried polymer. The explanation to these results could be either that no moisture has been removed (too strongly bound), or that such small amounts of water do not affect the degradation process. However, the fact that polyesters undergo hydrolysis in presence of water can not be neglected when the stability of PLLA in the melt is discussed.

With the mixing equipment used in this study the melt viscosity measurements could be performed in nitrogen or in oxygen atmospheres. Figure 4 shows the melt viscosity curve in nitrogen (a) and in oxygen (b) at 180 °C. During the first minutes, the melt viscosity of PLLA in the oxygen atmosphere decreased slower than in nitrogen atmosphere. Oxygen obviously retarded the melt degradation during the first minutes, but after about five minutes a drastic slope in the melt viscosity curve of PLLA occurred. This behavior can be explained by an initial stabilization followed by a dominant thermooxidative degradation after about five minutes.

Stabilization of Poly(L-lactide) in the Melt

By introducing oxygen in the melt zone, the melt viscosity curve was almost stable for several minutes. One theory was that the stabilization was due to an oxidation process, and this was tested by experiments where melt mixing in the presence of an other oxidant, dibenzoyl peroxide, was performed (*16*). Poly(L-lactide) and 0 - 3 wt-% dibenzoyl peroxide were mixed together at 180 °C, and melt viscosity curves were recorded (Figure 5). As much as an amount of 0.1 wt-% dibenzoyl peroxide significantly retarded the melt degradation, and this behavior was continued as the peroxide content was increased. However, an addition of more than 0.5 wt-% peroxide only slightly affected the torque further. Molar mass determinations of PLLA after peroxide addition and 5 minutes of mixing (Table II) showed that the peroxide stabilized the polymer.

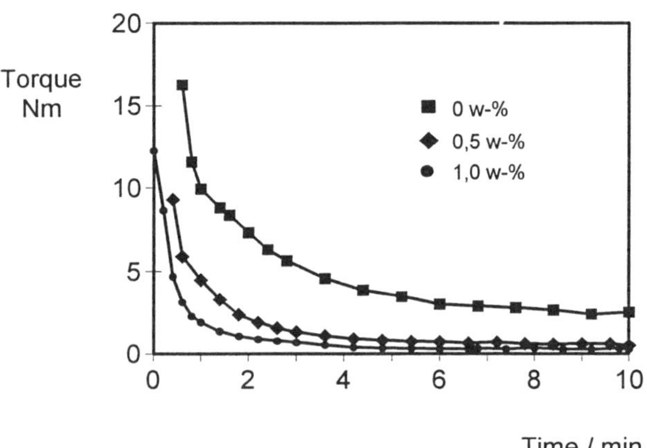

Figure 3. Melt viscosity of PLLA after addition of various amounts of Sn-octoate at 180 °C. (Reprinted with permission from reference 16. Copyright 1994, Elsevier Science Limited).

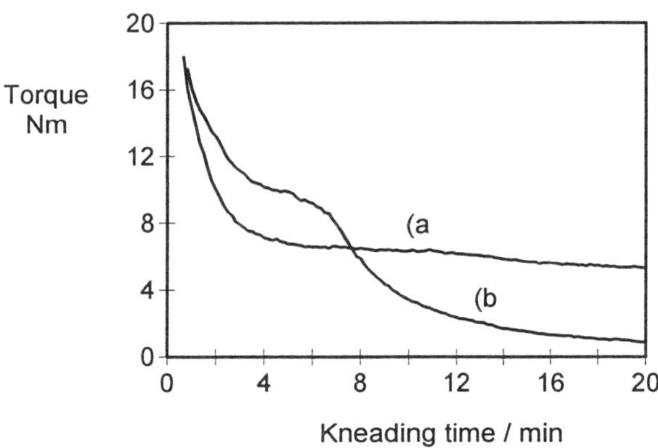

Figure 4. Melt viscosity behavior of PLLA in a) nitrogen b) oxygen atmospheres at 180 °C. (Reprinted with permission from reference 16. Copyright 1994, Elsevier Science Limited).

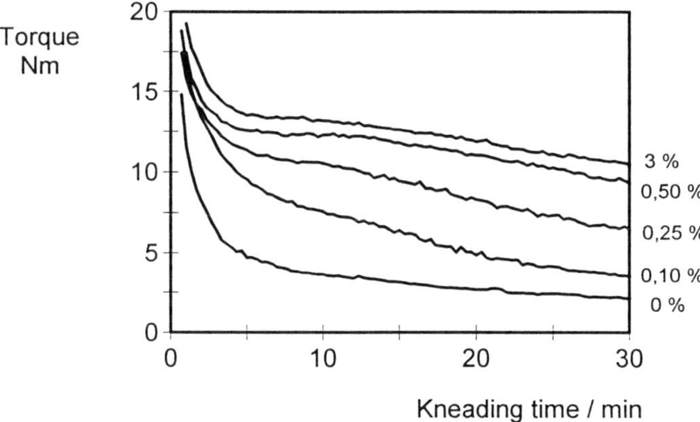

Figure 5. The influence of amount added peroxide on the melt viscosity of PLLA at 180 °C. (Reprinted with permission from reference 16. Copyright 1994, Elsevier Science Limited).

Table II. Weight average molar mass, gel content and particle size distribution of the crosslinked regions of PLLA as a function of dibenzoyl peroxide amount added after 5 minutes of melt mixing at 180 °C

Peroxide cont.	M_w	Gel cont.	Particle size distribution			
wt-%	g/mol	%	1.9-10 μm	10-50 μm	50-100 μm	100-188
0	104 000	0	0 %	0 %	0 %	0 %
0.10	173 000	0	0 %	0 %	0 %	0 %
0.25	159 000	0.9	86 %	13 %	0.2 %	0.8 %
0.50	168 000	1.8	50 %	31 %	3.7 %	15.5 %
3.0	102 000	5.1	0.2 %	1.3 %	5.1 %	93.4 %

An increase in the melt viscosity of polymers, when peroxides are added, is in general a result of crosslinking. In peroxide modified poly(L-lactide), however, no gel formation could be noticed at peroxide contents below 0.25 wt-%. After a dibenzoyl peroxide addition of 3 wt-%, the gel content was no more than 5 % (Table II). Verifying measurements were conducted with light scattering in order to detect crosslinked regions in the peroxide modified polymer. The size distribution in PLLA-$CHCl_3$ solutions proved that no agglomerates were formed after a peroxide addition less than 0.25 wt-% (Table II). Increased peroxide addition increased both the chain scissions and the crosslinking, which can be seen in Table II as a decreased molar mass and an increased gel formation. The observation of the increased amount of chain scissions was based on SEC measurements where the crosslinked regions were removed by filtering before the measurements.

The decrease in the melt viscosity of PLLA could be reduced by peroxide addition, and this stabilization was not caused by crosslinking. The most logical explanation is that the stabilization was due to a deactivation of the residual catalyst. XPS studies of the residual tin in the polymer proved that a change in the oxidation state occurred in the tin element after peroxide addition to the PLLA melt (Figure 6a and 6b) (*18*). The Sn 3d electron binding energy was shifted to higher binding energies as a result of the peroxide addition. This can been explained by a change in the oxidation state (*19*) and/or by in the coordination number and electronegativity of the tin compound (*20*).

Poly(L-lactide) was modified in the melt using various types of peroxides. The choice of peroxide type influenced the stabilization significantly. The peroxyester, the dialkyl peroxy dicarbonate, the diacyl peroxide and the diaroyl peroxide retarded the decrease in the molar mass of poly(L-lactide) most effectively. Hydroperoxides also caused a stabilization. Tert-butylperoxide and dicumylperoxide did not stabilize the melt. These types of peroxides have no carbonyl oxygen atoms present in the molecules, and apparently not able to deactivate the tin compound. Any differences in the stabilizing ability between aromatic and aliphatic hydroperoxides and between diaroyl- and diacylperoxids could not be noticed.

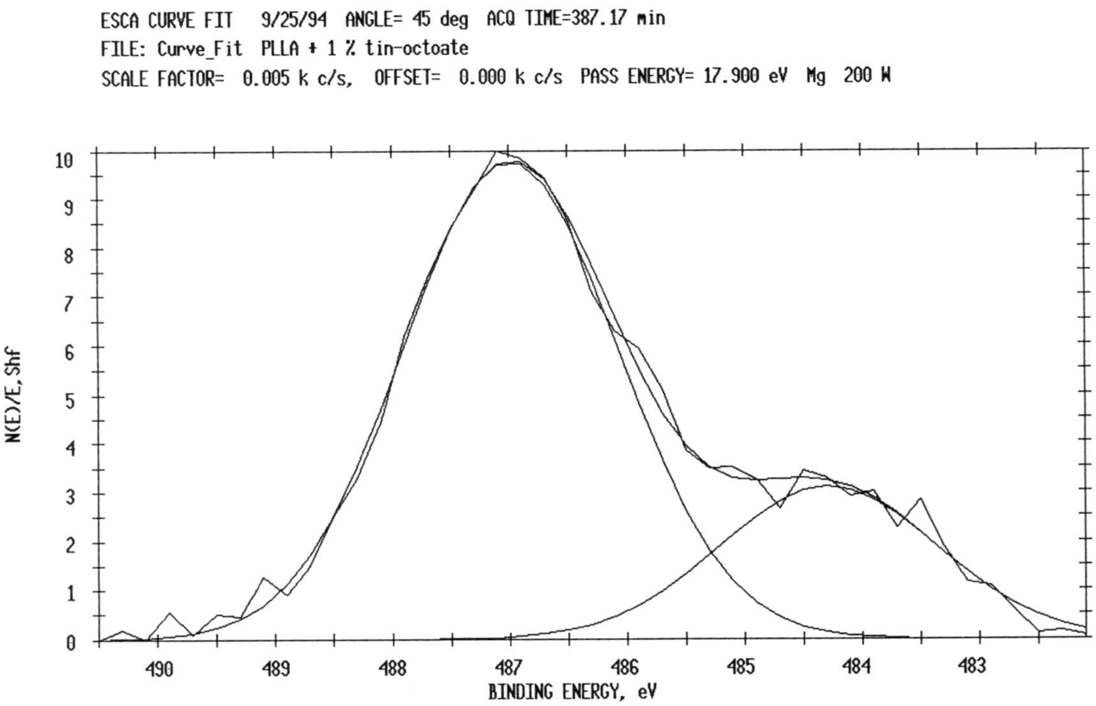

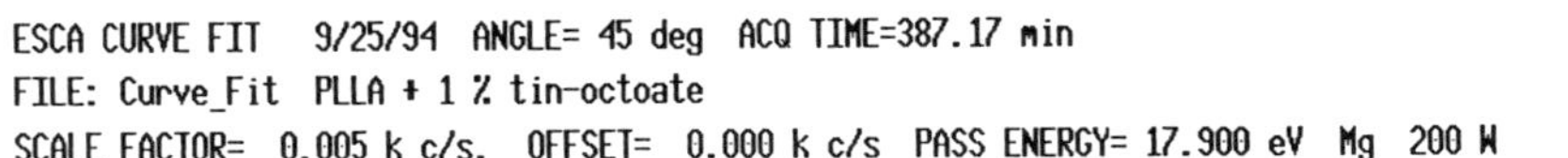

Figure 6a. XPS-spectra for the tin compound (Sn $3d_{5/2}$) in the PLLA sample after addition of 1 wt-% tin-octoate. (Reproduced from reference 15. Copyright 1995 American Chemical Society.)

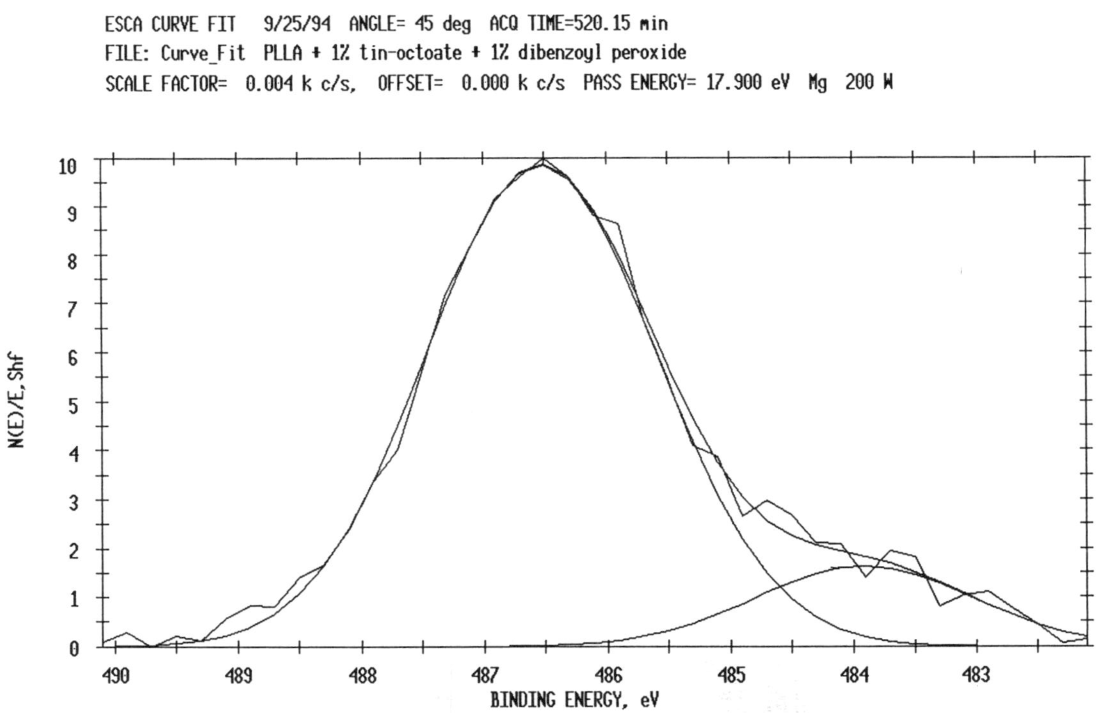

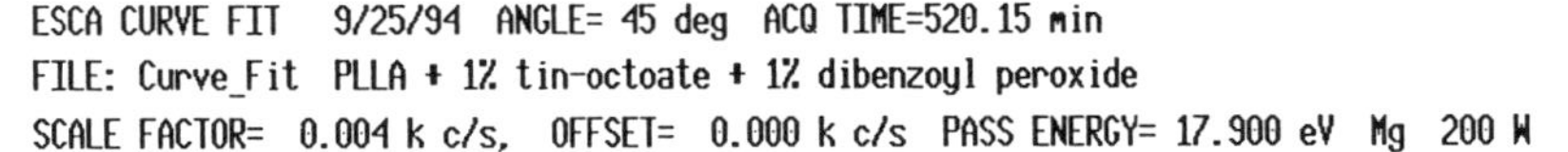

Figure 6b. XPS-spectra for the tin compound (Sn $3d_{5/2}$) in the PLLA sample after addition of 1 wt-% tin-octoate and 1 wt-% dibenzoyl peroxide.
(Reproduced from reference 15. Copyright 1995 American Chemical Society.)

Morphological Changes in the Stabilized Poly(L-lactide)

Calorimetric measurements of the stabilized polymer showed that the peroxide addition caused changes in the morphology. The degree of crystallinity, and also the melting point in some extent, decreased in the experiments when various amounts of dibenzoyl peroxide were added (Table III).

Table III. DSC data of the dibenzoyl peroxide modified PLLA

Peroxide content wt-%	T_m °C	X_c %
0	165.6	48
0.1	164.0	39
0.5	162.9	42
1.0	161.4	40
3.0	156.7	36

Microscopic studies of the crystals in the stabilized PLLA compared to unstabilized polymer showed that the crystal size decreased, but the amount of crystals increased when dibenzoyl peroxide was added (Figure 7). This explains the decrease in the degree of crystallinity for the stabilized poly(L-lactide) due to a slower nucleation rate. Further crystallisation studies were performed by annealing experiments using DSC (Figure 8). The degree of crystallinity was measured as a function of an annealing time between 5 minutes and 18 hours at 100 °C.

The crystallization rate of the samples drastically decreased when tert-butylperoxy benzoate was added. From these results it is apparent that the increased amount of side-chains, or the peroxide decomposition products, cause a slowing down effect on the crystallization. This can be explained by reduced molecular motion due to the introduced bulky groups or/and by effects from the peroxide residues. However, the degree of crystallinity for the peroxide stabilized poly(L-lactide) could with time reach the same level as the unstabilized PLLA. This is possible due to a slower reorganization of the polymer chains. No correlation between the crystallization rate and amount of peroxide added could be recorded. An addition of 0.25 wt-% tert-butylperoxy benzoate was enough to decrease the crystallization rate to the same extent as an addition of 0.5 wt-%.

In order to study the crystal thickening behavior, the lamellar thickness distribution was calculated from the DSC-measurements of the dibenzoyl peroxide stabilized poly(L-lactide). The lamellar thickness distribution proved to be independent of the peroxide amount added. This behavior is explained by the fact that the crystals are growing in the same manner in both unstabilized and stabilized PLLA, but the rate of the crystallization for the stabilized polymer is slower than that of the unstabilized due to the hindrance.

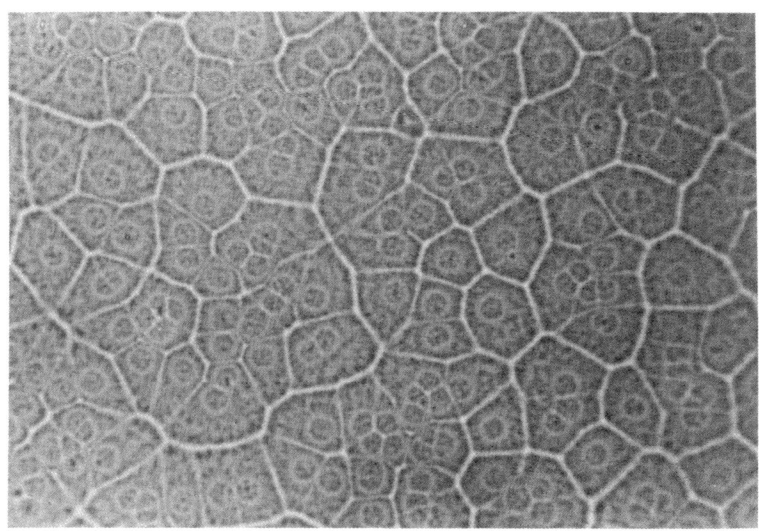

Figure 7a. Crystal size of PLLA basepolymer annealed 1 h at 130 °C (200x magnification).
(Reproduced from reference 15. Copyright 1995 American Chemical Society.)

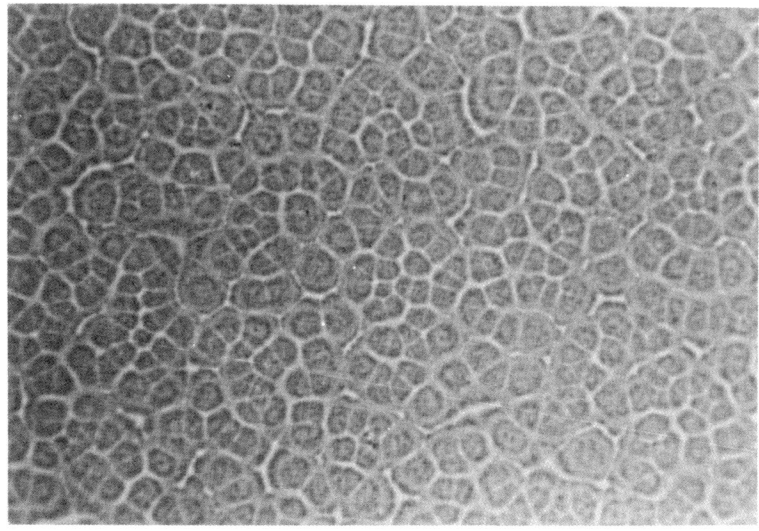

Figure 7b. Crystal size of PLLA + 0.5 wt-% dibenzoyl peroxide annealed 1 h at 130 °C (200x magnification).
(Reproduced from reference 15. Copyright 1995 American Chemical Society.)

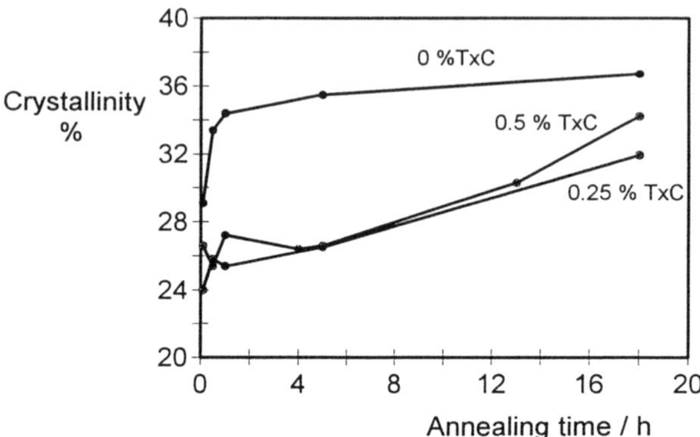

Figure 8. The degree of crystallinity as a function of annealing time for unstabilized and stabilized PLLA.
(Reproduced from reference 15. Copyright 1995 American Chemical Society.)

Conclusions

The following conclusions summarize the results of this study:

1) Poly(L-lactide) is a thermally unstable polymer, which undergoes a random main-scission melt degradation. Volatile product from the melt was identified as lactide, which obviously is produced by zipper-like depolymerisation

2) Melt mixing in oxygen atmosphere gives an initial stabilization of the melt, followed by a thermooxidative degradation. Addition of certain peroxide types reduces the melt degradation of poly(L-lactide).

3) The stabilization is achieved by a deactivation of the catalyst in the poly(L-lactide) by an oxidation of the tin compound.

4) The peroxide addition causes decreases in the degree of crystallinity and in the melt temperature in the peroxide modified PLLA due to a slower rate of crystallization caused by a decrease in crystal perfection.

Acknowledgments

The Fininsh Academy and Neste Säätiö are acknowledged for financial support and Neste Oy for materials. The author also wants to thank C. Eckerman for help with the microscopy and K. Ekman for valuable discussions concerning the XPS measurements.

Literature Cited

1. Williams, D.F., In *Comprehensive polymer science, Vol 6: Polymer Reactions*, Pergamon press, Oxford, Great Britain, 1989, 611.
2. Vert, M., Li, S. and Garreau, H., *J. Control. Rel.* **1991,** *16,* 15.
3. Nieuwenhuis, J., *Clin. Mater.* **1992,** *10,* 59.
4. Eenink, M.J.D, PhD Thesis, University of Twente, Holland, Feb. 1987.
5. Gogolewski, S., Jovanovic, M., Perren, S.M., Dillon, J.G. and Hughes, M.K., *Polym. Degr. Stab.* **1993,** *40,* 313.
6. von Oepen, R. and Michaeli, W., *Clin. Mater.* **1992,** *10,* 21.
7. Gupta, M.C. and Deshmukh, V.G., *Coll. Polym. Sci.*, **1982,** *260*, 308.
8. Törmälä, P., Presented at the Biopolymer Annual Seminar, Finnish Technology Center, Espoo, Finland, May 1993.
9. Jamshidi, K., Hyon, S.-H. and Ikada, Y., *Polymer* **1988,** *29,* 2229.
10. Zhang, X., Wyss, U.P., Pichora, D. and Goosen, M.F.A., *Polym. Bulletin* **1992,** *27*, 623.
11. McNeill, I.C. and Leiper, H.A., *Polym. Degr. Stab.* **1985,** *11,* 309.
12. Gupta, M.C. and Deshmukh, V.G., *Coll. Polym. Sci.*, **1982,** *260,* 514.
13. Migliaresi, C., De Lollis, A., Fambri, L., Cohn, D., *Clin. Mater.* **1991,** *8,* 111.
14. Alberlo, N., Cavaille, J.Y., Perez, J., *J. Poly. Sci.,: Part B: Polym. Phys.*, **1990**, *28*, 569.
15. Södergård, A., Niemi, M., Selin, J.-F., Näsman, J.H., *Ind. & Eng. Chem. Res.*, **1995,** *34*, 1203.
16. Södergård, A., Näsman, J.H., *Polym. Degr. Stab.*, **1994**, *46*, 25.
17. Kricheldorf, H.R., Berl, M. and Scharnagl, N., *Macromolecules* **1988**, *21,* 286.
18. Södergård, A. Näsman, J.H., *Polymer Preprints*, **1994**, *2*, 431.
19. Samorjai, G.A., *Introduction to Surface Chemistry and Catalysis*, John Wiley & Sons, Inc., New York, 1993.
20. Willemen, H., van de Vondel, D.F., van der Kerlen, *Inorg. Chim. Acta,* **1979,** *34,* 175.

RECEIVED January 2, 1996

Chapter 10

Biodegradability of Polycarboxylates: Structure–Activity Studies

M. B. Freeman, Y. H. Paik, G. Swift, R. Wilczynski, S. K. Wolk, and K. M. Yocom[1]

Rohm and Haas Company, 727 Norristown Road, Spring House, PA 19477

It is generally recognized that in order to improve the environmental acceptability of household laundry detergents, *biodegradble* replacements for the polycarboxylate components of these detergent formulations are needed. Although the poly(acrylic acid) homopolymers and copolymers currently in use are not considered harmful to the environment, they are not readily biodegradable; they enter environmental compartments beyond the sewage treatment plant. Poly(aspartic acid), on the other hand, has recently emerged as a benchmark for the biodegradability of synthetic polycarboxylates which exhibit acceptable detergent performance. Semi-continuous activated sludge (SCAS) removability tests and modified Sturm CO_2 production tests show that poly(α,β-D,L-aspartates) prepared via a phosphoric acid-catalyzed thermal polymerization process are rapidly and completely biodegraded by municipal treatment plant microorganisms. Polyaspartates prepared via uncatalyzed thermal polymerizations or maleic anhydride/ammonia-based processes are only partially biodegradable in these tests. Structural analysis indicates that a linear polyamide backbone is key to the total biodegradability of the acid-catalyzed polyaspartates. For the poly(acrylic acids) (C-C backbone), modified Sturm tests show that molecular weights must be in the oligomer range for total biodegradability.

Polycarboxylates - in particular, poly(acrylic acid) and copolymers of acrylic acid and maleic acid (Figure 1) - are widely used in low-phosphate and phosphate-free household laundry detergents. These polycarboxylates are generally found in zeolite- or soda ash-built detergent formulations at about 2-5 wt. % as builder assists (or cobuilders). As such they improve the cleaning performance of these powdered detergents by dispersing soils (thus improving soil removal and helping prevent soil redeposition) and by inhibiting the crystal growth of inorganic salts (thus preventing the incrustation of salts, e.g., calcium carbonate, on fabrics). Polycarboxylates are also known to function as process aids in the production of powdered laundry

[1]Corresponding author

0097–6156/96/0627–0118$15.00/0

detergents. To a lesser extent, polycarboxylates find utility as dispersants in such products as automatic dishwashing detergents and industrial and institutional cleaners.

The switch from phosphate-built laundry detergents to the no-P formulations has been driven primarily by environmental concerns, namely by the need to minimize the eutrophication of lakes and streams. Currently the use of phosphates in home laundry detergents is banned in many areas. The environmental fate and effect of *all* the components of laundry detergents (and other cleaning products as well) are coming under closer scrutiny. Polycarboxylates are no exception.

Recently reported studies on the environmental fate and ecotoxicity of the polycarboxylates currently used in household laundry detergents conclude that these materials have no adverse impact on the environment and are of extremely low toxicity (*1*). The necessity for these studies stems from the fact that the polycarboxylates commonly used in the U.S. (i.e., acrylic acid homopolymers, Mw ~4500) and in Europe (i.e., acrylic acid/maleic acid copolymers, Mw ~70,000), are not totally biodegradable. The favorable environmental assessment of these materials results primarily from the findings that they are, in large part, removed from wastewater by precipitation and/or adsorption on sewage sludge. (Some polymer does enter environmental compartments beyond the sewage treatment plant - in the aqueous effluent and/or with the sludge - thus the need for the toxicitiy testing.)

Although not critical from an environmental fate/toxicity standpoint, it is generally recognized that the environmental acceptability of household laundry detergents could be improved if totally biodegradable replacements for the polycarboxylate components of these detergent formulations were available. Ideally these polymers would be totally degraded to gaseous products, minerals, and biomass at a rapid rate, i.e., before they leave the treatment plant. Since it is unlikely that an effective detergent polymer will biodegrade within the hydraulic retention time of a treatment plant (4-8 hrs.) (*2*), it is advantageous to have a polymer which is highly adsorptive on sewage sludge so that biodegradation can occur within the sludge retention time (~5-15 days) (*2*) - a more realistic goal.

In the search for such a biodegradable water-soluble detergent polymer, poly(aspartic acid) has recently emerged as a benchmark (*3*). Poly(α,β-D,L-aspartates) (Figure 1) can be prepared which exhibit acceptable detergent performance, are totally removed in the presence of sewage sludge, and are rapidly and completely biodegraded by municipal treatment plant microorganisms. We have assessed the biodegradability of polyaspartates prepared via several different synthetic routes by carrying out semi-continuous activated sludge (SCAS) removability tests and modified Sturm (CO_2 production) tests. For comparison, we have also carried out some modified Sturm tests on low molecular weight poly(acrylic acids). In addition to the primary objective of synthesizing polyaspartates which are rapidly and completely biodegraded, an understanding of the structure-biodegradation relationships for the polyaspartates and the poly(acrylic acids) was also pursued.

Experimental

Materials. The sodium salts of poly(aspartic acid) used in the biodegradation studies were obtained by hydrolyzing polysuccinimide (poly(anhydroaspartic acid)). The polysuccinimides were prepared using several different synthetic routes (Figure 2). The synthetic methods used to prepare the polysuccinimides, and to hydrolyze them, are described below.

Synthesis of polysuccinimide without an acid catalyst. L-aspartic acid was heated at either 240°C or 270°C at ambient pressure in air. Reaction times were on the order of 5-9 hours, depending on the scale of the reaction and the type of

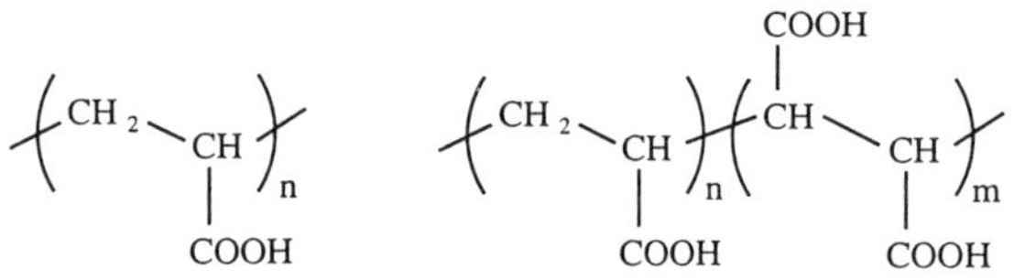

Figure 1. Detergent polycarboxylates.

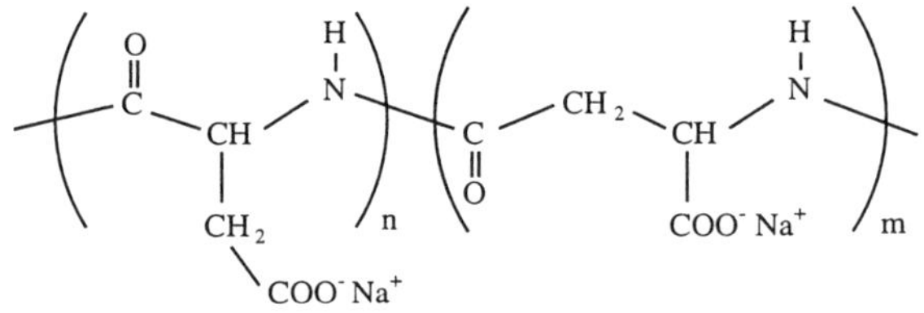

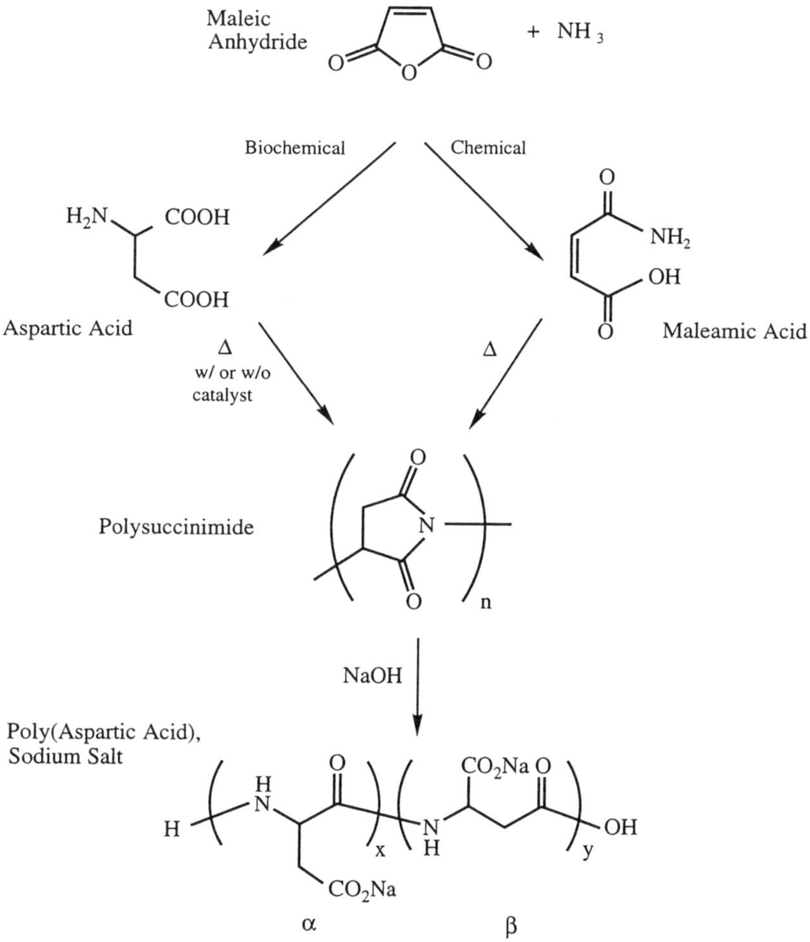
Maleic
Anhydride
+ NH 3
O
O
O
Biochemical
Chemical
H2N
COOH
COOH
Aspartic Acid
O
NH2
OH
O
Maleamic Acid
Δ
w/ or w/o
catalyst
Δ
Polysuccinimide
O
N
O
n
NaOH
Poly(Aspartic Acid),
Sodium Salt
H
N
O
H
x
CO2Na
CO2Na O
N
H
y
OH
α
β

Figure 2. Synthetic routes to sodium polyaspartate.

equipment used. Small lab-scale reactions were done in a beaker and the material was stirred at regular intervals. Larger scale reactions were done in a rotary tray dryer wherein the material was mixed continously throughout the reaction. Conversion to the desired product (a tan powder) was confirmed by NMR.

Synthesis of polysuccinimide with phosphoric acid as the catalyst. L-aspartic acid was mixed with 85% phosphoric acid and heated at 240°C at ambient pressure in air; reaction times ~2-7 hours. The amount of phosphoric acid used depended on the molecular weight desired. The higher molecular weight samples examined in this study were prepared by using up to 20 wt.% H_3PO_4 (85% aq. solution) in the reaction. These lab-scale reactions were generally carried out in beakers; the clumps which formed early in the reaction were ground with a mortar and pestle; thereafter the powder was stirred at regular intervals. A light tan powder was obtained. Complete conversion to polysuccinimide was confirmed by NMR.

Synthesis of polysuccinimide from maleic anhydride and ammonia. Maleamic acid was prepared in toluene from maleic anhydride and ammonia gas, similarly to known methods (*4*). The isolated and purified maleamic acid was converted to polysuccinimide by heating it to 190-200°C for one hour. These reactions were carried out either in a plow mixer or a double planetary mixer. Complete conversion of maleamic acid was confirmed by NMR. Polysuccinimide was also prepared on a laboratory scale by passing molten maleic anhydride and anhydrous ammonia through a stainless steel tube at ~235°C.

Hydrolysis of polysuccinimide to poly(aspartic acid), sodium salt. Typically, polysuccinimide powder in water was heated to 90°C and an aqueous solution of sodium hydroxide (50 wt. %) was added dropwise while maintaining the pH of the solution at about 11. After all the polysuccinimide was dissolved and the NaOH addition was complete, the solution was held at 90°C for an additional 0.5 hr. Some hydrolyses were done at lower temperatures and/or slightly higher pH. For biodegradation testing, all solutions were lyophilized to give solid sodium polyaspartate.

In some cases, removal of the catalyst salts from samples prepared with phosphoric acid was desired. This was accomplished by thoroughly washing the insoluble polysuccinimide with water to remove the soluble phosphate species. This was done prior to hydrolysis. Complete removal of the phosphates was confirmed by elemental analysis.

The poly(acrylic acid) homopolymers used in this study were laboratory versions of commercially available polymers. They were prepared by the free radical polymerization of acrylic acid in aqueous solution. The acrylic acid oligomer samples were obtained from a mixture of poly(acrylic acid) homopolymer solutions (Mw's ~500 and 1000) by chromatographic isolation of oligomer-containing fractions (TosoHaas TSK Gel G2500PW column). The glacial acrylic acid was obtained internally (Rohm and Haas product); 2-methyl glutaric acid was obtained from Aldrich Chemical Company.

Analytical Methods. All polyaspartate samples tested for biodegradability were characterized by nuclear magnetic resonance spectroscopy (NMR), gel permeation chromatography (GPC), and elemental analysis. Selected samples were also subjected to additional analyses (e.g., infrared spectoscopy, mass spectrometry) for structural characterization. Poly(acrylic acids) were characterized primarily by GPC and NMR.

Gel Permeation Chromatography. Aqueous GPC was used to measure the weight average molecular weights (Mw) and number average molecular weights (Mn) of both the sodium polyaspartates and the sodium salts of the poly(acrylic acid) homopolymers. A Progel TSK GMPWXL gel column, 30 cm x 7.8 mm (Supelco, Inc., Bellefonte, PA) was used, with a 0.1 M NaSO4 mobile phase and a flow rate of

1.0 mL/min. The Mw and Mn values reported are relative to a Mw 4500 poly(acrylic acid) in-house standard, and were calculated using proprietary software. The composition of the pAA oligomer solutions was determined using six GPC columns with higher resolution for low molecular weight species, i.e., AllTech Macrosphere GPC 60 columns, 25 cm x 4.6 mm each, with a pH 7 phosphate buffer mobile phase and a flow rate of1.0 mL/min. Detection was by refractive index in all cases.

Nuclear Magnetic Resonance Spectroscopy. Both proton (^1H) and carbon-13 (^{13}C) NMR spectroscopy were used to extensively characterize the poly(aspartic acid) samples. The levels of residual aspartic acid present during synthesis, the levels of residual succinimide following hydrolysis, the ratio of α to β linkages in the polyaspartates, and the extent of branching were all determined by NMR. All samples were analyzed on a Bruker AMX500 MHz spectrometer. Polysuccinimide samples were dissolved in DMSO-d_6, and sodium polyaspartate samples in D_2O. The details of these analyses are described in Reference 5.

Elemental Analysis. Lypholized samples of sodium polyaspartate were analyzed for C, H, N, Na (and P, where applicable) prior to biodegradation testing. Analyses were done by Galbraith Laboratories, Inc. Good agreement of observed values with theory was obtained. Good agreement was also always observed between the carbon values obtained by elemental analysis on the lyophilized samples and those obtained by TOC analysis on the biodegradation stock solutions prepared from these samples.

Biodegradation Test Methods. All biodegradation tests were run at Roy F. Weston, Inc. (West Chester, PA) using standard test procedures and analytical protocols. The sludge for the tests was obtained from local residential/light industrial sewage treatment plants.

Semi-Continuous Activated Sludge (SCAS) Removability Test. The SCAS test procedure as carried out on the sodium polyaspartate samples was similar to the OECD Guidelines Test 302A (*6*). The suspended solids level in this test was ~2500 mg/L. Acclimation was carried out over seven days. Duplicate units were run on each samples with a polymer concentration of 40 mg active/L. The test itself was run for seven days - each day the sludge is allowed to settle after 23.5 hrs. of aeration, effluent is drained off and replaced with influent containing fresh polymer, and SOC measurements are made on the effluent collected. The numbers reported here are 7-day averages of the percent carbon removal, and are averages of the two SCAS units. Rarely was the unit-to-unit, or day-to-day, variability statistically significant.

Mini-Continuous Activated Sludge (Mini-CAS) Acclimation. Mini-CAS units were used for the acclimation phase of the biodegradation studies carried out on the AA/pAA samples. The mini-CAS system was designed so that its operation would model a conventional activated sludge treatment process. The design and operation of the mini-CAS units are described in Reference 7. Eleven mini-CAS units were acclimated to various mixtures of low molecular weight pAA's and poly(ethylene glycols) for about one month i.e., approx. 3 SRT's (sludge retention times). Total active polymer concentration in the units was 20 mg/L. The sludge from these units was combined and homogenized. After settling, supernatant from the composited sludge was used as inoculum for the subsequent modified Sturm tests carried out on the monomeric, oligomeric, and polymeric AA samples.

Modified Sturm (CO_2 Production) Test. The modified Sturm test procedure used here was similar to the OECD Guidelines Test 301B (*6*) with the addition of periodic soluble organic carbon analyses. The acclimated inocula for the polyaspartate tests and the AA/pAA tests were obtained from the SCAS units (sludge supernatant) and from the mini-CAS units, respectively. The CO_2 test flasks, containing inoculum, polymer, and test medium (buffer, salts) were aerated, and the CO_2 evolved was trapped by reaction with $Ba(OH)_2$. The excess $Ba(OH)_2$ is titrated with standardized HCl solution to determine the amount of CO_2 produced. The extent of carbon removal was determined by SOC measurements taken at the beginning and end of the test, and periodically during the test. Generaly, two test flasks were run per sample. For the polyaspartates, one flask was run at 20 mg active/L and one at 40 mg active/L; a blank control (no polymer) and a glucose control were run with each test. For the AA/pAA samples, both flasks were run at 20 mg active/L (monomer, dimer, polymers) or at 10 mg/L (oligomers); blank controls and glucose controls were also run. The CO_2 production tests were run until the CO_2 evolution and carbon removal values leveled out; test duration was a minimum of 28 days in all cases.

Results and Discussion

Poly(Aspartic Acid) Biodegradation - SCAS Tests. Following a one-week acclimation period, the SCAS test measures polymer removability by SOC measurements. The bar chart in Figure 3 summarizes the SCAS removability results on poly(aspartic acid) samples prepared via the three different synthetic routes, i.e., the acid-catalyzed thermal polymerization of aspartic acid, the uncatalyzed thermal polymerization of aspartic acid, and the maleic anhydride/ammonia-based polymerizations. Each bar in the chart represents a unique sample, i.e., no replicate tests are shown here. Within each synthetic process group, the samples differ slightly from one another in process conditions used and/or catalyst level and/or hydrolysis conditions. As a result, within the H_3PO_4-catalyzed samples, the Mw's vary across a range of about 7,000-17,000 (GPC values relative to pAA standards). All the uncatalyzed samples, however, are Mw ~4000-5000, and all the maleic/ammonia-based samples are Mw ~2000. For each sample tested, the number above the bar in Figure 3 represents the average removability over the seven-day test period, and is the average over the two replicate SCAS units. (Averages are reported because the day-to-day and unit-to-unit variability were not statistically significant, with the exception of the last maleic/ammonia sample where the difference between the two replicate units was 5%.)

Figure 3 shows that the catalyzed poly(aspartic acid) samples are essentialy completely removed in this high biomass SCAS test, followed in order by the uncatalyzed and the maleic/ammonia-based samples, which are partially removed. The averages across the samples for each process are 97%, 79%, and 67% removal, respectively (Table 1). Since removal in this system can be attributed to adsorption on sludge, biodegradation, or a combination of the two, the subsequent modified Sturm (CO_2 evolution) tests are needed to determine to what extent (and how rapidly) the polymers are actually mineralized (biodegraded to CO_2, N oxides, water, and mineral salts).

Poly(Aspartic Acid) Biodegradation - Modified Sturm (CO_2 Evolution) Tests. The modified Sturm test was used to determine the rate and extent of biodegradation of the different poly(aspartic acids) under aerobic conditions. Supernatant from the SCAS test units was used as the source of acclimated inoculum for the Sturm test flasks. Thus, the biomass in these tests was low relative to the SCAS tests. The effect of polymer concentration was also examined in most of these tests by looking at two test concentrations, i.e., 20 and 40 mg active polymer (as sodium polyaspartate) per liter.

The results of a modified Sturm test carried out at 20 mg active/L on a H_3PO_4-catalyzed sodium polyaspartate sample, Mw 16,000, are shown in Figure 4. (The phosphate catalyst salts were washed out of this sample prior to hydrolysis of the polysuccinimide to sodium polyaspartate.) CO_2 evolution (as a % of the theoretical CO_2 evolution possible, given total mineralization) as a function of time is shown in the top of Figure 4; % carbon removal (100 - % of the initial carbon remaining) is

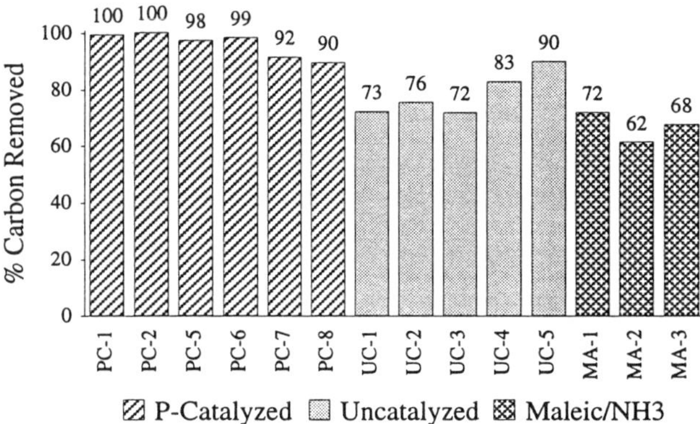

Figure 3. Summary of semi-continuous activated sludge (SCAS) removability results on sodium polyaspartates at 40 mg active/L. Each bar represents the mean removability over seven days and is the mean of two replicate units.

Table 1. Summary of Sodium Polyaspartate Biodegradation Data

Biodegradation Test [a]	*No. of Tests*	*Range of Results*	*Mean Value ± Std. Dev.*
H_3PO_4-Catalyzed p(Asp), Na salt:			
SCAS @ 40 mg/L	6	90-100%	96 ± 4%
CO_2 Evolution @ 20 mg/L	9[b]	78-99%	90 ± 7%
Rate of CO_2 Evol. @ 20 mg/L	9[b]	0.06-0.29 day^{-1} 0.0-3.0 day lag	0.17 ± 0.06 day^{-1} 2.1 ± 1.1 day lag
SOC Removal @ 20 mg/L	9[b]	92-100%	98 ± 3%
CO_2 Evolution @ 40 mg/L	6	70-101%	82 ± 12%
Rate of CO_2 Evol. @ 40 mg/L	6	0.17-0.22 day^{-1} 1.9-3.7 day lag	0.21 ± 0.04 day^{-1} 2.8 ± 0.6 day lag
SOC Removal @ 40 mg/L	6	83-99%	94 ± 7%
Uncatalyzed p(Asp), Na salt:			
SCAS @ 40 mg/L	5	72-90%	79 ± 8%
CO_2 Evolution @ 20 mg/L	5	63-92%	73 ± 13%
Rate of CO_2 Evol. @ 20 mg/L	5	0.11-0.20 day^{-1} 1.4-2.8 day lag	0.15 ±0.04 day^{-1} 2.0 ± 0.6 day lag
SOC Removal @ 20 mg/L	5	65-80%	73 ± 7%
CO_2 Evolution @ 40 mg/L	5	48-68%	59 ± 8%
Rate of CO_2 Evol. @ 40 mg/L	5	0.10-0.45 day^{-1} 1.5-6.5 day lag	0.21 ± 0.15 day^{-1} 3.1 ± 2.0 day lag
SOC Removal @ 40 mg/L	5	65-80%	71 ± 6%
Maleic/NH_3-Based p(Asp), Na salt:			
SCAS @ 40 mg/L	3	62-72%	67 ± 5%
CO_2 Evolution @ 20 mg/L	3	44-80%	57 ± 20%
Rate of CO_2 Evol. @ 20 mg/L	3	0.20-0.49 day^{-1} 1.7-5.2 day lag	0.39 ± 0.16 day^{-1} 3.6 ± 1.8 day lag
SOC Removal @ 20 mg/L	3	55-67%	60 ± 6%
CO_2 Evolution @ 40 mg/L	3	42-67%	54 ± 13%
Rate of CO_2 Evol. @ 40 mg/L	3	0.16-0.24 day^{-1} 1.7-3.7 day lag	0.20 ± 0.04 day^{-1} 2.6 ± 1.0 day lag
SOC Removal @ 40 mg/L	3	56-70%	61 ± 8%

[a]Units on results: SCAS - % carbon removal (ave. of 7 days, 2 replicate units); CO_2 evolution - % of theoretical CO_2, final value reached after leveling out; SOC removal - % soluble organic carbon removed, final value.

[b]Represents 8 unique samples.

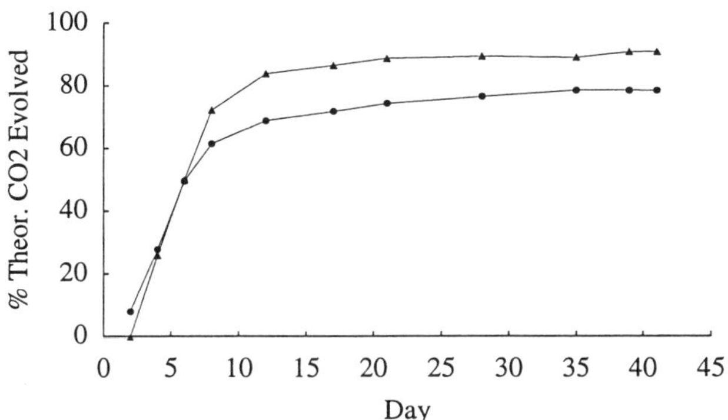

Figure 4a. Kinetics of CO_2 evolution during biodegradation of phosphoric acid-catalyzed sodium polyaspartate, Mw 16,000, (▲, Sample PC-8) and a glucose control (●) at 20 mg active/L using SCAS inoculum.

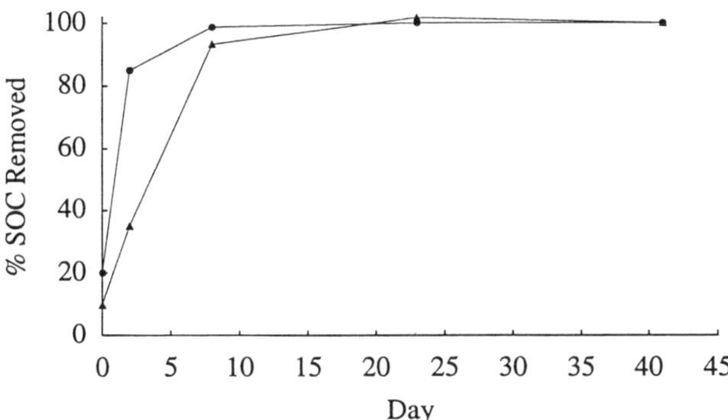

Figure 4b. Kinetics of soluble organic carbon removal during biodegradation of phosphoric acid-catalyzed sodium polyaspartate, Mw 16,000, (▲, Sample PC-8) and a glucose control (●) at 20 mg active/L using SCAS inoculum.

shown in the bottom of Figure 4. The carbon removal data shows that the polymer is completely removed within 2-3 weeks. The corresponding CO_2 evolution data shows that the primary degradation pathway here is mineralization (91% of the carbon removed is evolved as CO_2). CO_2 evolution also levels out in about 2-3 weeks; the calculated rate constant for CO_2 evolution in this case is ~0.3 day^{-1}, following a 3-day lag period. It is assumed that the carbon that is removed, but not converted to CO_2, is converted to biomass. (These studies were not designed to address this issue directly, however.) Data at 20 mg/L for the reference compound, glucose, are also shown in Figure 4. The degradation profile of glucose is similar to that of the catalyzed poly(aspartic acid), with a final value of 79% CO_2 evolution and 100% carbon removal (k=0.2 day^{-1}, 1.6 day lag).

Most of the catalyzed polyaspartate samples were tested after first removing the phosphate salts. However, to confirm that the presence of the catalyst salts would not affect the biodegradation results, two H_3PO_4-catalyzed polyaspartate samples, Mw's 8000 and 16,000, were tested both with and without removal of the catalyst salts. The results are shown in Figure 5. (The "washed" Mw 16,000 sample is the same one shown in Figure 4.) For both the lower and higher Mw species, the biodegradation behavior is unaffected by the presence of the catalyst salts. All these samples were highly biodegradable with final carbon removals of >92% and final CO_2 evolution values of >78% of theory.

CO_2 evolution and carbon removal profiles (at 20 mg/L) for an uncatalyzed polyaspartate sample and a maleic/NH_3 polyaspartate sample are shown in Figure 6. Although the rates of biodegradation for these uncatalyzed and maleic/NH_3 samples as measured by CO_2 evolution (0.20 day^{-1} and 0.47 day^{-1}, respectively) are similar to the rates of biodegradation seen for the catalyzed samples (Table 1, Figures 4 and 5), the extent of biodegradation is significantly less. In Figure 6, the CO_2 evolution of the uncatalyzed sample reached only ~63% of theory with 71% carbon removal; the maleic/NH_3 sample values are 49% CO_2 and 55% carbon removal.

The bar chart in Figure 7 summarizes the final CO_2 evolution and corresponding carbon removal data obtained on numerous samples of sodium polyaspartate prepared by the three different synthetic processes. The bars marked with an asterisk indicate the samples discussed above (Figures 4-6). One replicate test on a catalyzed sample is depicted (PC-2); the remaining bars represent unique samples. With the exception of samples PC-3 and PC-4, the inoculum used in each modified Sturm test was obtained from the SCAS unit used for acclimating and testing the removability of the same polymer. The inoculum for PC-3 and PC-4 was obtained from the SCAS unit used to acclimate sample PC-2.

These summary charts (Figure 7) show that the H_3PO_4-catalyzed polyaspartate samples are essentially totally biodegradable - the mean carbon removal for these samples is 98%. Mineralization of the polymeric carbon to CO_2 accounts for most of the carbon removal observed - the CO_2 evolution mean for these samples is 89%. In contrast, the uncatalyzed and maleic/NH_3 polyaspartate samples are only partially biodegradable in this test. The mean carbon removal and CO_2 evolution values for the uncatalyzed samples are 73% and 73%, respectively; for the maleic/NH_3 samples, 60% carbon removal and 57% CO_2. This data is summarized in Table 1 along with the mean CO_2 evolution rate data for these samples.

Within each group of samples, the variability of the CO_2 evolution data was found to be higher than the variability of the carbon removal data. It is postulated that this may merely be a reflection of greater sample-to-sample variability in the balance of carbon converted to CO_2 versus the carbon converted to biomass, while the total amount of carbon converted (i.e., removed) is relatively invariant across the samples within each group.

Most of the samples tested at 20 mg/L in the modified Sturm test were also tested at 40 mg/L. At 40 mg/L the relative biodegradation behavior of the different

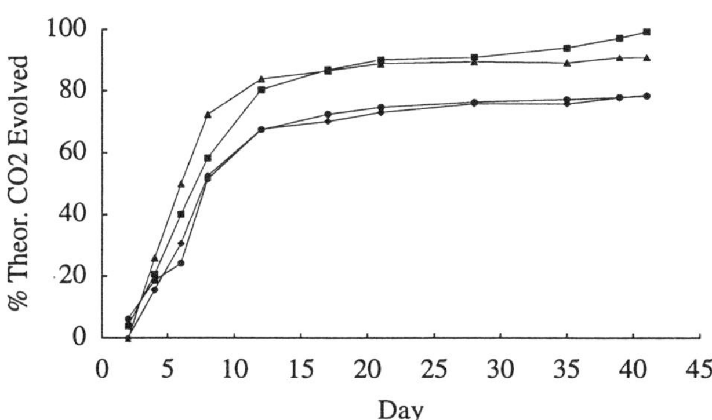

Figure 5a. Kinetics of CO_2 evolution during biodegradation of phosphoric acid-catalyzed sodium polyaspartates with and without the catalyst salts removed: Mw 8000 - with catalyst salts (●, Sample PC-5); Mw 8000 - catalyst salts removed (◆, Sample PC-6); Mw 16,000 - with catalyst salts (■, Sample PC-7); Mw 16,000 - catalyst salts removed (▲, Sample PC-8). Test concentration = 20 mg active/L; SCAS inoculum.

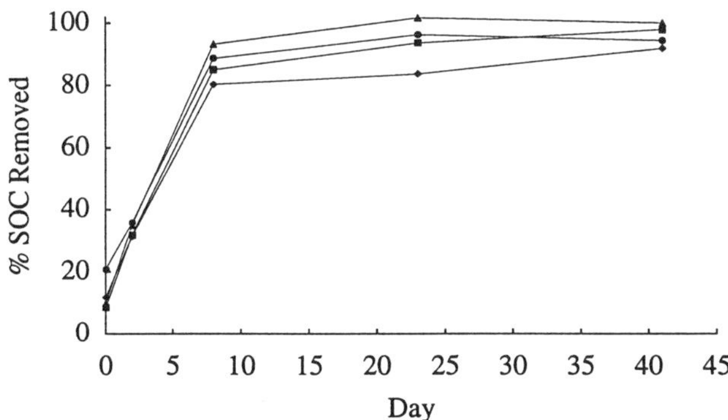

Figure 5b. Kinetics of soluble organic carbon removal during biodegradation of phosphoric acid-catalyzed sodium polyaspartates with and without the catalyst salts removed: Mw 8000 - with catalyst salts (●, Sample PC-5); Mw 8000 - catalyst salts removed (◆, Sample PC-6); Mw 16,000 - with catalyst salts (■, Sample PC-7); Mw 16,000 - catalyst salts removed (▲, Sample PC-8). Test concentration = 20 mg active/L; SCAS inoculum.

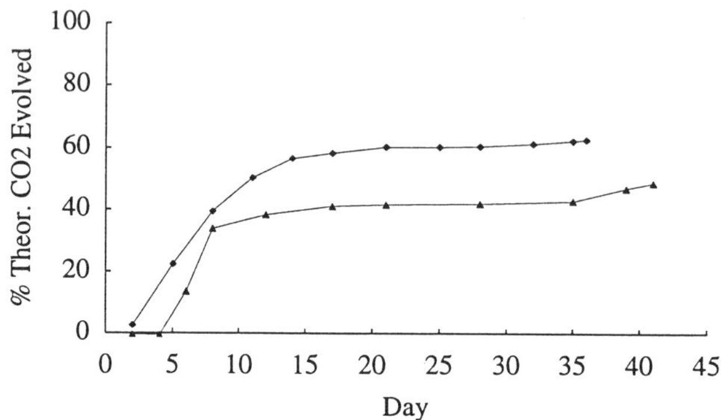

Figure 6a. Kinetics of CO_2 evolution during biodegradation of uncatalyzed sodium polyaspartate, Mw 4000, (◆, Sample UC-1) and maleic anhydride/ ammonia-based sodium polyaspartate, Mw 2000 (▲, Sample MA-3) at 20 mg active/L using SCAS inoculum.

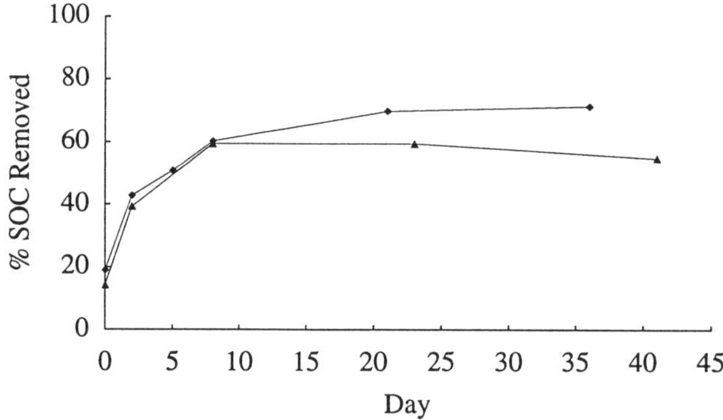

Figure 6b. Kinetics of soluble organic carbon removal during biodegradation of uncatalyzed sodium polyaspartate, Mw 4000, (◆, Sample UC-1) and maleic anhydride/ammonia-based sodium polyaspartate, Mw 2000 (▲, Sample MA-3) at 20 mg active/L using SCAS inoculum.

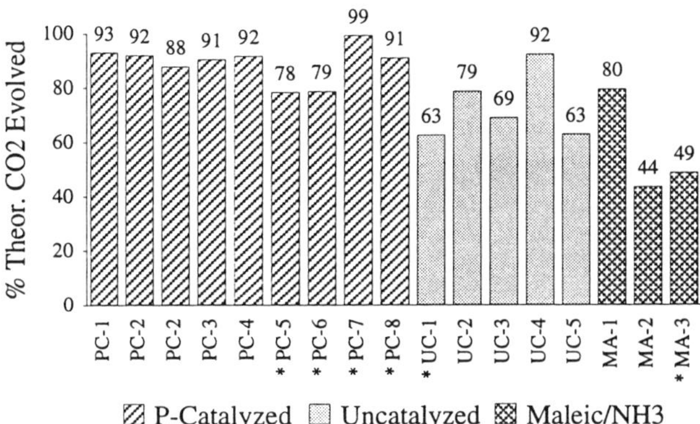

Figure 7a. Summary of final CO_2 evolution values from modified Sturm tests on sodium polyaspartates at 20 mg active/L using SCAS inoculum.

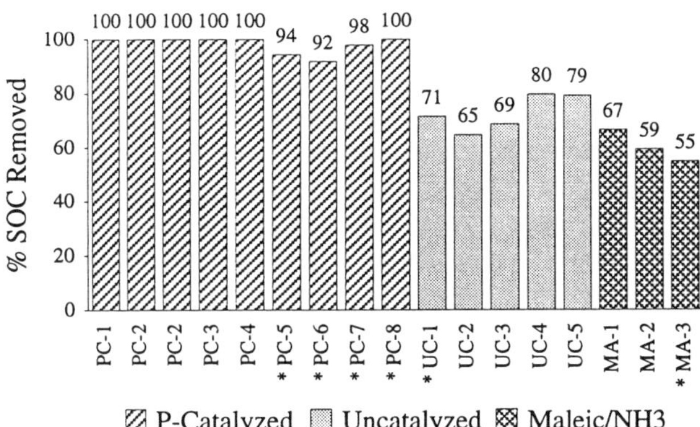

Figure 7b. Summary of final soluble organic carbon removal values from modified Sturm tests on sodium polyaspartates at 20 mg active/L using SCAS inoculum.

polyaspartates was the same as it was at 20 mg/L, i.e., catalyzed>uncatalyzed> maleic/ammonia, however, the extent of biodegradation was slightly less at 40 mg/l than at 20 mg/L (Table 1). From a practical standpoint, this observed concentration dependence is of little concern since environmentally realistic concentrations of detergent polycarboxylates in U.S. sewage treatment influent is <1 mg/L (*1*).

In order to show that polymer decomposition to CO_2 is due to biodegradation by the microorganisms present in the flask, i.e., degradation does not occur under the modified Sturm test conditions in the absence of active inoculum, a sample of uncatalyzed polyaspartate was tested at 40 mg/L using SCAS inoculum which had been "killed" by the addition of $HgCl_2$. As expected, no CO_2 evolution was observed in this test.

Poly(Aspartic Acid) Biodegradation - Analysis of Polymer Structure. Initially, in order to determine why the acid-catalyzed polyaspartates were superior in biodegradability to the polyaspartates prepared by the other processes, a sample of the undegraded residue from biodegradation tests carried out on an uncatalyzed poly(aspartic acid) sample was analyzed. Following a modified Sturm test that gave about 60% CO_2 evolution and 70% carbon removal, the test solutions were concentrated and the residual undegraded polymer was isolated. By NMR it was determined that the isolated polymer was sodium polyaspartate and that the α/β ratio was essentially the same, i.e., about 1/3, as in the starting polymer. Hydrolysis of the isolated polymer to aspartic acid monomer, followed by analysis by chiral HPLC, showed that the D/L ratio was 50/50, as in the starting polymer. Thus, these analyses did not identify the unique structural feature(s) of the uncatalyzed polymer which would explain its resistance to biodegradation.

To test the hypothesis that the extent of branching correlates with the observed differences in biodegradation between the catalyzed, uncatalyzed, and maleic/ammonia polyaspartates, it was necessary to examine the polymers in their unhydrolyzed polysuccinimide form. The nature of the proposed branch site (Figure 8) is such that the NMR and IR spectral features attributable to the structure of these sites are observable when the polymer is in the polysuccinimide form, but not when it is in the hydrolyzed form. In the IR spectra, bands consistent with the amide structure of the proposed branch sites are observed at 1529 cm^{-1} and 3400 cm^{-1}; their intensities decrease as the level of H_3PO_4 catalyst increases. In the NMR spectra, the methine proton peak at 4.6 ppm, which decreases with increasing H_3PO_4 level, is also consistent with the branch site structure. Additional NMR data (e.g., COSY, $^1H/^{13}C$ HMQC, and $^1H/^{15}N$ HMQC-TOCSY data) were obtained which further support these assignments (*5*). In short, the spectral data on the polysuccinimides prepared by the different synthetic processes are all consistent with branching in the uncatalyzed and maleic/ammonia samples, and increasing linearity with increasing catalyst level in the H_3PO_4-catalyzed samples. Thus, taken together with the biodegradation results , the extent of branching is inversely correlated with the extent of biodegradation.

(It should be noted here that the spectral features consistent with the proposed branched polysuccinimide structure are also consistent with a partially ring-opened form of polysuccinimide (Figure 8). However, if the presence of a few ring-opened sites per chain were the only structural feature which distinquished the uncatalyzed and maleic/ammonia polysuccinimides from the catalyzed polysuccinimides, the differences in biodegradation would remain unexplained since all the polymers would be structurally indistinguishable once hydrolyzed to the totally ring-opened form (poly(aspartic acid)) for testing.)

Poly(Acrylic Acid) Biodegradation - Modified Sturm (CO_2 Evolution) Tests. The results of a study done on the biodegradation of poly(acrylic acid) as a function of molecular weight are shown in Figure 9. This figure shows the extent of

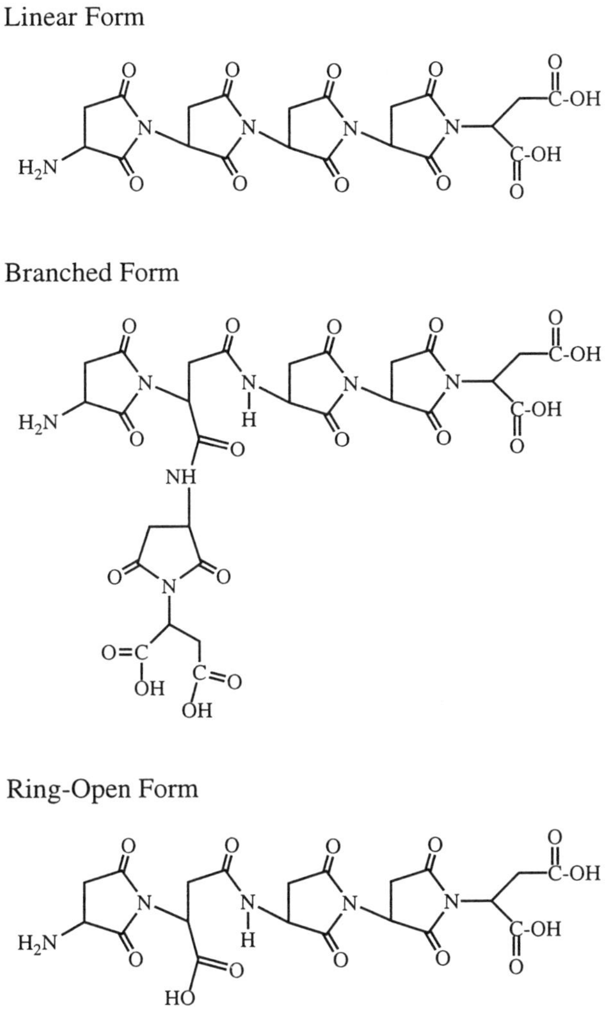

Linear Form

Branched Form

Ring-Open Form

Figure 8. Proposed structures of polysuccinimide consistent with the spectroscopic data.

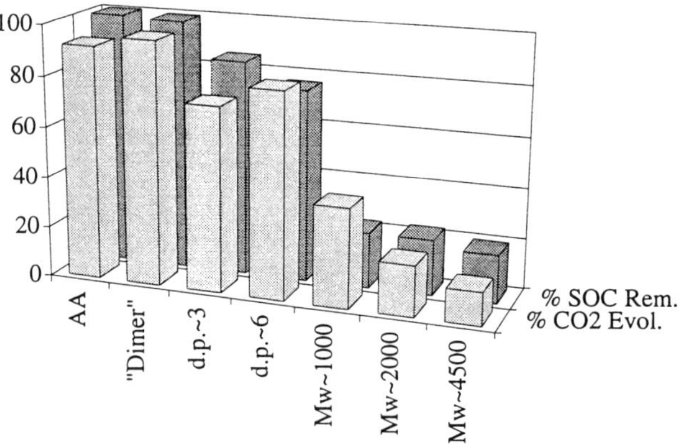

Figure 9. Summary of final CO_2 evolution and soluble organic carbon removal values from modified Sturm tests on acrylic acid monomer (at 20 mg/L), AA "dimer" (at 20 mg/L), AA oligomers (at 10 mg/L), and AA homopolymers (at 20 mg/L) using mini-CAS inoculum.

biodegradation as measured by the final CO_2 evolution and carbon removal values obtained in modified Sturm tests. The source of inoculum for these tests was sludge which had been acclimated to low molecular weight poly(acrylic acids) and poly(ethylene glycols). The AA oligomer samples were run at 10 mg/L; the remaining samples were run at 20 mg/L.

Clearly, the pure acrylic acid monomer and the acrylic acid dimer "model compound" (α-methyl glutaric acid) were completely biodegraded. The AA oligomer samples tested were isolated by chromatographic separation. Each of these samples contained a distribution of oligomers, the low molecular weight sample being centered at about a degree of polymerization (d.p.) of 3, the higher molecular weight sample at about d.p. 6. As shown in Figure 9, these samples were found to be highly, although not totally, biodegradable. And, as expected, the extent of biodegradation of the higher molecular weight pAA's was quite low. These results show that the biodegradability of AA oligomers and homopolymers is clearly a function of molecular weight and that the cutoff for rapid and complete biodegradation of poly(acrylic acids) in a sewage treatment environment is probably in the oligomeric molecular weight range. Additional work is needed, however, to be able to clearly define the molecular weight (d.p.) cutoff for pAA biodegradation with sewage treatment plant inoculum; individual oligomers should be isolated, purified, and tested in SCAS and modified Sturm biodegradation tests.

Consistent with the biodegradation results reported here are several recent reports in the literature (*8,9*)on the biodegradation of AA oligomers and homopolymers suggesting that acrylic acid oligomers/polymers above about molecular weight 500 are not easily biodegraded, even with AA oligomer-utilizing bacteria isolated from sludge and/or soil by enrichment culture techniques. Interestingly, the recently published work of Kawai suggests that trimer-utilizing bacteria can be isolated, the resting cells of which will degrade higher molecular weight (Mw 1000-4500) poly(acrylic acids) (*10*).

Conclusions

Poly(aspartic acid) represents a new benchmark for totally biodegradable water-soluble polymers. Poly(α,β-D,L-aspartates) can be prepared via an acid-catalyzed, thermal polymerization of L-aspartic acid which are highly removed on sewage sludge and are rapidly and completely biodegraded by municipal treatment plant microorganisms. From an environmental perspective, these biodegradation characteristics, coupled with good detergent performance, make the phosphoric acid-catalyzed sodium polyaspartates attractive replacements for the currently used poly(acrylic acid)-based detergent polymers. Given that the current raw material cost of aspartic acid is significantly higher than that of acrylic acid, cost-performance considerations will undoubtedly dictate the commercial attractiveness of these polymers.

Based on these studies, a linear polyamide backbone appears to be the key to the total biodegradability of the phosphoric acid-catalyzed polyaspartates. Polyaspartates prepared by uncatalyzed thermal polymerizations and by maleic anhydride/ammonia routes are only partially biodegradable in a sewage treatment plant, consistent with other previously reported results (*11-13*). The partial biodegradability of these polyaspartates is attributable to their branched structures. Additional studies aimed at elucidating the mechanisms of biodegradation for these different polyaspartates (e.g., susceptibility to endo- vs. exopeptidases) would be of interest.

In contrast to the amide bonds which form the backbone of the polyaspartates, polymers with carbon-carbon backbones are generally not found to be readily biodegradable, with the possible exception of poly(vinyl alcohol) (*14*). In the case of poly(acrylic acid), molecular weights must be in the oligomer range for these materials to be highly biodegradable in a sewage treatment plant environment. Since the pAA's used as detergent additives are generally much higher than this in molecular weight, achieving *both* biodegradability and detergent performance with poly(acrylic acids) remains a challenge.

Literature Cited

1. Freeman, M. B.; Bender, T. M. *Environ. Technol.* **1993**, *14,* 101-112.
2. Metcalf & Eddy, Inc. *Wastewater Engineering: Treatment, Desposal, Reuse*; McGraw-Hill: New York, 1979.
3. Freeman, M. B.; Paik, Y. H.; Swift, G.; Wolk, S. K.; Yocom, K. M., presented at *Polyamino Acids, The Emergence of Life and Industrial Applications*, ACS Symposium - Division of Industrial and Engineering Chemistry; SanDiego, CA, 1994.
4. Robinson, R. S.; Humburger, E. L. U.S. Patent 2,459,964, 1949 (assigned to Beck, Koller and Company).
5. Wolk, S. K.; Swift, G.; Paik, Y. H.; Yocom, K. M.; Smith, R. L.; Simon, E. S. *Macromolecules* **1994**, *27*, 7613-7620.
6. OECD *OECD Guidelines for Testing of Chemicals*; Organization for Economic Cooperation and Development: Paris, 1981.
7. Larson, R. J.; Williams, R. T.; Swift, G. *Proceedings of the American Chemical Society - Division of Polymeric Materials and Engineering* (Fall Meeting, Washington, D.C.) **1992**, *67*, 348-349.
8. Matsumara, S.; Maeda, S.; Yoshikawa, S.; Chikazumi, N., *J. Jpn. Oil Chem. Soc. (Yukagaku)*, **1990**, *39* , 245-249.
9. Hayashi, T.; Mukouyana, M.; Sakano, K.; Tani, Y., *Appl. Environ. Microbiol.*, **1993**, *59* , 1555-1559.
10. *Kawai, F. Appl. Microbiol. Biotechnol.* **1993**, *39*, 382-385.
11. Schornick, G.; Kroner, M.; Lungershausen, R., presented at *Polyamino Acids, The Emergence of Life and Industrial Applications*, ACS Symposium - Division of Industrial and Engineering Chemistry; SanDiego, CA, 1994.
12. Wheeler, A. P.; Alford, D. D.; Koskan, L. P., presented at *Polyamino Acids, The Emergence of Life and Industrial Applications*, ACS Symposium - Division of Industrial and Engineering Chemistry; SanDiego, CA, 1994.
13. Alford, D. D.; Wheeler, A. P.; Pettigrew, C. A. *J. Environ. Polym. Degrad.* **1994**, *2*, 225-236.
14. Suzuki, T.; et al. *Agric. Biol. Chem.* **1978**, *42*, 1217.

RECEIVED November 28, 1995

Chapter 11

Biodegradation of Poly(vinyl alcohol) and Vinyl Alcohol Block as Biodegradable Segment

Shuichi Matsumura and Kazunobu Toshima

Department of Applied Chemistry, Keio University, 3–14–1 Hiyoshi, Kohoku-ku, Yokohama 223, Japan

The biodegradation of poly(vinyl alcohol) (PVA) and the vinyl alcohol block as a biodegradable segment was evaluated for the design of biodegradable water-soluble functional polymers using PVA-assimilating aerobic and anaerobic microbes. PVA having a number-average molecular weight ($\overline{Mn}$) between 90000 and 530 (corresponds to the vinyl alcohol heptamer of the PVA oligomer) was equally and substantially biodegraded by the aerobic microbes. No difference in the biodegradability of PVA was observed for molecular weights above the heptamer of the vinyl alcohol monomer units. The isotactic moieties of PVA were preferentially biodegraded compared to the atactic or syndiotactic moieties. That is, the biodegradability was increased with increasing isotactic content of the PVA. PVA having a $\overline{Mn}$ of 2200 and 14000 was also biodegraded by the anaerobic microbes. However, the anaerobic biodegradation rate was slower than that under aerobic conditions. Also, the lower molecular weight fractions of PVA tended to biodegrade more rapidly under anaerobic conditions. The minimum structure of the vinyl alcohol block, which is accepted as a substrate for PVA-assimilating microbes and PVA-dehydrogenase, is estimated to be an isotactic-type vinyl alcohol block having a chain length of 3 monomer units. A significant increase in the biodegradability of the poly(sodium carboxylate) containing vinyl alcohol blocks was observed for a vinyl alcohol block length of more than about 3.

Poly(vinyl alcohol) (PVA) is the only carbon-carbon backbone polymer that is biodegradable (*1-5*). PVA has attracted attention as a biodegradable segment for the design of biodegradable water-soluble functional polymers. There may be two ways to develop biodegradable functional polymers using PVA as a biodegradable segment; incorporation of a vinyl alcohol block into a functional polymer chain, and grafting functional oligomers into the PVA chain. The introduction of biodegradable vinyl alcohol blocks into the main chain of a nonbiodegradable functional polymer is one way to design a biodegradable functional polymer (*6-9*). The functional polymer

0097–6156/96/0627–0137$15.00/0

containing vinyl alcohol blocks is cleaved to produce low molecular weight fractions which are further assimilated by the environmental microbes. In order to act as a biodegradable segment in the polymer chain, the vinyl alcohol blocks must be incorporated into a polymer chain in such a manner that they are accepted as a substrate for the degrading microbes or enzymes. It is considered that there are two major factors, the block length and its stereochemistry, which are responsible for the enzymatic reaction at the biodegradable segment. However, the minimum required structure of a vinyl alcohol block necessary to react with a PVA-degrading microbe or enzyme has not yet been clarified. Grafting a functional oligomer chain into water-soluble biodegradable polymers, such as PVA, is the other potential route to biodegradable functional polymers, provided that the molecular weight of the graft chain can be controlled *(10)*. It is important to know the effects of molecular weight and stereoregularity on the biodegradation of PVA for the design of a graft-type biodegradable polymer.

This report summarizes the biodegradation of PVA and the vinyl alcohol block with respect to their chain length and stereochemistry using the isolated PVA-assimilating microbes and a PVA-degrading enzyme for the design of biodegradable functional polymers.

Structures of PVA, Vinyl Alcohol Blocks and Polycarboxylates

Figure 1 summarizes the molecular structures of PVA, vinyl alcohol blocks and polycarboxylates containing the vinyl alcohol groups discussed in this paper. PVA (100% hydrolyzed) having $\overline{Mn}$=14000 and triad isotacticity=22.3% (PVA-14000) was purchased from Aldrich Chem. Co. (Milwaukee, WI, USA). PVA having $\overline{Mn}$=3700 and triad isotacticity=20.6% (PVA-3700) was prepared by the radical polymerization of vinyl acetate in a phenol solution using 2,2'-azobis(isobutyronitrile) (AIBN) as an initiator with subsequent hydrolysis according to the literature *(11,12)*. PVA having $\overline{Mn}$=35000 and triad isotacticity=15.9% (PVA-35000) was prepared by the radical polymerization of vinyl pivalate in the presence of AIBN with subsequent hydrolysis *(11,13)*. PVA (100% hydrolyzed) having $\overline{Mn}$=2200 (PVA-2200), 50000 and 90000 were supplied by Kuraray Co. Ltd. (Osaka, Japan). Vinyl alcohol oligomers having a $\overline{Mn}$ of 530 as shown in Figure 1 were prepared by the group transfer polymerization method as described in the literature *(11,14,15)*.

The isotactic-type vinyl alcohol pentamer (*i*-5VA), (*2S,4S,6S,8S,10S*)-1,12-di-*O*-benzyl-1,2,4,6,8,10,12-dodecaneheptaol, isotactic-type vinyl alcohol trimer (*i*-3VA), (*2S,4S,6S*)-1,8-di-*O*-benzyl-1,2,4,6,8-octanepentaol, and atactic-type vinyl alcohol trimer (*a*-3VA), (*2S,4S,6R*)-1,8-di-*O*-benzyl-1,2,4,6,8-octanepentaol were synthesized from L-malic acid according to the literature *(9,16-21)*. The isotactic-type vinyl alcohol dimer (*i*-2VA), (*S,S*)-2,4-pentanediol, and the atactic-type vinyl alcohol dimer (*a*-2VA), *meso*-2,4-pentanediol, were purchased from Tokyo Kasei Kogyo Co., Ltd. (Tokyo, Japan). Poly[(disodium methylene malonate)-*co*-(vinyl alcohol)] [P(DSMM-VA)], poly[(sodium acrylate)-*co*-(vinyl alcohol) [P(SA-VA)], poly[(disodium fumarate)-*co*-(vinyl alcohol) [P(DSF-VA)] and poly[(disodium maleate)-*co*-(vinyl alcohol)] [P(DSM-VA)] with varying amounts of vinyl alcohol groups were prepared by the copolymerization of diethyl methylene malonate, methyl acrylate, dimethyl fumarate and dimethyl maleate with vinyl acetate, respectively, with subsequent saponification *(6-8)*.

Microorganisms and Aerobic Biodegradation Test

Symbiotic PVA-assimilating microbes (*PVA-IMX*), comprised mainly of *Arthrobacter* sp., were obtained by the acclimation of river water using PVA-14000

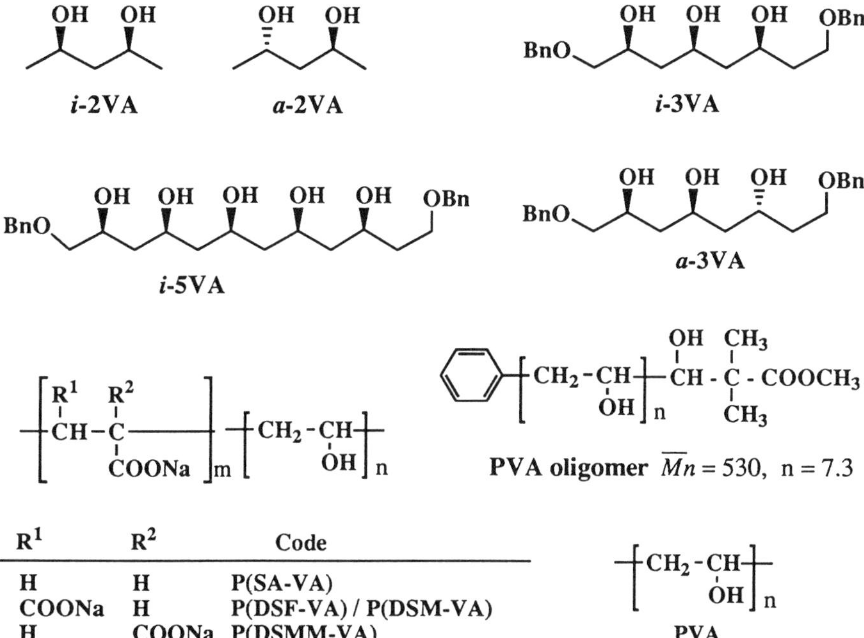

R^1	R^2	Code
H	H	P(SA-VA)
COONa	H	P(DSF-VA) / P(DSM-VA)
H	COONa	P(DSMM-VA)

Figure 1. Molecular structure and abbreviation of vinyl alcohol block, poly(vinyl alcohol) (PVA) and polycarboxylate copolymer.

as the sole carbon source and stored in our laboratory. The isolated PVA-assimilating strain, *Alcaligenes faecalis* KK314 *(11)*, was maintained on agar slants using a 0.1% PVA concentration of the culture medium and was used throughout this work.

Aerobic biodegradability was evaluated using an isolated PVA-assimilating microbial strain and acclimated microbes for better reproducibility. Biochemical oxygen demand (BOD) was measured with a BOD Tester (Model 200F, TAITEC Corp., Koshigaya-shi, Japan) using the oxygen consumption method, basically according to the OECD Guidelines for Testing of Chemicals (301C, Modified MITI Test) *(22)* at 25°C using PVA-assimilating microbes. The concentration of the polymer in the incubation media was 25 mg/L, and 0.001 mg/L PQQ (pyrroloquinoline quinone) was added to facilitate the biodegradation *(23)*. The BOD measurements were repeated at least two times. The residual polymers present in the culture broth before and after the biodegradation test (BOD test) were analyzed by gel permeation chromatography (GPC). That is, 200 mL BOD test solution was evaporated to dryness under reduced pressure and the residue was dissolved in the GPC elution buffer (1~5 mL). After ultrasonic treatment with a small amount of a nonionic surfactant to avoid any adsorption loss of the polymer onto the microbial cells *(24)* the residual polymer was analyzed directly by GPC. The detectable limits of GPC were below 0.5 mg/L polymer fractions (molecular weight, MW>800) in 200 mL BOD test solution. Biodegradability was also evaluated by the total organic carbon (TOC) of the incubation media using a TOC analyzer in order to analyze the assimilation of low-molecular weight biodegradation fragments by the cells.

Biodegradation of PVA

Influence of Molecular Weight on Biodegradation. The aerobic biodegradability of PVA having of several molecular weights, as determined by the 21-day BOD values and the theoretical oxygen demand (ThOD), is shown in Figure 2 as a function of molecular weight using two different strains of PVA-assimilating microbes *(9,11)*. Tacticities of these polymers were between 21-23% isotactic in triad basis. It was found that PVA having a $\overline{Mn}$ between 90000 and 530 (corresponds to the heptamer of the vinyl alcohol monomer units of PVA oligomer as shown in Figure 1) was equally and substantially biodegraded by *Alcaligenes faecalis* KK314 and the symbiotic PVA-assimilating microbes, *PVA-IMX*. Over 90% of the TOC was removed from the incubation media after the 21-day incubation, suggesting that the biodegradation of PVA had certainly occurred. No difference in the biodegradability of PVA with molecular weights above the heptamer was observed. The slight difference in the extent of biodegradation as determined by the BOD value will be ascribed to the producibility of the microbial cells using PVA as a carbon source according to the kind of microbial strains. Similar results were obtained using the other PVA-assimilating microbes.

Influence of Stereoregularity on Biodegradation. PVA is biodegraded by the conversion of the 1,3-diol moiety into the corresponding 1,3-diketone by PVA-dehydrogenase with subsequent hydrolytic cleavage by hydratase to yield low molecular weight fractions. Because these reactions are carried out enzymatically, biodegradation of PVA may be influenced by the stereochemical configuration of the hydroxyl groups of PVA. A triad tacticity of PVA was estimated by comparing the peak area of the hydroxyl proton in the ^1H NMR spectrum. Three peaks of 4.35, 4.56 and 4.77 ppm (δ values in DMSO-d_6) correspond to the hydroxyl proton of the syndiotactic, atactic and isotactic moieties of PVA, respectively. A triad tacticity composition of the PVA can be obtained from the relative peak area of these three peaks. Figure 3 shows the triad tacticity change in PVA-14000 after a 2-day aerobic biodegradation by *A. faecalis* and after 10-day biodegradation by *PVA-IMX* *(11)*. It

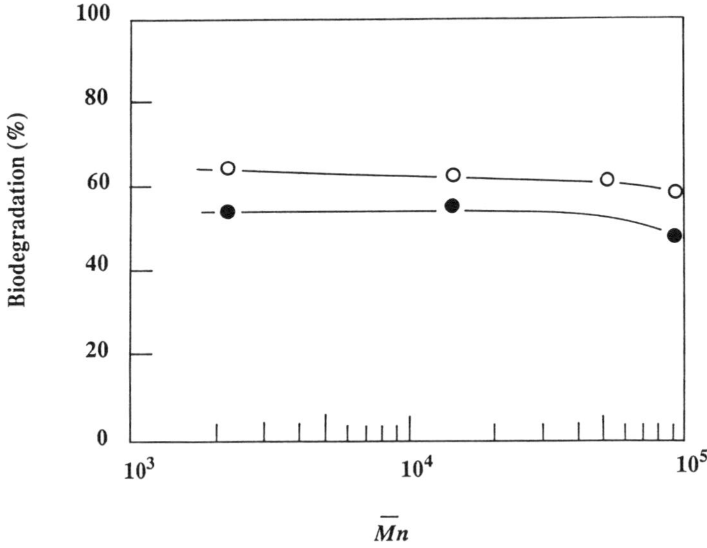

Figure 2. 21-Day biodegradation of PVA (triad *i* : 21~23%) as a function of number-average molecular weight ($\overline{Mn}$).
Biodegradation (%) = (1 - BOD/ThOD) x 100

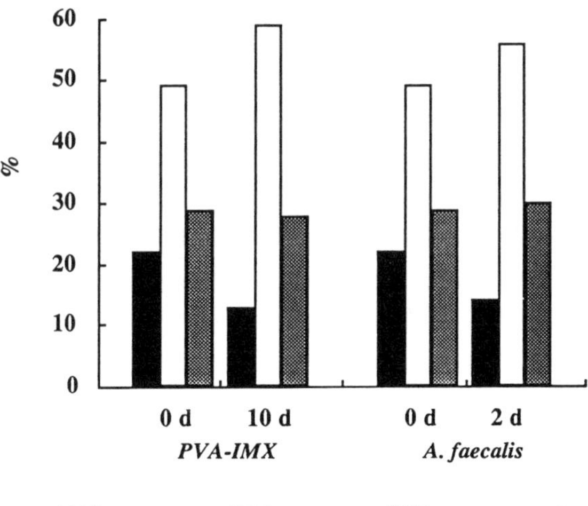

Figure 3. Tacticity (triad) change of PVA-14000 during the biodegradation in shake flask. (Adapted from ref. 11.)

was found that the isotactic moieties of PVA-14000 were preferentially degraded by both *A. faecalis* and *PVA-IMX*. These results indicate that the biodegradability is influenced by its configuration of the hydroxyl groups of PVA. Table I shows the biodegradability of PVA having a varying amount of isotactic hydroxyl groups as measured by TOC using *PVA-IMX* (*11*). Biodegradability was increased markedly with increasing isotactic content of PVA, suggesting the stereochemical preference of the enzymes involved. The 5-day biodegradation of PVA was increased with increasing isotactic moiety; however, the extent of biodegradation after a 28-day incubation was the same. Similar tendencies were also observed using *A. faecalis* (*9,11*).

Anaerobic Biodegradation of PVA. Water-soluble polymeric compounds may be widely diffused in anaerobic environments as well as in the aerobic environments of the earth's soil, river water, or seawater after their use. It will be important to estimate the biodegradability of these water-soluble polymers under anaerobic conditions in addition to aerobic conditions. The anaerobic biodegradation test was carried out basically according to the literature (*25,26*) in a 3 L glass bottle equipped with a sampling tube into the medium and also having a serum cap for gas sampling. The inoculum consisted of black river sediments obtained from the river mouth in an industrial area or anaerobically treated sludge from municipal sewage treatment plant. Each biodegradation vessel containing the mineral solutions, the anaerobically preincubated microbial source and the test polymers (0.01%) was purged with nitrogen gas and incubated at 27°C in the dark with a one-hour stirring period twice a day. Biodegradability was evaluated by the GPC and TOC of the incubation media.

PVA having $\overline{Mn}$ of 2200 and 14000 was biodegraded under anaerobic conditions (*26*). It was confirmed that the biodegradability of PVA-2200 was comparable to that of D-glucose as a control by the anaerobic microbes in river sediments. Though the anaerobic biodegradation of PVA-14000 by the river sediments was slower than that of PVA-2200, the polymer was gradually biodegraded, and after 125 days' incubation the extent of biodegradation exceeded 60%. In general, a longer time was needed to biodegrade PVA under anaerobic conditions compared to the aerobic conditions. It was also found that under anaerobic conditions, low molecular weight fractions tended to biodegrade more rapidly than the high molecular weight fractions. On the other hand, under aerobic conditions no difference in the biodegradability of PVA was observed according to the molecular weight. Figure 4 shows the GPC profiles of PVA-14000 and PVA-2200 before and after the anaerobic and aerobic conditions (*11,26*). It was found that PVA of low molecular weight as well as high molecular weight was biodegraded under anaerobic conditions. However, the low molecular weight fractions of PVA were biodegraded more rapidly than the high molecular weight fractions in the biodegradation media. These facts mean that PVA of any molecular weight was certainly biodegraded under both aerobic and anaerobic conditions. That is, PVA is potentially applicable as a biodegradable segment for the design of a biodegradable polymer.

Biodegradation of Vinyl Alcohol Blocks as Biodegradable Segments

Microbial Degradability of Vinyl Alcohol Blocks. Because the vinyl alcohol block is readily incorporated into a functional polymer chain by the copolymerization of functional vinyl monomers with vinyl acetate and subsequent hydrolysis, PVA has attracted attention as biodegradable segments in the polymer chain for the design of functional polymers. It is important to clarify the minimum structure of the vinyl alcohol block needed to react with a PVA-assimilating microbe, or PVA-degrading enzyme, with respect to its stereochemical configuration and vinyl alcohol block length. We previously reported that the minimum chain length of the vinyl alcohol

Table I. Five-day Biodegradation of PVA Having Varying Tacticities by Symbiotic PVA-Assimilating Microbes, *PVA-IMX* Biodegradation (%) = (1-TOC_5 / TOC_0) x 100

Polymer code	Tacticity (triad *i* %)	Five-Day Biodegradation (%)
PVA-35000	15.9	38
PVA-3700	20.6	50
PVA-14000	22.3	69

SOURCE: Adapted from ref. 11.

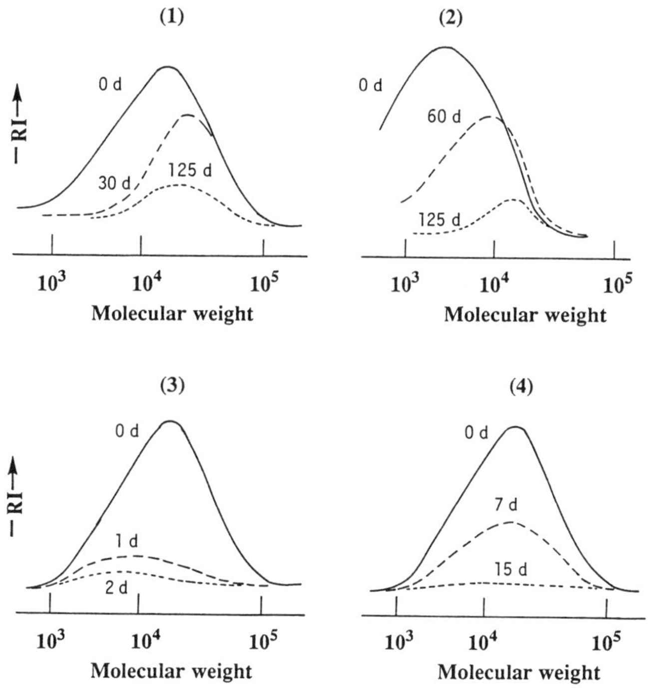

Figure 4. GPC profiles of PVA before and after the biodegradation. (1) Anaerobic biodegradation of PVA-14000 with anaerobic river sediments. (2) Anaerobic biodegradation of PVA-2200 with anaerobic river sediments. (3) Aerobic biodegradation of PVA-14000 with *Alcaligenes faecalis* KK314. (4) Aerobic biodegradation of PVA-14000 with *PVA-IMX*. (Adapted from ref. 26.)

block, which acts as a biodegradable segment, was statistically estimated to be about 3 and 5 monomer units for P(DSMM-VA) and P(DSF-VA), respectively, when PVA-assimilating microbes (*PVA-IMX*) were used (*6,7*). Based on the fact that the isotactic moiety of PVA was preferentially biodegraded by the PVA-assimilating microbes, the minimum required structure, which is necessary to form a substrate for the PVA-degrading enzymes, is indicated to be a trimer to pentamer of vinyl alcohol units having isotactic form. For further analyses, the biodegradabilities of the dimer to the pentamer of the vinyl alcohol units having different configurations as indicated in Figure 1 were evaluated using PVA-assimilating microbes and enzyme. Figure 5 shows the biodegradability of vinyl alcohol blocks as measured by the BOD tester using *A. faecalis* and *PVA-IMX* (*9*). It was confirmed that the isotactic-type pentamer of the vinyl alcohol (*i*-5VA) showed excellent biodegradability similar to that of PVA-2200. The isotactic-type trimer (*i*-3VA) was also substantially biodegraded by the PVA-assimilating microbes. On the other hand, the biodegradability of the atactic-type trimer (*a*-3VA) was significantly decreased. The dimer was not biodegraded. The minimum required structure for the microbes was estimated to be an isotactic-type 3-5mer of vinyl alcohol monomer units. Similar results were obtained using the other PVA-assimilating microbes.

PVA-Dehydrogenase Activity Measurements. PVA-dehydrogenase (PVA-DHase) was obtained from the PVA-assimilating strains according to the literature (*27*). The outline of the procedure is as follows. PVA-assimilating microbial cells were harvested by centrifugation. The cells were re-suspended in distilled water containing EDTA, phenylmethylsulfonyl fluoride and Tween 80, and were disrupted the cells with an ultrasonic oscillator. The cell debris was removed by centrifugation and the supernatant was used as the cell-free extract. After streptomycin treatment and ammonium sulfate precipitation, the partially purified PVA-DHase was collected using DEAE-Sepharose FF column fractionation. PVA-DHase activity was assayed by measuring the rate of reduction of phenazine ethosulfate as determined by the decrease in A_{660} resulting from the coupled reduction of 2,6-dichlorophenolindophenol according to the method of Shimao *et al.* (*28*). The protein concentration was determined by the method of Lowry *et al.* (*29*) with bovine serum albumin as a standard.

Enzymatic Degradation of Vinyl Alcohol Block by PVA-DHase. It is well known that the 1,3-diols of PVA are first converted into the corresponding 1,3-diketone by PVA-DHase. In order to elucidate the minimum required structure of the vinyl alcohol block which is accepted as a substrate for PVA-assimilating microbes, the enzyme activity of the vinyl alcohol block was measured using PVA-DHase (*27*). Table II shows the substrate specificity of the PVA-DHase obtained from *A. faecalis* (*9*). The isotactic-type pentamer (*i*-5VA) showed the strongest PVA-DHase activity among the PVA and vinyl alcohol blocks tested. The isotactic-type trimer (*i*-3VA) showed activity similar to that of PVA-14000. However, the PVA-DHase activity of the atactic-type trimer (*a*-3VA) was significantly decreased. No activity was observed with the dimer of the vinyl alcohol units. These results suggest that an isotactic-type vinyl alcohol block having a chain length of more than 3 monomer units is needed as a biodegradable segment to be a substrate for PVA-DHase. Table II also shows the substrate specificity of PVA-DHase obtained from the PVA-assimilating microbes, *Pseudomonas* sp. 113 P3 (*27*) in order to compare the difference in the substrate specificity of PVA-DHase for the microbial strain. It was found that the substrate specificity of PVA-DHase obtained from *A. faecalis* was quite similar, supporting similar biodegradation results but using the different degrading microbes.

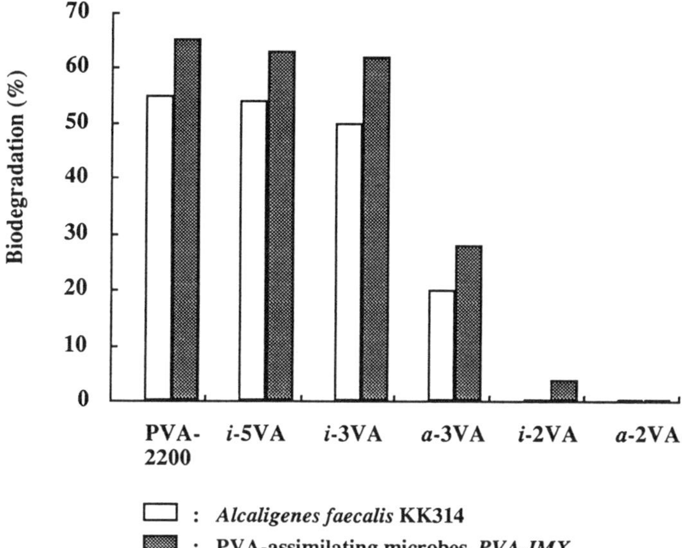

Figure 5. 21-Day biodegradation of PVA and vinyl alcohol blocks. Biodegradation (%) = (1 - BOD/ThOD) x 100

Table II. Substrate Specificity of PVA-Dehydrogenase Obtained from *Alcaligenes Faecalis* KK314 and *Pseudomonas* sp. 113 P3

Substrate	Relative activity (%)	
	A. faecalis KK314	*Pseudomonas* sp. 113 P3
PVA-14000	57	55
PVA-2200	100	100
i-5VA	216	202
i-3VA	78	64
a-3VA	10	10
i-2VA	0	0
a-2VA	0	0

SOURCE: Adapted from ref. 9.

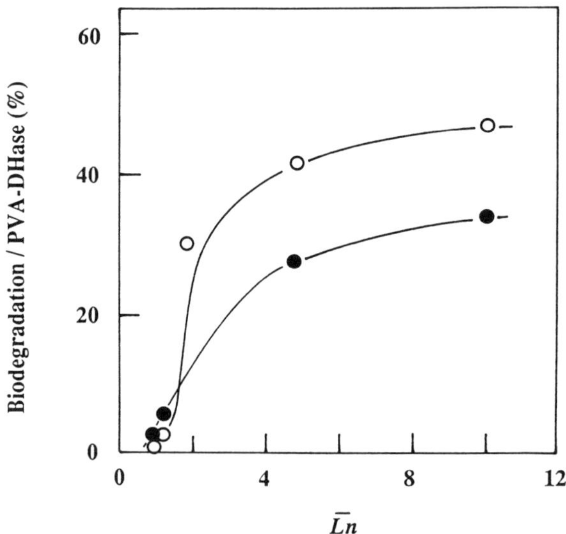

O : Biodegradation (%), ● : Relative activity of PVA-DHase

Figure 6. 21-Day biodegradation of P(DSMM-VA) and relative activity of PVA-dehydrogenase as a function of vinyl alcohol block length ($\overline{Ln}$). Biodegradation (%) = (1 - BOD/ThOD) x 100
(Adapted from ref. 9.)

Vinyl Alcohol Blocks in a Polycarboxylate Chain. In order to demonstrate the biodegradability of such vinyl alcohol blocks in the polycarboxylate chain, the enzymatic degradability of P(DSMM-VA) containing vinyl alcohol blocks was evaluated as a function of calculated vinyl alcohol block length. Figure 6 shows the relation between the relative activity of PVA-dehydrogenase from *A. faecalis* and the aerobic biodegradability determined by BOD values as a function of the number-average vinyl alcohol block length ($\overline{Ln}$) for P(DSMM-VA). The $\overline{Ln}$ showing a rapid increase in activity corresponds to the minimum chain length of a vinyl alcohol block which acts as a biodegradable segment in the polymer chain. This tendency was in good agreement with the results obtained from the BOD values (*7*). From these facts, it was estimated that the minimum chain length of the vinyl alcohol block, which acts as a biodegradable segment, was about 3 vinyl alcohol monomer units for P(DSMM-VA).

In our previous report, the microbial degradabilities of poly(sodium carboxylate)s containing vinyl alcohol groups as biodegradable segments were significantly increased when the vinyl alcohol block length was larger than 5-6, 5, and 7 monomer units for P(SA-VA), P(DSF-VA) and P(DSM-VA), respectively (*6,8*). In the copolymer chain, a slightly longer vinyl alcohol block length was needed to form a substrate for the PVA-degrading enzyme than the minimum required chain length as determined using model compounds. This longer block length may be attributable to the steric and electrostatic hindrance caused by the entire copolymer chain. The expanding polymer chain in aqueous solution may be more accessible to the enzymatic reaction at the biodegradable segment in the polymer chain compared to the more tightly coiled polymer chain in an aqueous solution (*6,7*).

Conclusions

PVA was biodegraded by both aerobic and anaerobic microbes. Isotactic moieties of PVA were preferentially biodegraded. The minimum required structure of the vinyl alcohol block, which is accepted as a substrate for PVA-assimilating microbes and PVA-DHase, is an isotactic-type vinyl alcohol block with a chain length of 3 monomer units.

Literature Cited

1. Suzuki, T.; Ichihara, Y,; Yamada, M.; Tonomura, K. *Agric. Biol. Chem.* **1973**, *37*, 747.
2. Watanabe, Y.; Hamada, N.; Morita, M.; Tsujisaka, Y. *Archiv. Biochem. Biophys.* **1976**, *174*, 575.
3. Sakazawa, C.; Shimao, M.; Taniguchi, Y.; Kato, N. *Appl. Environ. Microbiol.* **1981**, *41*, 261.
4. Shimao, M.; Saimoto, H.; Kato, N.; Sakazawa, C. *Appl. Environ. Microbiol.* **1983**, *46*, 605.
5. Matsumura, S.; Maeda, S.; Takahashi, J.; Yoshikawa, S. *Kobunshi Ronbunshu* **1988**, *45*, 317.
6. Matsumura, S.; Shigeno, H.; Tanaka, T. *J. Am. Oil Chem. Soc.* **1993**, *70*, 659.
7. Matsumura, S.; Tanaka, T. *J. Environ. Polym. Degrad.* **1994**, *2*, 89.
8. Matsumura, S.; Ii, S.; Shigeno, H.; Tanaka, T.; Okuda, F.; Shimura, Y.; Toshima, K. *Makromol. Chem.* **1993**, *194*, 3237.
9. Matsumura, S.; Shimura, Y.; Toshima, K. *Polym. Prepr. (Am. Chem. Soc., Div. Polym. Chem.)* **1994**, *35(2)*, 429.
10. Swift, G. *Polym. Degrad. Stab.* **1994**, *45*, 215.

11. Matsumura, S.; Shimura, Y.; Terayama, K.; Kiyohara, T. *Biotech. Lett.* **1994**, *16*, 1205.
12. Imai, K.; Shiotani, T.; Oda, N.; Otsuka, H. *J. Polym. Sci.* **1986**, *A24*, 3225.
13. Yamamoto, T,; Yoda, S.; Sangen, O.; Fukae, R.; Kamati, M. *Polym. J.* **1989**, *21*, 1053.
14. Sogah, D.Y.; Webster, O.W. *Macromolecules* **1986**, *19*, 1775.
15. Risse, W.; Grubbs, R.H. *Macromolecules* **1989**, *22*, 1558.
16. Hanessian, S.; Ugolini, A.; Dube, D.; Glamyan, A. *Can. J. Chem.* **1984**, *62*, 2146.
17. Meyers, A.I.; Lawson, J.P. *Tetrahedron Lett.* **1982**, *23*, 4883.
18. Mori, Y.; Kuhara, M.; Takeuchi, A.; Suzuki, M. *Tetrahedron Lett.* **1988**, *29*, 5419.
19. Mori, Y.; Takeuchi, A.; Kageyama, H.; Suzuki, M. *Tetrahedron Lett.* **1988**, *29*, 5423.
20. Mori, Y.; Suzuki, M. *J. Chem. Soc., Perkin Trans. 1*, **1990**, 1809.
21. Mori, Y.; Suzuki, M. *Tetrahedron Lett.* **1989**, *30*, 4383.
22. *OECD Guidelines for Testing of Chemicals, 301C, Modified MITI Test, Organization for Economic Cooperation and Development (OECD)*, Paris, **1981**.
23. Shimao, M.; Yamamoto, H.; Ninomiya, K.; Kato, N.; Adachi, O.; Ameyama, M.; Sakazawa, C. *Agric. Biol. Chem.* **1984**, *48*, 2873.
24. Matsumura, S.; Nishioka, M.; Yoshikawa, S. *Kobunshi Ronbunshu* **1991**, *48*, 231.
25. Ito, S.; Naito, S.; Unemoto, T. *J. Jpn. Oil Chem. Soc. (Yukagaku)* **1988**, *37*, 1006.
26. Matsumura, S.; Kurita, H.; Shimokobe, H. *Biotech. Lett.* **1993**, *15*, 749.
27. Hatanaka, T.; Asahi, N.; Tsuji, M. *Abstracts of the annual meetings of Japan Society for Bioscience, Biotechnology, and Agrochemistry,* 3D2p5, *Nippon Nogeikagaku Kaishi* **1993**, *67*, 311.
28. Shimao, M.; Ninomiya, K.; Kuno, O.; Kato, N.; Sakazawa, C. *Appl. Environ. Microbiol.* **1986**, *51*, 268.
29. Lowry, O.H.; Rosenbrough, N.J.; Farr, A.L.; Randall, R.J. *J. Biol. Chem.* **1951**, *193*, 265.

RECEIVED October 4, 1995

Chapter 12

Soil and Marine Biodegradation of Protein–Starch Plastics

K. E. Spence[1], A. L. Allen[2], S. Wang[1], and J. Jane[1,3]

[1]Center for Crops Utilization Research and Department of Food Science and Human Nutrition, Iowa State University, Ames, IA 50011
[2]U.S. Army Natick Research, Development, and Engineering Center, Natick, MA 01760–5020

Plastics produced from soy protein and cornstarch were determined to be biodegradable in marine and soil environments. Dialdehyde starch and zein plastics were less readily biodegradable in soil. The plastics were manufactured by extrusion and injection-molding, and by compression-molding the protein/starch mixtures into standard tensile articles. Respirometric studies, measuring CO_2 evolution, showed the ground molded protein/starch plastics to have faster biodegradation rates than that of the individual raw materials. This was attributed to the denatured protein and gelatinized starch of the molded specimens, which were more susceptible to microbial biodegradation.

Over 21.9 billion pounds of plastic materials were discarded in 1992 [1]. The loss of landfill space and a change in the public perception of acceptable waste, waste reduction, and waste elimination has increased interest in biodegradable plastics. It is estimated that the degradable plastics market will exceed 3.2 billion pounds by the year 2000 [2]. Recent research efforts to develop polymeric alternatives to petroleum-based products have centered on biopolymers as starting materials.

Protein and starch are two major biopolymers in crops. Soybean seeds consist of ca. 30-45% protein [3]. Soy proteins are classified as globulins, being soluble in water or salt solutions above or below the pH (4.5) of their isoelectric point [3]. The properties of the protein are largely defined by the higher number of acidic and basic amino acids in the proteins and the disulfide bonds which bind the polypeptide subunits together. The proteins have a molecular weight range of 181,000 to 350,000 [3]. Soy protein isolate contains >90% protein [3].

Corn grains contain ca. 72% starch and 10% protein [4]. Corn starch consists of two polymers, amylose and amylopectin. Amylose is a linear polymer

[3]Corresponding author

0097–6156/96/0627–0149$15.00/0

of 1,4-α-linked gluco-pyranose units with few branches [5]. Amylopectin is a highly branched polymer defined as α-1,4-glucopyranose chains linked by α-1,6-branches. Normal corn starch typically consists of 28% amylose and 72% amylopectin. Zein is the major protein in corn (ca. 50% of total protein)[4] and is categorized as a prolamin, soluble in aqueous alcohol. Zein is low in ionized amino acids and high in nonpolar amino acids; and the high proportion of hydrophobic side chains results in its hydrophobicity [6,7].

Dialdehyde starch is a unique product produced by the oxidation of native corn starch with periodate. The reaction involves a high degree of specificity, with the periodate cleaving the bond between C2 and C3 of the glucopyranose ring [8]. This bond cleavage generates an aldehyde on each of the two carbons.

Research using vegetable protein in plastics dates to the early part of this century. In the 1930's, formaldehyde was used as a hardening agent with water-plasticized soy protein to produce plastics [9,10]; however, the soy plastics were more hydrophilic and costly than their petroleum-based counterparts. With growing environmental pollution concerns and increasing waste-disposal problems, the use of agricultural biopolymers for plastics is once again being investigated.

Gennadios et al. [11] have shown that vegetable proteins can be used to cast films. Paetau et al. [12,13] demonstrated the effects of molding temperature, acid treatment, crosslinking, and fillers on the mechanical properties and water absorption of compression- molded soy protein. Soy protein isolate (11.7% moisture content) displayed a maximum tensile strength (39 MPa) when compression molded at 140°C [12]. Acid-treatment (HCl, propionic acid, acetic acid) of the soy protein, to the isoelectric point, reduced the 26 h water absorption of the plastics from >100% to 30% and did not effect the tensile strength [12]. Crosslinking with formaldehyde (5% w/w) significantly increased the tensile strength (48 Mpa) and reduced the water absorption (23%) of the molded plastics. Crosslinking with glyoxal or adipic/acetic anhydride decreased the extensibility and tensile strength of the plastics [13]. Incorporating short-fiber cellulose (up to 20% w/w) into the soy protein plastic increased the tensile strength and Young's modulus, but long-fiber cellulose decreased the plastic's mechanical properties [13].

Corn starch has also been compounded with petroleum-based polymers in the production of plastics [14,15,16]. Lim and Jane [17] demonstrate that corn starch and the hydrophobic corn protein, zein, can be cross-linked and molded into a plastic. A modified corn starch, dialdehyde starch, forms crosslinks with zein in compression molding to produce a plastic that has good tensile strength (49 MPa) and is water resistant (2.5% after 24 h soaking)[18].

For many years the terms degradable and biodegradable were used interchangeably. The 1988 ISO 472 definition reads: Biodegradable plastic - A degradable plastic in which the degradation results from the action of naturally-occurring microorganisms such as bacteria, fungi, and algae [19]. Tests for biodegradability have included studies of weight loss, mechanical property changes, O_2 consumption, and CO_2 evolution. Weight loss is of limited use for polymers that fragment during the test. A test of CO_2 evolution is considered a good criterion for mineralization and hence ultimate biodegradation of an organic molecule [20]. In the test, a portion of the carbon is incorporated into microbial biomass and the percentage of CO_2 produced will never reach 100% of that theoretically to be produced from the total carbon of the substrate. Therefore, regulations, such as those adopted for surfactants, regard chemical substances that produce 60% of their theoretical total mass within a defined time period, (28 days), as "readily biodegradable" [21]. It is assumed the time period for other materials, particularly insoluble materials such as plastics, would need to be longer.

Soy protein isolate, native corn starch, modified starch, and zein have been investigated for manufacturing biodegradable plastics. Various combinations of the protein and starch were molded into shaped specimens by suitable processing methods. The objective of this study was to investigate the biodegradability of soy protein/starch plastics and dialdehyde starch/zein plastics, by respirometry, in simulated soil and marine environments.

Experimental

Materials. Soy protein isolate (PRO-FAM) (Archer Daniels Midland, Decatur, IL), zein (Freeman Industries, Tuckahoe, NY), corn starch (American Maize, Hammond, IN), and polymeric dialdehyde starch (PDS, 90% oxidized starch)(Sigma Chemical Company, St. Louis, MO) were used. Other chemicals were reagent grade and used without further purification.

Preparation of Plastics. Native soy protein isolate was compression molded (45 MPa, 140°C, 10 min) to form standard tensile bars using a Wabash compression molding machine (Wabash Metal Products, Wabash, IN). Mixtures of soy protein isolate (3 parts) and corn starch (2 parts) with glycerol (14%, and 18%) as a plasticizer were mixed in a heavy duty mixer (KitchenAid, St. Joseph, MI). The mixture was equilibrated to the desired moisture content and extruded on a twin-screw extruder (Micro 18 twin screw, American Liestritz Extruder Corp., Somerville, NJ) and pelletized. The pellets were injection molded using a Boy injection-molding machine (Boy 22S, Boy Machine Inc., Berwyn, IL) at 1300 bar to form tensile specimens.

Zein (1 part) was dissolved in 75% aqueous methanol and mixed with dialdehyde starch (3 parts) of various degrees of oxidation (1%, 5%, and 90% oxidation) [22]. After evaporating the methanol in a fume hood, the mixture was dried at 50°C in a forced air oven. The mixture was ground in a cyclone mill (UDY Corp., Fort Collins, CO). The moisture content was adjusted to the desired level in a humidity chamber. Tensile bars were produced using the Wabash compression molding equipment under the same conditions as described in the preceding paragraph.

Biodegradation

Soil/Sand/Composted Manure Environment. A respirometer was designed to measure mineralization of polymers in a soil environment. The basic design of the system consisted of air conditioning pretraps, the soil reactor, and a CO_2 posttrap. CO_2-free humidified air was generated by bubbling air through four 2-L pretraps containing 1-L of 2N NaOH, 2N NaOH, 2N H_2SO_4, and distilled water, respectively. Air flow, at 15 mL/min, was directed through a leakproof barbed bulkhead fitting in the base of one side of each 500 mL polycarbonate square bottle (Figure 1). Each reactor contained an all-purpose potting soil (Hyponex All-Purpose potting soil, Hyponex Corp., Marysville, OH), sand, and compost and manure (Fertalife Compost and Manure, Hyponex Corp., Marysville, OH) (1:1:1) mixture (200 g total dry weight basis) with a moisture content of 30%. The reactor bottles were kept in a water bath at 25°C. Temperature and water circulation were controlled by an Immersion Circulator Model 70 (Fischer Scientific, Pittsburgh, PA). Gases exited each reactor to the posttrap through the stoppered top. Five replicates (2 g each) of each sample of raw materials and processed plastics and the controls (soil/sand/compost and manure) were analyzed for comparison. Evolved CO_2 gas was trapped in 10 mL NaOH solution and titrated using HCl with a Mettler DL 12 autotitrator (Mettler Instruments, Hightstown, NJ). The degree of polymer biodegradation was calculated as the difference between the average CO_2 value of the reactors with samples and the controls. These values were then converted to percent biodegradation based on the amount of carbon evolved divided by the theoretical carbon amount of each sample. Native soy protein isolate, native corn starch, soy protein plastic, soy protein/starch (3/2, w/w) plastics with 18% glycerol, and the controls were monitored over 456 h.

Simulated Marine Environment. Mineralization of ground soy protein/starch plastic specimens was carried out by using respirometry procedures in a simulated marine environment developed by Allen et al. [23]. Erlenmeyer flasks (250 mL) were used with a working volume of 100 mL. In each flask, 10 mg of sample substrate were used as the sole carbon source. The composition of the defined marine medium was as follows, in g/L: NH_4, 2; $MgSO_4$, 2; K_2HPO_4, 0.1; KNO_3, 0.5; and "Instant Ocean" salts (Aquarium System Inc., Mentor, OH), 17.5. The media were inoculated with a suspension (100 μL) of eleven marine cultures previously characterized for polymer degradation [24]. A Microoxymax respirometer (Columbus Instruments International Corporation, Columbus, OH) was used for the study. Gas sensors were calibrated using two flasks with 100 mL of 0.1% phosphoric acid to which 90 mg of Na_2SO_3 and 50 mg of Na_2CO_3 were added. Oxygen consumption (sulfite oxidation) and CO_2 evolution for the calibration solutions were monitored along with the sample flasks. A three-minute sampling time was used for each flask, followed by a three-minute purge of the sensor cell. This cycle repeated every three hours. All the flasks were kept in a light-tight incubator at 30°C and agitated with magnetic stirrer bars.

Raw materials and processed plastics were analyzed in parallel for comparison. Samples of ground soy protein plastic, ground soy protein/starch (3/2, w/w) plastics with glycerol (14%), native soy protein isolate, and native corn starch were analyzed along with controls (no substrate) and empty flasks. Triplicate analyses were run for each sample. Moles of CO_2 evolved, derived by

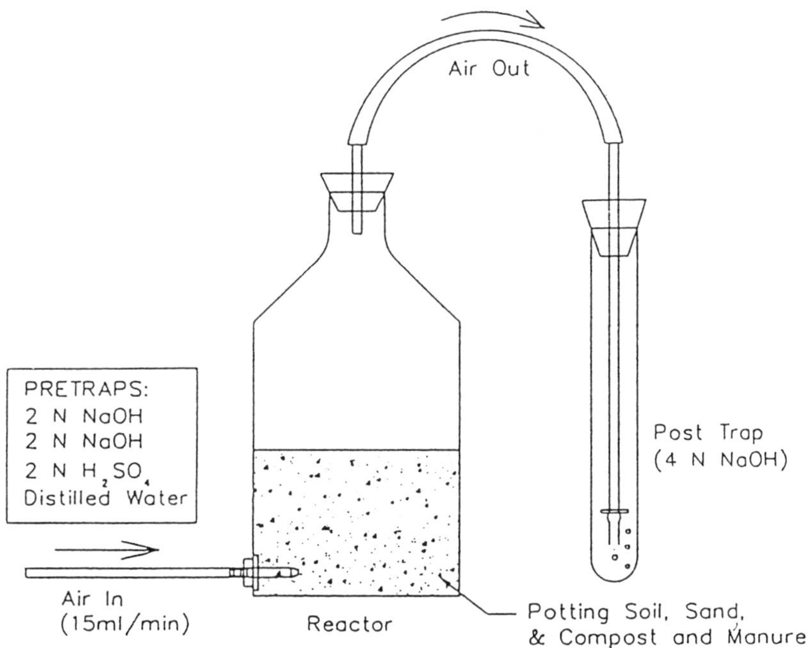

Figure 1. Respirometry reactor showing the details of CO_2-free air entering the bottom of the reactor and exiting to the posttrap through the top.

subtracting CO_2 from controls, were averaged for each sample and divided by the theoretical carbon ($ThCO_2$) in the sample to calculate percentage mineralization.

Soil/Sand Environment. Biodegradation of the dialdehyde starch and zein plastics was conducted in potting soil and sand (1:1) under the same conditions as described for soil/sand/compost and manure [18]. However, in this set-up both the incoming air and exiting gases passed through the top of a 400 mL glass bottle above the soil line and the water pretrap was placed at beginning rather than the end. The analysis was conducted over 180 days. Corn starch, zein, polymeric dialdehyde starch, and ground samples of plastics made from dialdehdye starch (various degrees of oxidation) and zein (3:1) were studied. Each reactor contained 10 g of sample in 100 g soil.

Results and Discussion

Respirometry using potting soil, sand, and compost and manure demonstrated that the soy, soy plastics, and soy and starch plastics were readily biodegraded (Figure 2). Soy protein isolate and starch (3/2, w/w) (18% glycerol) and compression molded soy protein isolate plastics reached 62% and 60% mineralization, respectively. The percentage of carbon evolved as CO_2 for native soy protein was 58% after 456 h. Corn starch displayed a limited carbon evolution of 28%. The enhanced degradation rates of the plastics (CM Soy, Inj. SCG) were attributed to the molding process, which denatured proteins and gelatinized starch; thus, the material became more susceptible to microbial degradation. Mixtures of protein and starch provided a balanced nitrogen and carbon source for microorganism growth which also enhanced degradation. The slow degradation of the native starch sample was attributed to its granular structure. Native starch is present in a semicrystalline, granular form, which is more resistant to enzymatic and chemical degradation.

After 174 h of simulated marine-environment biodegradation, soy protein/starch injection-molded plastics (14% glycerol) displayed 41% mineralization, and compression molded soy protein showed 37% (Figure 3). Native soy protein isolate displayed 30%, whereas native corn starch displayed only 4% mineralization. The results showed that soy protein/starch plastics were degraded faster than the raw materials from which they were made. The results also showed that the soy protein/starch plastics were promptly degraded in a marine environment. Oxygen consumption (Figure 4) was indicative of respiratory activity of biologics in the reactors. It is evident that the microbial activity was greatest in reactors containing protein-based plastics, which had enhanced nitrogen availability. Oxygen consumption in the reactor containing starch was no greater than the control.

Potting soil and sand as media, without microbial inoculation via compost and manure, was the initial experiment for biodegradation studies and demonstrated a slow rate of substrate degradation (Figure 5). Aeration of the reactor from the top, rather than through the soil, and the arrangement of the pretraps with 2N H_2SO_4 at the end had a dehydrating effect and likely slowed

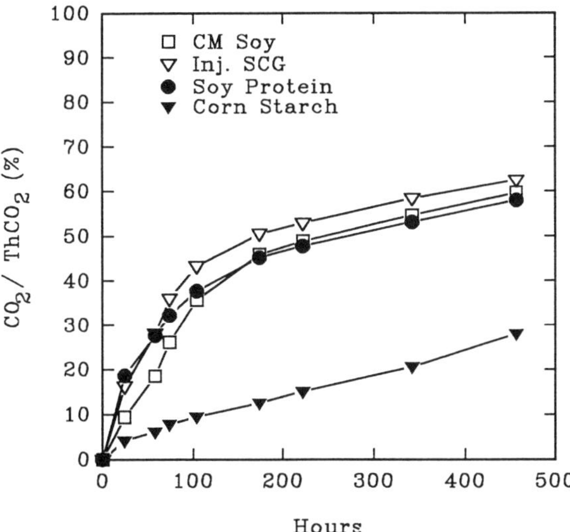

Figure 2. CO_2 evolved represented as percent of theoretical CO_2 each sample could produce in a soil/sand/compost and manure (1/1/1, w/w) environment. Samples are compression molded soy protein (CM Soy), injection molded soy protein/starch (3/2, w/w) with 18% glycerol (Inj. SCG), native soy protein, and corn starch.

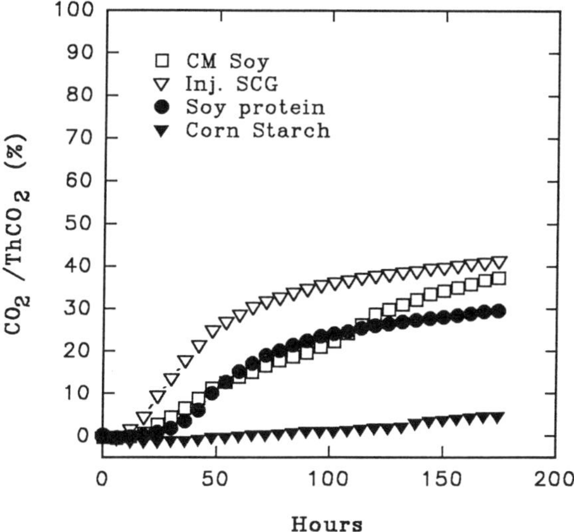

Figure 3. CO_2 evolved represented as percent of theoretical CO_2 each sample could produce in a simulated marine environment. Samples are compression molded soy protein (CM Soy), injection molded soy protein/starch (3/2, w/w) with 14% glycerol (Inj. SCG), native soy protein, and corn starch.

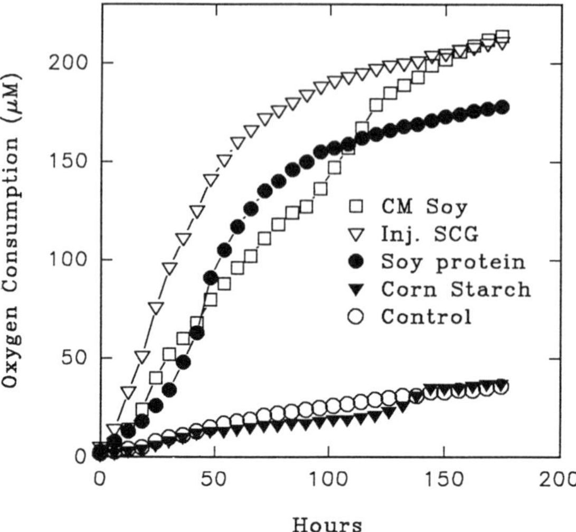

Figure 4. O_2 consumption of samples in simulated marine environment. Legend the same as Figure 3.

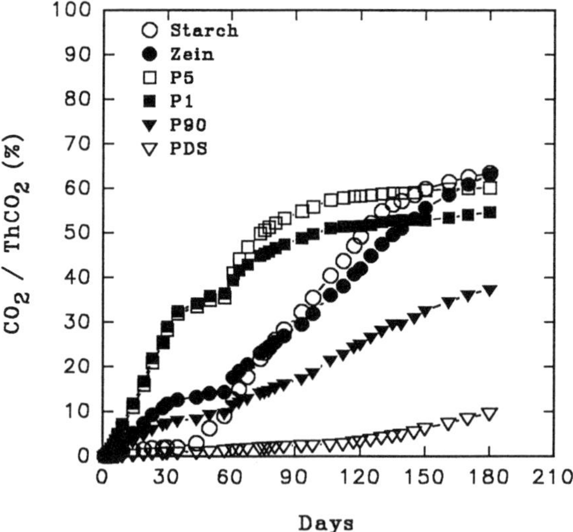

Figure 5. CO_2 evolution from soil/sand respirometer represented as percentage of theoretical CO_2 which a sample could produce. Samples are native corn starch (starch), corn zein protein (zein), 90% oxidized dialdehyde starch (PDS), and dialdehyde starch and zein (3/1, w/w) plastic in which the number following "P" represents the percentage of starch oxidation. (Reproduced with permission from Ref. 18. Copyright 1995 Journal of Environmental Polymer Degradation.)

down mineralization. After 180 days, both corn starch and zein reached about 63% mineralization. Polymeric dialdehyde starch of ca. 90% oxidation showed little mineralization (9.6%). Plastics made from zein and dialdehyde starch of 5% and 1% oxidation showed 60% and 55% mineralization, respectively. The values were fairly close to those of zein and cornstarch. The plastic samples made from zein and polymeric dialdehyde starch (90% oxidation) showed a retarded mineralization of 37%. Degradation curves showed that plastic samples made from zein and dialdehyde starch (1% and 5% oxidation) had much greater initial degradation rates than raw materials (i.e., zein and corn starch). This further demonstrated that processing made the starch and protein more susceptible to microbial attack. The degradation rates, however, slowed down after the first 100 days. Corn starch showed almost no CO_2 evolution during the first 40 days. After the lag period, mineralization of corn starch increased and reached 63% at the end of 180 days. Replacement of this respirometer design with airflow through the bottom and the H_2O pretrap at the end helped maintain a humid environment for faster biodegradation in the other experiments.

Conclusions

In conclusion, plastics made from soy protein and starch mixtures were determined to show "ready biodegradability" in both marine and soil environments under the conditions of these experiments. Plastics made from highly oxidized dialdehyde starch and protein, however, showed retarded biodegradation.

Acknowledgments

The authors would like to thank Anthony L. Pometto III for his advise and assistance, and P. Puechner for her advise, on the biodegradation studies. We would also like to thank D.L. Kaplan and J. Mayer, U.S. Army Natick Research Center, for suggestions and discussions. Financial support for this research was provided by the American Soybean Association and the Corn Refiners Association.

Journal Paper J-16246 of Iowa Agriculture and Home Economics Experiment Station, Ames, Iowa, Project No. 3258.

Literature Cited

1. Plastic Durables: Environmental impact. *Plastics Eng.* **1994**, 50:34.
2. North American Demand for Degradable Plastics. *Polymer News.* **1992**, 17, 150.
3. Nielsen, N.C. In *New Protein Foods*; Altschul, A.M.; Wilcke, H.L., Eds.; Academic Press: Orlando, FL, **1985**, Vol. 5; pp. 27-64.
4. Watson, S.A. in *Corn, Chemistry and Technology*; Watson, S.A.; Ramstad, P.E., Eds.; AACC; St. Paul, MN, **1987**, pp. 53-82.
5. Takeda, Y.; Kanako, S.; Hizukuri, S. *Carbohydr. Res.* **1984**, 132, 83-92.
6. Paulis, J.W.J. *Agric. Food Chem.* **1982**, 30, 14.

7. Pomes, A.F. *Encycl. Polym. Sci. Technol.* **1971**, 15, 125.
8. Jackson, E.L.; Hudson, C.S. *J. Am. Chem. Soc.* **1937**, 59, 2049.
9. Brother, G.H.; McKinney, L.L. *Ind. Eng. Chem.* **1940**, 32, 1002-1005.
10. McKinney, L.L.; Deanin, R.; Babcock, G.; Smith, A.K. *Ind. Eng. Chem.* **1943**, 35, 905-908.
11. Gennadios, A.; Weller, C.L. *Food Technol.* **1990**, 44, 63-69.
12. Paetau, I.; Chen, C.-Z.; Jane, J. *Ind. Eng. Chem. Res.* **1994**, 33, 1821-1827.
13. Paetau, I.; Chen, C.-Z.; Jane, J. *J. Environ. Polym. Degrad.* **1994**, 2, 211-217.
14. Otey, F.H.; Westhoff, R.P. U.S. Patent 4,337,181, **1992**.
15. Griffin, G.J.L. U.S. Patent 4,016,177, **1977**.
16. Lay, G.; Rehm, J.; Stepto, R.F.; Thoma, M.; Sachetto, J.-P.; Lentz, D.J.; Silbiger, J.; U.S. Patent 5,095,054, **1992**.
17. Lim, S.; Jane, J. in *Carbohydrates and Carbohydrate Polymers*; Yalpani, M. Ed.; ATL Press: Mount Prospect, IL, **1993**, pp. 288-297.
18. Spence, K.E.; Pometto, III, A.L.; Jane, J. *J. Environ. Polym. Degrad.* (In press)
19. Narayan, R. In *Biodegradable Polymers and Plastics*; Vert, M.; Feijen, J.; Albertsson, A.; Scott, G.; Chiellini, E., Eds.; Royal Society of Chemistry, Cambridge, UK, **1992**, pp. 176-187.
20. Seal, K.J. In *Chemistry and Technology of Biodegradable Polymers*; Griffin, G.J.L., Ed.; Blackie Academic and Professional: Glasgow, UK, **1994**, pp. 116-134.
21. Mehltretter, C.L. In *Methods in Carbohydrate Chemistry*; Whistler, R.L.; Ed., Academic Press: Orlando, FL, **1964**, Vol. 4; pp. 316-317.
22. OECD *Guidelines for testing of Chemicals*; OECD, Paris, FR, **1992**, 1-62.
23. Allen, A.L.; Mayer, J.M.; Stote, R.E.; Kaplan, D.L. *J. Polym. Degrad.* (In press).
24. McCassie, J.E.; Mayer, J.M.; Stote, R.E.; Shupe, A.E.; Stenhouse, P.J.; Dell, P.A.; Kaplan, D.L. in *Biodegradable Polymers and Packaging*; Ching, C.; Kaplan, D.; Thomas, E., Eds.; Technomic Publishing, Co.; Lancaster, PA, **1993**, pp. 247-256.

RECEIVED November 16, 1995

Chapter 13

Biodegradation of Polymer Films in Marine and Soil Environments

J. M. Mayer, D. L. Kaplan, R. E. Stote, K. L. Dixon, A. E. Shupe, A. L. Allen, and J. E. McCassie

Biotechnology Division, U.S. Army Natick Research, Development, and Engineering Center, Natick, MA 01760–5020

The emergence of biodegradable plastics has necessitated the development of standard methods to determine biodegradation rates in various environments. Standardized accelerated marine and soil laboratory biodegradation test systems were developed in which comparative polymer biodegradation rates could be determined by quantifying and plotting the weight loss/surface area of each sample over time and determining the maximum slopes of the curves generated. The results indicate that, in general and depending on the environment, biodegradation rates for unblended polymers were: polyhydroxybutyrate-co-valerate > cellophane > chitosan > polycaprolactone. Results from blends are more difficult to interpret since different biodegradation rates of the component polymers and leaching of plasticizers and additives can impact the data.

The assessment of biodegradability of polymer films in natural environments is a difficult problem because of the inherent variations in environmental conditions from one site to another. This is further confounded by the need to balance material performance vs. rates of biodegradation *(1,2)*. Therefore, many laboratory approaches have been developed to simulate natural biodegradation processes, but in a more controlled setting to try and predict natural environmental susceptibility of materials to biodegradation. These methods recently have been summarized and include enzyme assays, plate tests, clearing zones or changes in optical absorbance, biological oxygen demand, changes in carbon isotope ratios, release of radioactive products from radioactively labeled polymers, automated respirometry in biometer flasks, and acceleratated simulated laboratory systems or mesocosms *(3-5)*. Many of these methods are coupled to assessments of changes in weight, molecular weight, mechanical properties, morphological appearance, or chemical functionalities of the

Published 1996 American Chemical Society

polymer evaluated. In addition, new practices and methods to assess biodegradability are being developed by many organizations including the American Society for Testing and Materials (ASTM). All of the above methods are potentially useful but limited when used by themselves. If the test results are properly interpreted, important data to enhance the understanding of the environmental fate of the subject polymer film can be developed. It is clear that a battery of tests is usually required to fully assess the biodegradability of a polymer coupled with environmental impact and risk assessment. Tests on individual polymers must be factored against the effects of polymer processing and blending, consideration of the disposal environment where the material may end it's lifecycle, and other constraints such as avoiding entrapment and ingestion hazards to organisms in the environment.

We define biodegradation as a process carried out primarily by bacteria or fungi in which a polymer chain is cleaved or modified by hydrolytic or oxidative enzymatic activity *(6)*. Related terms or subsets of biodegradation include biotransformation and biomineralization. Biotransformation is the biologically-mediated change in chemical structure of the target polymer. Mineralization is the biologically-mediated complete breakdown of the polymer generating simple gases like carbon dioxide, methane and nitrogen, water and biomass, so that all elements from the polymer re-enter natural geochemical and microbial cycles.

With a goal of developing biodegradable polymers useful for a range of applications, it is important to use appropriate test methods to provide data on biodegradability and environmental impact. Our primary concerns are with soil and marine environments, thus we focused on developing appropriate test methods to assess candidate materials for disposal in these environments. We have used a three tier system to characterize biodegradability and assure that environmental impact is negligable. In the first tier, susceptibility of the individual polymers to mineralization is assessed using automated respirometry *(7)*. If the polymer is mineralizable, then in the second tier, the processed polymer blend or formulation is assessed for evidence of biodegradability in the laboratory by exposure in simulated environments (marine or soil). After exposure for varying periods of time, the materials are characterized for changes in mechanical properties or chemical structure. In the third tier, the material is exposed in natural environments to corroborate the laboratory results. In all tiers, we also assess the toxicity of the polymer, the blends after processing, and the residues subsequent to biodegradation and exposure in the various systems *(8)*. Toxicological impact is a critical component in the assessment process since rates of biodegradation will differ significantly in different environments. For example, if no toxicity is observed, then rates of biodegradation in marine environments become less of a factor as long as the more acute effects, such as entrapment or ingestion hazards to marine animals, are addressed, or the asthetics of litter are tolerable. This may be handled by demonstrating rapid loss of mechanical properties in the target disposal environment, even if rates of mineralization are very low. All three tiers need to be considered for a complete environmental risk assessment before a new biodegradable polymer formulation should be considered for use. In addition, if the new formulation will be in contact with food, the U.S. Food and Drug Administration regulations also will have to be considered.

In the process of carrying out our program goals as described above, we found it necessary to develop new test systems and methods for some of the evaluations we needed to perform on biodegradable materials. We describe our efforts to develop laboratory-scale simulated soil and marine biodegradation test systems and some of the biodegradation kinetics observed for candidate materials. Some aspects of the development of these methods have been previously published *(6,9)*. Our test methods for automated respirometric analysis of the polymers have already been described *(7)* and the toxicological studies have also, in part, been published *(8)*.

Materials and Methods

Polymer Films Polyhydroxybutryrate-co-valerate (PHBV), containing 8, 16 and 24% valerate, were obtained from Imperial Chemical Industries (Zeneca), Billingham, UK. Uncoated- and nitrocellulose-coated cellophane films were supplied by DuPont, Wilmington, DE. Crosslinked chitosan (Protan Laboratories, Redmond, WA) films were produced by reaction with epichlorohydrin *(10)*. Starch/ethylene vinyl alcohol (St/EVOH) blend films and pure EVOH film (38 mole percent ethylene) were obtained from Novamont (Novara, Italy) and EVALCo (Lisle, IL), respectively. Polycaprolactone (PCL), molecular weight 80,000 Daltons, in film form, was received from Union Carbide (Bound Brook, NJ).

Cellulose Acetate/Starch (CA/St) Blends for Tensile Bars CA/St blends were produced by mixing cellulose acetate (degree of substitution 2.1) (Eastman Chemical, Kingsport, TN) with 30% amylose starch (Melogel, National Starch, Inc., Bridgewater, NJ) and propylene glycol (Dow Chemical, Midland, MI) at a ratio of 59:19:22 (weight precent) *(11)*. Mixing was in a high intensity Henschel mixer (Purnell International, Houston, TX) at 85°C, 3000 rpm for approximately two minutes. Pellets were then produced using a Brabender (Hackensack, NJ) single screw extruder (1.9 cm diameter X 47.6 cm length) operating at 60 rpm with the following temperature profile: Zone 1 = 120°C, Zone 2 = 140°C, Zone 3 = 160°C, and die (4 hole, 4 mm) = 170°C. Pellets were converted into dogbone shaped tensile bars (3.7 cm X 0.3 cm) using a bench top miniature injection molder (Custom Scientific Instruments, Inc., Cedar Knolls, NJ) heated to 185°C.

Marine Simulator Sample exposures were conducted in 76 liter aquaria maintained at 30°C as previously described *(9)*. Aquaria were filled with 50 mm thick sediment (natural or defined - see later) topped with standardized simulated seawater (Aquarium Systems, Mentor, OH). The components (weight percent as provided by the manufacturer) included: chlorine, 54.2; sodium, 31.5; sulfate, 8.1; magnesium, 3.9; calcium, 1.1; potassium, 1.1; in mg/L, included: boron, 32.1; strontium, 10.0; phosphorus, 2.4; lithium, 2.24; tin, 2.22; aluminum, 0.87; vanadium, 0.57; molybdenum, 0.42; silicon, 0.42; iron, 0.21; barium, 0.13; chromium, 0.13; nickel, 0.1; cobalt, 0.06; maganese, 0.03; zinc, 0.02. This composition is designed to mimic natural seawater *(12)*. The marine water was aerated by aquaria air pumps (Willinger Bros., Oakland, NJ) at approximately one liter per minute. The water in each aquarium was continuously replaced by a peristaltic pump (Rainin Instrument Co., Woburn, MA) with fresh simulated sea water at a weekly exchange rate of 15% of the total aquarium volume to minimize accumulation of potentially inhibitory biodegradation metabolites. Twenty watt fluorescent lights were used at 12 hour on/off cycles to simulate light effects on marine organisms. Triplicate samples of films (72 mm X 25 mm) or tensile bars (3.7 cm X 0.3 cm) were placed in fiberglass screening (18 X 16 mesh, opening size of approximately 1 mm). Screens were placed both vertically in the water phase and horizontally in the sediment, 25 mm below the surface. Total maximum loading of polymer in the system was approximately 1.3 g/L.

Marine Sediments Natural marine sediment was collected in the tidal zone along the beach at Gloucester, MA. In the defined marine sediment, commercial grade sand, 0.45 - 0.55 mm particle size, served as the matrix. Marine agar 2216 (Difco Laboratories, Detroit, MI) was mixed with the sand at 0.2, 0.4 and 0.6 percent (wt/wt) to determine the levels of the agar required to support equivalent numbers of marine microorganisms in comparison to the levels present in the natural sediments. Organic carbon contents of natural sediment and the commercial sand were determined by heating samples at 560°C for 18-24 hours in a furnace (Thermolyne, Dubuque, IA).

Marine Inoculum and Counts For the natural sediment marine test systems, water containing local flora were obtained from marine waters off a beach in Gloucester, MA. For the defined sediment marine systems, the inoculum contained nine marine microorganisms that we had previously isolated based on their ability to utilize a series of polymer substrates as carbon sources *(9)*. Some of the organisms were identified using the Biolog Microstation System, Release 3.5 (Biolog, Inc., Hayward, CA). Table I identifies the marine isolates used to inoculate the defined marine systems and the substrates they were able to utilize as sole carbon sources in liquid minimal media. During the exposures of polymer samples in the marine test systems, microbial counts in the marine water and sediment were determined by serial dilutions in sterile simulated seawater and using spread plates with 0.1 mL of each dilution bottle onto marine agar. Plates were incubated 1-2 days at 30°C.

Soil Simulators A water retentive, aerated soil was produced by mixing equal parts by weight topsoil (Earthgrow, Inc., Lebanon, CT), composted cow manure (1881 brand, Earthgrow, Inc., Lebanon, CT) and sand (0.45 - 0.55 mm particle size) with 30% (wt/wt) water. Moisture content and pH were 30% and 7.1, respectively. The natural flora in the soil served as the inoculum. The counts of microorganisms in the sand, topsoil and composted cow manure were determined by placing one gram of each medium into phosphate buffer, pH 7.0, diluting, spread plating 0.1 ml onto nutrient agar, and incubating the plates at 30°C for 1-2 days. The soil mixture was placed in soil boxes (33 cm long X 22 cm wide X 8 cm high) covered with sliding plexiglass sheets and turned periodically. Film and tensile bar samples were placed into the soil boxes without fiberglass screens and incubated at 30°C. The soil and composted cow manure were analyzed for pH, nitrate and ammonia.

Sample Retrieval and Analysis. Samples were retrieved from marine and soil simulators at selected weekly intervals, washed with distilled water to remove debris, dried to constant weight at 70°C and weighed. Weight loss data are presented as weight loss/surface area ($\mu g/mm^2$) and plotted vs. time. Maximum rates of biodegradation for each polymer or blend were determined using the Origin software program (Version 2.67, Microcal Software, Inc, Northampton, MA) to calculate the maximum slope for each weight loss/surface area vs. time curve.

Results and Discussion

Weight loss/surface area is equated here with biodegradation. We recognize this is an extrapolation due to dissolution and hydrolytic effects being confounded with biodegradation as defined earlier. However, in refering back to the context of our three tier system, we will be discussing polymers that we have already studied for mineralizability, and therefore those polymers lost from the samples, by biodegradation, solubilization or hydrolysis, will be mineralized over time, and thus we consider them part of the biodegradable fraction. In addition, we have examined several of these polymers in sterile control systems and found negligible effects of hydrolysis alone on weight loss at 30°C. Finally, there are many reports in the literature detailing rates of hydrolytic degradation or rates of solubilization of these polymers. In many cases these rates of hydrolysis are very low *(see later 12)*, unless high temperatures or pH extremes were used in the studies.

Table I. Substrates Utilized by Microorganisms in the Defined Marine Systems

	Polymer Substrate				
Organism	Starch	EVOH	PHBV(16%V)	PCL	Cellulose
Vibrio halophlanktis	X[1]	---[2]	---	---	X
Vibrio proteolyticus	X	---	---	---	---
Bacillus megaterium	X	---	X	X	X
Pseudomonas sp. 1	X	---	---	---	X
Zoogloea	X	---	X	---	---
Pseudomonas sp. 2	X	---	---	---	X
Actinomycete sp.	---	X	X	X	X
Bacillus sp.	X	---	---	---	X
Unknown	X	---	---	---	X

[1] substrate utilized, [2] substrate not utilized

In developing the defined marine simulators, we assessed the effect of adding various amounts of marine agar to commercial sand on the biodegradation rate of PHBV(8%V) in comparison to the natural marine sediment (Figure 1). The results with 0.2% marine agar most closely mimicked the rates found in natural sediment, based on comparisons of rates of weight loss/surface area. The organic analysis of the natural sediment and the commercial sand showed 0.18% and 0.04% organic content, respectively. These data correlate well with the degradation results in that approximately 0.2% organic content correlates to maximum activity in both the defined and native marine simulators.

The counts of microorganisms from marine and soil test simulators are presented in Table II. The marine counts varied by approximately two logs in magnitude in the natural population due to seasonal variations. Topsoil contained the highest microbial counts, followed by the composted cow manure and then the sand. Based on morphology of colony types, the topsoil appeared to contain the most diverse populations.

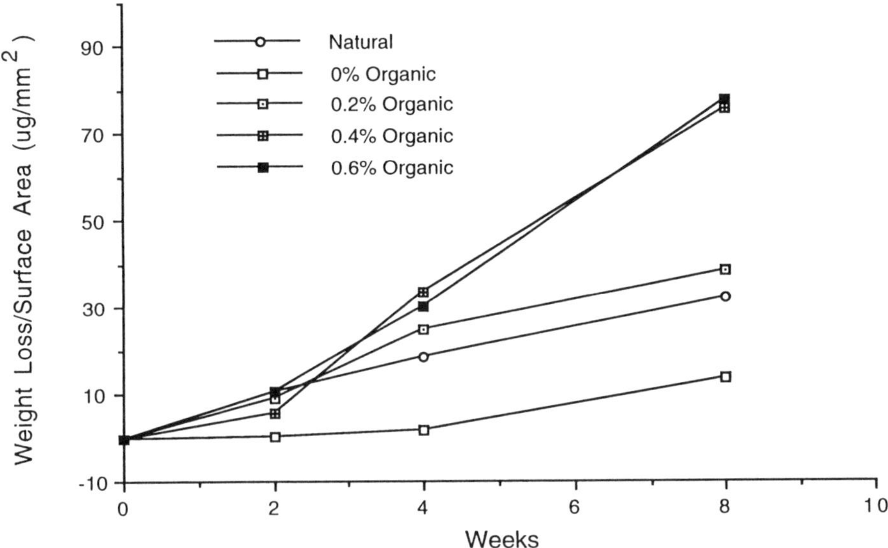

Figure 1. Influence of marine agar on rates of biodegradation of PHBV(8%V).

Table II. Counts of Microorganisms in the Simulator Components

Source	Microbial Counts/ml or gram
Natural Marine Sediment	10^3-10^5
Natural Marine Water	10^3-10^5
Defined Marine Sediment (0.2% marine agar)	10^6
Topsoil	10^6
Composted Cow Manure	10^5
Sand	10^4

The analysis of the nutrient content of some of the components used in the soil simulators indicates that the composted cow manure contributes a large amount of ammonia nitrogen which should facilitate the biodegradation of the carbon-rich, nitrogen-poor polymers. The pH, nitrate (mg/L), ammonia (mg/L) and total kjeldahl nitrogen contents of topsoil were 6.3, 110, 167, and 815, respectively, while the corresponding numbers for the composted cow manure were 8.6, 0.8, 2570 and 6130.

Graphs of weight loss/surface area vs. time of PHBV(8%V) following marine and soil exposures are presented in Figure 2, and a graph for the different polymers in soil is shown in Figure 3. It is often difficult to extrapolate relative biodegradation rates by comparison of the lines on the graphs, therefore, we used maximum slope calculated by the Origin software. Table III presents the data for maximum biodegradation rates in μg/mm^2/week for several polymers or blends in the soil simulators, the defined marine water and sediment simulators, and the natural marine water and sediment simulators. The results indicate, in general, that biodegradation in soil is more rapid

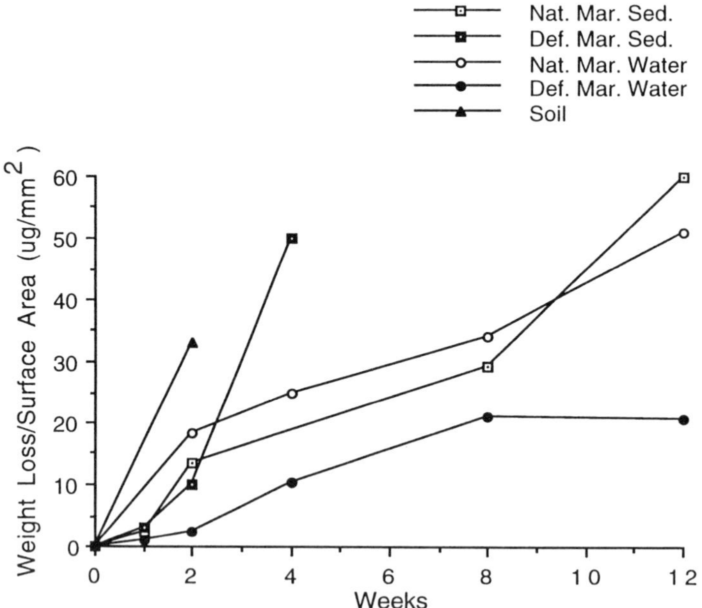

Figure 2. Weight loss/surface area vs. time of PHBV(8%V) in marine and soil simulators.

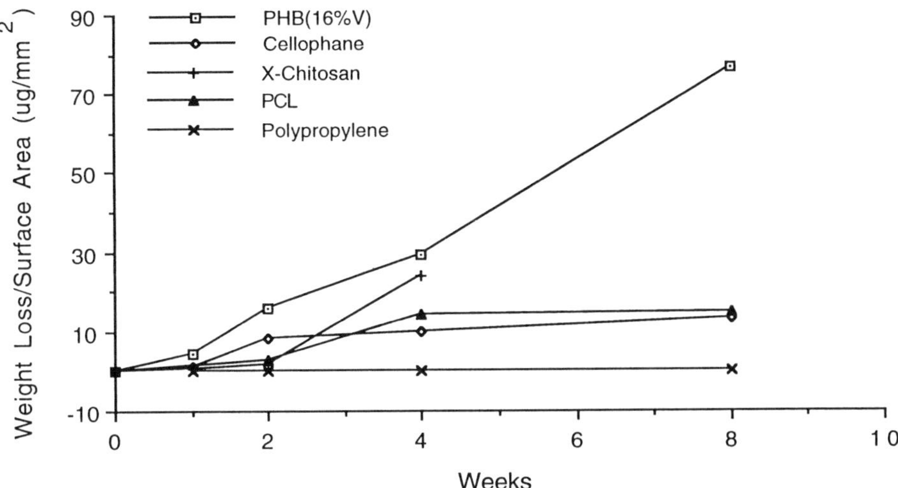

Figure 3. Weight loss/surface area vs. time of the different polymers in soil simulators.

than in marine water and sediment, except for the PCL which appears to biodegrade at about the same rate in all environments (all of which are at 30°C). PCL exhibits slow initial weight loss in all the test environments, followed by rapid weight loss after approximately eight weeks incubation time. This effect could be due to a requirement for initial nonbiological hydrolysis of the amorphous domains to lower molecular weight to a more suitable range for microbial degradation. Therefore, rates of biodegradation of PCL and related polyesters may be greatly influenced by temperature based on the correlation between temperature and hydrolytic attack and dissolution. This may a critical factor to consider in designing biodegradable materials that may be disposed in a marine environment since the temperature on the deep ocean floor often ranges between 5 - 15°C (personal communication - C. Wirsen).

Of the homopolymers evaluated, PHBV, uncoated cellophane and crosslinked chitosan were readily biodegraded based on rates generally above around 5 $\mu g/mm^2$/week (Table III). PCL was moderately biodegradable at rates around 3 to 4 $\mu g/mm^2$/week, and polypropylene and EVOH were recalcitrant under the simulator test conditions. A marine actinomycete was isolated that appeared to grow well on EVOH as the sole carbon source in preliminary experiments; however, rates appear to be too low to detect in the simulators or in respirometry. St/EVOH blends show some evidence they are susceptible to biological activity, but total degradation does not exceed that expected from the starch alone. The data for the maximum rates of biodegradation for many of the polymers are higher than the rate observed for bond paper. This is a particularly important consideration when designing materials to degrade in a municipal compost environment with well prescribed cycle times.

Table III. Maximum biodegradation rates (μg/mm²/week) for polymers and blends

Polymer/Blend	Soil	Defined Marine Sediment	Natural Marine Sediment	Defined Marine Water	Natural Marine Water
PHB(8%V)	17.0 (n=3)	15.6 (n=27)	6.2 (n=6)	3.7 (n=18)	11.4 (n=6)
PHB(16%V)	12.0 (n=6)	ND[1]	6.8 (n=12)	ND	22.2 (n=9)
PHB(24%V)	25.5 (n=6)	12.6 (n=12)	5.0 (n=12)	11.2 (n=12)	10.5 (n=15)
PCL	3.5 (n=3)	3.0 (n=18)	3.2 (n=6)	3.2 (n=12)	3.9 (n=4)
St/EVOH	2.0 (n=6)	3.0 (n=9)	4.2 (n=6)	1.2 (n=3)	2.8 (n=6)
EVOH	0 (n=3)	0 (n=3)	0 (n=3)	0 (n=3)	0 (n=3)
PP[2]	0 (n=3)	0 (n=3)	0 (n=3)	0 (n=3)	0 (n=3)
Uncoated Cellophane	12.0 (n=3)	2.0 (n=3)	3.1 (n=3)	9.4 (n=3)	4.7 (n=3)
Coated Cellophane	4.0 (n=3)	ND	1.6 (n=9)	ND	1.5 (n=9)
CA[2]/St/PG[2]	ND	43.7 (n=9)	42.5 (n=9)	47.3 (n=9)	39.0 (n=9)
PE[2]/Paper	12.7 (n=3)	ND	2.0 (n=3)	ND	7.8 (n=3)
Bond Paper	ND	8.0 (n=3)	4.7 (n=3)	10.8 (n=3)	4.5 (n=3)
X-linked Chitosan	10.1 (n=3)	ND	11.6 (n=3)	ND	7.5 (n=3)

[1] ND = no data

[2] PP = polypropylene; CA = cellulose acetate; PG = propylene glycol; PE = polyethylene

Mergaert *et al. (13)* reported 0.03 to 0.64% weight loss/day upon exposure of PHB or PHBV (10%V) in soils. Rates increased from 15°C up to 40°C. They also found no loss in weight in sterile buffers up to 55°C over 98 days, although some loss in molecular weight was attributed to abiotic hydrolysis. Doi *et al. (14)* reported the biodegradation of various copolymers of PHBV in film and fiber forms in natural marine environments over one year of exposure. Surface erosion was the primary effect observed and this was found to correlate directly with sea water temperature and not polymer composition. Changes in gravimetric weight were more pronounced than polymer molecular weight due to the surface effects. Two actinomycetes were also isolated and found capable of using PHB as a sole carbon source.

The three tier system we have described (only a part of these assessments was included in this paper), provides for a reasonable assessment of new biodegradable polymers. Many other methods to assess biodegradation are being developed which can be integrated into this tier system as they develop. For example, Yabannavar and Bartha *(15)* recently presented soil degradation data for polymer films and found GPC-coupled with carbon dioxide evolution useful to distinguish between biodegradation due to plasticizer and additives vs. the polymers themselves. However, the author's recognized that even this approach can lead to problems in interpretations when polymer cross-linking occurs or in cases where there are surface errosion effects.

The rationale for using a limited set of previously isolated microorganisms in the defined marine simulators was to overcome the high degree of variability observed when using natural populations collected at different sites at different times. The goal was to 'standardize' everything in the simulators, the water, organisms and sediment, so that reproducibility would be optimal. The use of a set of defined microorganisms is inherently prone to possible omissions of key microorganisms; however, this is addressed with the inclusion of new isolates to the mix that are selected for their ability to biodegrade new types of polymers.

Literature Cited

1. Kaplan, D. L.; J. M. Mayer; D. Ball; J. McCassie; A. L. Allen; P. Stenhouse. In *Biodegradable Polymers and Packaging*, C. Ching, D. L. Kaplan, E. Thomas, Eds.; Technomic Publishing Co., Lancaster, PA, 1994, pp 1-42.
2. Mayer, J. M.; D. L. Kaplan.*Trends Polym.Sci.* **1994**, *2*, 227-235.
3. Mayer, J. M.; D. L. Kaplan. In *Biodegradable Polymers and Packaging*, C. Ching, D. L. Kaplan, E. Thomas, Eds.; Technomic Publishing Co., Lancaster, PA, 1994, pp 233-246.
4. Muller, R.-J.; J. Augusta; T. Walter; H. Widdecke. In *Biodegradable Plastics and Polymers*, Y. Doi and K. Fukuda, Eds., Elsevier Science Publishers, 1994, pp 237-248.

5. Andrady, A. L. *J. Macromolecular Sci. - Rev. Macromol. Chem. Phys.* **1994**, *C34*, 25-76.
6. Kaplan, D. L.; J. M. Mayer; M. Greenberger; R. Gross; S. McCarthy. *Polym. Degradation Stability* **1994**, *45*, 165-172.
7. Allen, A.; J. M. Mayer; R. Stote, D. L. Kaplan. *J. Environ. Polymer Degn.* **1994**, *2*, 237-244.
8. Sullivan, B. K.; C. A. Oviatt; G. Klein-MacPhee. In *Biodegradable Polymers and Packaging*, C. Ching, D. L. Kaplan, E. Thomas, Eds.; Technomic Publishing Co., Lancaster, PA, 1994, pp 281-296.
9. McCassie, J. E.; J. M. Mayer; R. E. Stote; A. E. Shupe; P. J. Stenhouse; P. A. Dell; D. L. Kaplan. In *Biodegradable Polymers and Packaging*, C. Ching, D. L. Kaplan, E. Thomas, Eds.; Technomic Publishing Co., Lancaster, PA, 1994, pp 247-256.
10. Mayer, J. M.; D. L. Kaplan. **1991**. U. S. Patent 5,015,293.
11. Mayer, J. M.; G. R. Elion. **1994**. U. S. Patent 5,288,318.
12. Pytkowicz, R. M.; E. Atlas; C. H. Culberson. In Marine Chem. Coastal Env., Ed. T. M. Church, Am. Chem. Soc. p.2.
13. Mergaert, J.; A. Webb; C. Anderson; A. Wouters; J. Swings. *Appl. Envionm. Microbiol.* **1993**, *59*, 3233-3238.
14. Doi, Y.; Y. Kanesawa; N. Tanahasi; Y. Kumagai. *Polym. Degradation Stability* **1992**, *36*, 173-177.
15. Yabannavar, A. V.; R. Bartha. *Appl. Environ. Microbiol.* **1994**, *60*, 3608-3614.

RECEIVED November 28, 1995

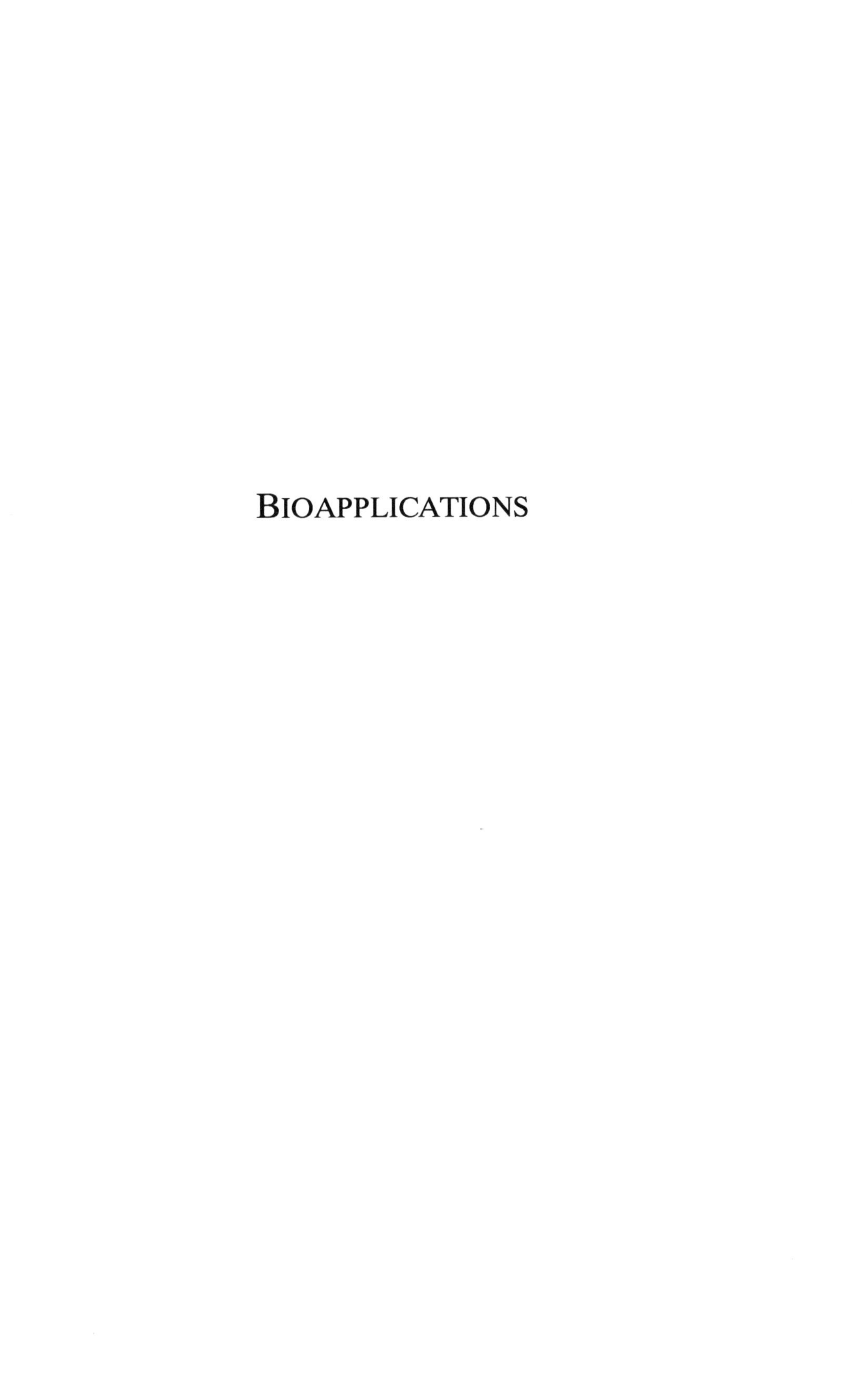

BIOAPPLICATIONS

Chapter 14

Two-Dimensional Latex Assemblies for Biosensors

S. Slomkowski, M. Kowalczyk, M. Trznadel, and M. Kryszewski

Center of Molecular and Macromolecular Studies, Polish Academy of Sciences, 90–363 Lodz, Poland

Materials with surfaces allowing the controlled attachment of biomolecules that are suitable for recognition of the desired analyte, are often required for manufacturing of biosensors. There are many kinds of latex particles which, in principle, could be used for this purpose. In this paper there are described the results of our studies on the formation of two-dimensional latex assemblies covalently immobilized on the surface of solids. Computer simulations illustrate the general properties of the two-dimensional latex assemblies and provide a theoretical tool for discriminating between assemblies of particles with repulsion or attraction during deposition. An example of real assemblies is presented for the composite poly(styrene/acrolein) latex on the surface of quartz modified with γ-aminopropyltriethoxysilane. The main characteristics of the morphology of two-dimensional assemblies are discussed. Immobilization of human serum albumin (HSA) and proteins from the goat anti-HSA serum on the two-dimensional latex assemblies as well as the potential suitability of materials with latex modified surfaces for the diagnostic applications are discussed.

The typical biosensor consists of detector and transducer. The changes in a detector, due to interactions with an analyte, induce a signal (e.g. electrical or optical) which is transmitted from the transducer to the recording elements of the analytical device *(1)*. Detectors usually contain biomolecules, often proteins, immobilized on the surface and ready to bind selectively the molecules to be detected. Thus, one of the crucial steps in the fabrication of a biosensor involves the immobilization of appropriate proteins so as to provide sufficient protein concentration of selected sensor elements. Immobilized proteins should retain a maximum of their biologic activity. Very often the adsorption and/or covalent immobilization of proteins on solid surfaces is accompanied with a considerable degree of protein denaturation and loss of biological function *(2-10)*. On the other hand, there are known several kinds of latex particles on which proteins could

0097–6156/96/0627–0172$15.00/0

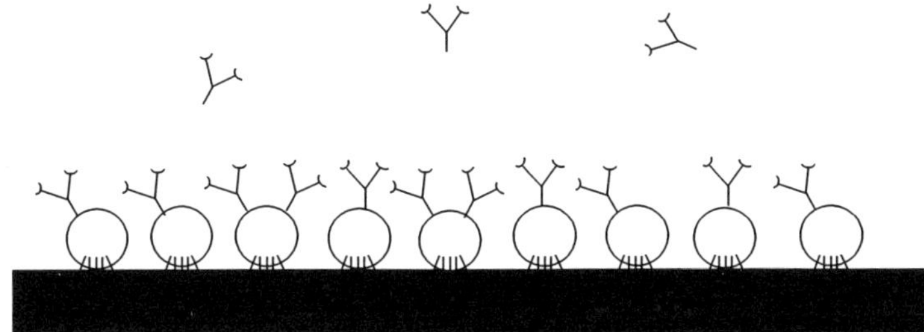

Figure 1. Surface of solid material modified by the covalent binding of latex particles suitable for immobilization of protein macromolecules.

be immobilized with retention of a considerable part of the biologic activity which allows the use of these particles for diagnostic purposes *(11-21)*. Thus, it seems to be reasonable to develop procedures for the controlled immobilization of these latexes on the surface of solids to yield the new composite materials assuring immobilization of proteins in the active form. The schematic structure of such material is illustrated in Figure 1.

The two-dimensional latex assemblies are interesting also from pure theoretical reasons. Suspensions of latex particles involved in Brownian motion were for a long time considered as excellent examples of unordered assemblies, similar to molecules in the gas state. However, recently it has been found that in certain systems long range interparticle interactions result in the formation of crystal like structures made of latex particles *(22-24)*. The formation of ordered latex assemblies was observed on the liquid-gas *(25)* and liquid-solid *(26)* interfaces. Evaporation of liquid from diluted suspensions placed on glass surfaces was found to produce densely packed two-dimensional latex assemblies *(27)*. However, for the modification of solid materials surfaces, the covalent immobilization of latex particles is required.

Although the kinetics of adsorption and desorption of latex particles onto the surfaces of solids (collector surfaces) was investigated in many laboratories *(8-37)* very little is known about the influence of the interparticle and particle-support interaction on the morphology of covalently immobilized latex assemblies. We expected that computer simulation would be especially suitable for the fast, convenient, and inexpensive estimation of the basic features of the possible morphologies of latex assemblies. Some results of such simulations will be discussed later in this paper.

Recently, we developed the synthetic procedures for the production of composite poly(styrene/acrolein) particles with the controlled fraction of polyacrolein in the surface layer *(38)*. These particles were found to be suitable for the controlled adsorption and/or covalent immobilization of protein molecules *(39)* (in reaction between aldehyde groups located on the surface of particles and amino groups of proteins). Gamma globulins (γG) and horseradish peroxidase (HRP) attached onto the P(S/A) latexes retained considerable part of their biological activity making the protein-latex systems suitable for diagnostic *(39)* and/or immobilized enzyme (Basinska, T., Slomkowski, S. *Colloid Polym.Sci.,*) applications. Human serum albumin (HSA) adsorbed and/or covalently bound to the P(S/A) latex particles, in spite of at least partial denaturation, preserved some epitopes intact and thus, could be recognized by antibodies against HSA (anti-HSA) *(39,40)*. Due to the properties mentioned above, the P(S/A) latexes could be considered as good candidates for the studies of solid surface modification, leading to new materials for the diagnostic applications. The aldehyde groups on the surface of P(S/A) latexes could be used also for the covalent immobilization of latex particles on superficially functionalized quartz. Quartz elements, modified by covalent binding of the P(S/A) latexes, could be tailored for the controlled immobilization of proteins, and thus could became suitable as parts of biosensors with optical (e.g. involving fluorescence) detection.

Basic Features of the Two-Dimensional Latex Assemblies Determined by the Computer Simulations

The density of packing and the arrangement of latex particles depend on the nature of the latex, structure of the surface, and on the procedures of attachment. The main set of parameters characterizing this process should include: parameters determining the interparticle interactions (attraction and/or repulsion), probabilities of the particle lateral movement on the surface, probabilities of desorption, and of the covalent immobilization due to the chemical reaction between the reactive groups on the surface of latex particles and on the surface of solid support onto which the deposition of particles takes place.

The computer program, written in our laboratory for the simulation of the formation of two-dimensional latex assemblies, included all parameters mentioned above. Simulations could be carried out for surfaces allowing accommodation from 500 to 10 000 particles at the maximum packing density. Attraction between particles was assumed to reduce the probabilities of lateral movement (P_L) and of desorption (P_D) according to equations:

$$P_L(n) = P_L(0) \cdot A^n \qquad (1)$$

$$P_D(n) = P_D(0) \cdot A^n \qquad (2)$$

where $P_L(n)$ and $P_D(n)$ denote the probability of lateral movement and desorption, respectively, for every particle being in contact with n other particles, A ($0 < A \leq 1$) characterizes attraction between particles (for strong attracting force A is close to 0, in the absence of attraction A = 1). The short range repulsive forces were taken into account by introducing parameter increasing the probability of back scattering and decreasing the probability of particle adsorption in an event which consists on "sliding" of the incoming particle on the particle being already on the surface.

The detailed description of the program and analysis of the morphology of two-dimensional latex assemblies simulated for various conditions requires a full separate paper and will be published elsewhere (Trznadel, M., Slomkowski, S. *Colloid Polym.Sci.*,). Here we would like to present only one important result.

The two-dimensional assemblies generated by computer were characterized by two parameters: the degree of coverage (DC) and the average number of neighbors (ANN). The degree of coverage was defined as the ratio of the number of particles actually present on the surface to the number of particles present at the same surface at the maximal packing conditions (hexagonal lattice). The average number of neighbors was defined as the average number of latex particles being in contact with every particle from the ensemble. ANN could vary from 0, for the system of isolated particles, to 6, for particles packed in the hexagonal lattice. Computer simulations revealed that it is possible to make the diagram, similar to the phase diagrams, illustrating the relation between ANN and DC for systems with various interparticle interactions. This diagram is shown in Figure 2.

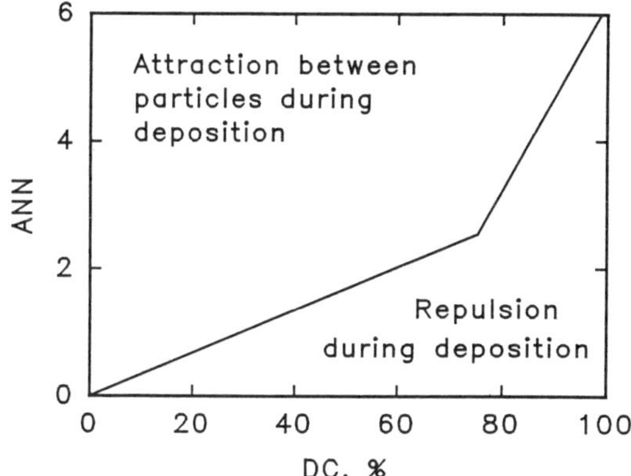

Figure 2. ANN and DC diagram for two-dimensional latex assemblies with attraction and repulsion between particles. Line corresponds to assemblies of not interacting particles. (Reproduced with permission. Copyright 1995 Elsevier.)

The line on this diagram corresponds to all systems, regardless of the relative probabilities of desorption, lateral movement on the surface, and covalent immobilization of latex particles. The only limitation was that two particles can not occupy the same space. Points corresponding to the ANN and DC pairs, for systems with the interparticle attracting forces, are always located above this line. For all systems with the repulsion between particles the points defined by values of ANN and DC are below this line. It is important to stress that this observation was valid at any moment of the formation of the two-dimensional latex assemblies. Thus, for any real system, the determination of the character (repulsive or attractive) of the interparticle interactions in the nearest proximity of the surface of deposition is possible by the system, the determination of the character (repulsive or attractive) of the interparticle interactions in the nearest proximity of the surface of deposition is possible by the simple measurement of ANN and DC, on the electron micrographs of the two-dimensional assemblies.

Preparation of the Two-Dimensional Assemblies of Poly(styrene/acrolein) Latex Immobilized Covalently on Quartz

Poly(styrene/acrolein) latex (P(S/A)) was synthesized and characterized according to the procedure described earlier *(38)*. Polyacrolein content in the surface layer of latex particles (f) was determined from the XPS calibration and was found equal to f= 0.65. The number averaged diameter $\bar{D}_n$ and the polydispersity factor $\bar{D}_v/\bar{D}_n$ (D_v denotes the volume averaged diameter), determined from the scanning electron microphotograms were equal 0.44 μm and 1.008, respectively. Concentration of aldehyde groups on the surface of P(S/A) latex, available for chemical reaction, was determined using 2,4-dinitrophenylhydrazine as the reference reagent *(38)* and was found equal $1.42 \cdot 10^{-6}$ mol/m^2. Concentration of anionic groups on the surface of latex particles, determined by conductometric titration, was found equal $1.36 \cdot 10^{-6}$ mol/m^2.

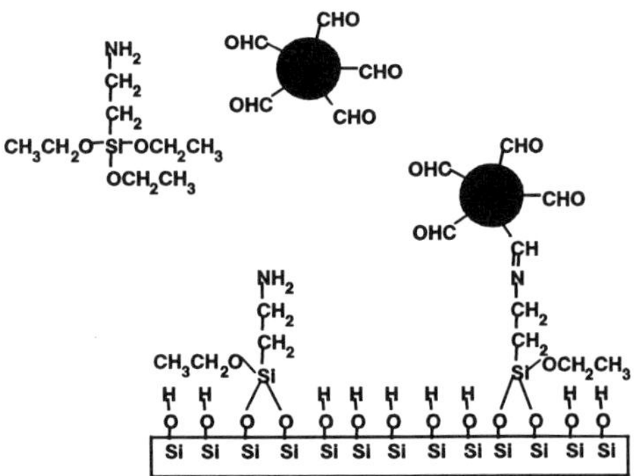

Scheme I.

Surface of quartz slides, used for the covalent immobilization of P(S/A) latex, was modified with solutions of γ-aminopropyltriethoxysilane (APTS) similarly as it was described by Weetal *(41)*. Before immobilization quartz slides were washed for 10 min. with 5 mol/L solution of KOH, and subsequently for 30 min with running distilled water. Finally, slides were dried at 70 °C for 2h. Modification was carried out by immersion of the purified slides for 18 h into solutions of APTS in toluene. Concentrations of APTS were varied from 2% to 12% and the temperature of the solution was equal 20 °C. Presence of amino groups on the surface of modified quartz slides was detected by the XPS method. The background level of nitrogen atoms (presumably from air), detected on the quartz slides not treated with APTS gave the nitrogen to silicon ratio equal to 0.022. Ratio of nitrogen and silicon atoms in the surface layer of modified slides varied from 0.049 to 0.055 (after correction for background varied from 0.027 to 0.033), for slides modified with toluene solutions containing from 2% and 12% of APTS, respectively. Assuming that XPS provides information on the composition of the layer 30 Å thick one could estimate the ratio of nitrogen and silicon atoms for quartz covered with monolayer of APTS. A monolayer of APTS molecules, stretching perpendicularly from the surface of the quartz, would be ca 6 Å thick. Calculations of the composition of the 30 Å thick slice of quartz covered with APTS (6 Å APTS and 24 Å quartz) lead to the nitrogen to silicon ratio of ca 0.025. Thus, the results of the XPS analysis conform to the hypothesis that the quartz slides, after modification, are covered with the monolayer of the densely packed molecules of APTS.

Prior to the immobilization of P(S/A) latex particles, the modified slides were conditioned for 5 h in the phosphate buffer saline (PBS, pH = 7.4). Thereafter, the immobilization was carried out by incubation of quartz slides with suspensions of P(S/A) latex (0.2 wt/vol%) in PBS. Incubation time was varied from 30 min to 72 h. During this process particles were immobilized via the imine linkages (Schiff base) in the reaction of amino groups on the modified quartz surface and aldehyde groups of the surface of latex particles. Slides with attached P(S/A) latex were washed with water and then with 2% sodium dodecyl sulfate (SDS). In this step the adsorbed particles, not immobilized covalently, were removed. Finally, particles were mechanically removed from one side of the slides and the covered side of the slides were investigated by various techniques.

Basic Properties of The Two-Dimensional Assemblies of P(S/A) Latex Immobilized on Quartz

An example of the SEM microphotogram of the two-dimensional assembly of P(S/A) latex covalently immobilized on quartz is shown in Figure 3. The degree of coverage (DC) of the quartz slides with latex particles depends on the concentration of APTS in solution used for the quartz modification and on the time of incubation with latex particles. For example, for slides modified with solutions containing 2, 6 and 8 % of APTS and then incubated with P(S/A) latex for 3 h DC was equal to 30.8, 37.8, and 41.0 %, respectively.

The dependence of DC on time of incubation with P(S/A) latex, for quartz slides modified with 8 % solutions of APTS, is illustrated in Figure 4. From the plot in Figure 4 it follows that after rapid immobilization within the first five hours the subsequent increase of DC becomes very slow. Inspection of SEM microphotographs indicated that for DC exceeding 40 % it is impossible to immobilize new latex particles. The distances between the already immobilized particles are to small (less than the diameter of particle) to accommodate the new ones. Covalent immobilization prevents desorption and/or lateral movement of particles on the surface which could lead to the denser packing of latex.

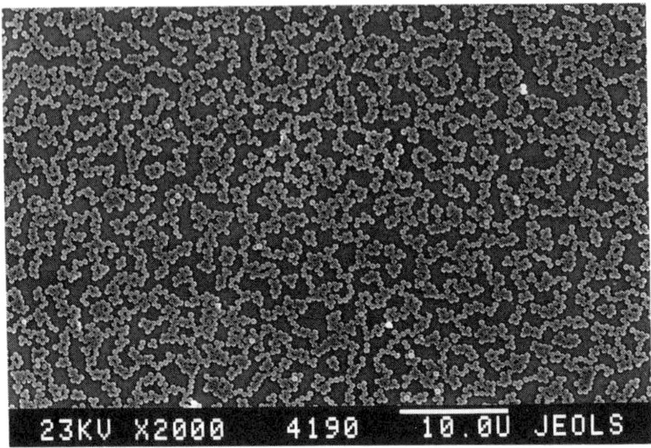

Figure 3 SEM microphotogram of P(S/A) latex immobilized covalently on the surface of quartz modified with 8 % APTS. Time of incubation with latex 5 h.

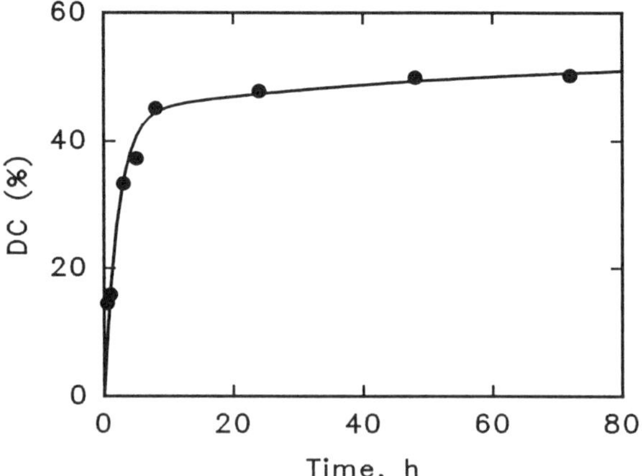

Figure 4 Dependence of the degree of coverage of quartz slides on time of incubation with P(S/A) latex. Quartz slides modified with 8 % APTS.

Based on the analysis of the SEM pictures the values of DC and ANN (average number of neighbors) were calculated for various two-dimensional assemblies. The results are presented in Table 1.

Table 1 Values of DC and ANN for assemblies of P(S/A) latexes on quartz slides

DC (%)	23.2	27.3	49.3	49.8	69.0	76.2	76.5
ANN	2.47	2.42	2.62	2.75	2.92	3.28	3.29

It is worth to note that points defined by pairs of DC and ANN collected in Table 1 are located in the diagram in Figure 1 in the area corresponding to latex systems with interparticle attraction during immobilization. On the other hand, it was established earlier *(38)* that the P(S/A) latex particles in the suspension are stabilized electrostatically, due to the presence of negatively charged $-OSO_3^-$ groups on the surface. The repulsive forces between particles are eliminated at low pH, due to protonation of the surface anionic groups. Apparently, as in the case of quartz slides immersed in buffer at pH = 7.4 (conditions for the immobilization of latex), the actual local pH value, at a distance from the quartz comparable with the latex particle diameter ($\bar{D}_n$ = 0.44 μm), is considerably lower.

The two-dimensional latex assemblies could be characterized also by light scattering. The angular distribution of the intensity of light (He-Ne laser, λ = 632.8 nm) scattered from the P(S/A) latex assemblies immobilized on quartz is shown in Figures 5a and 5b.

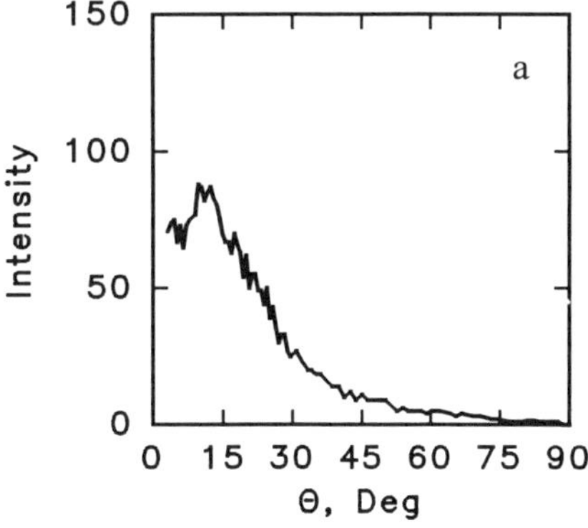

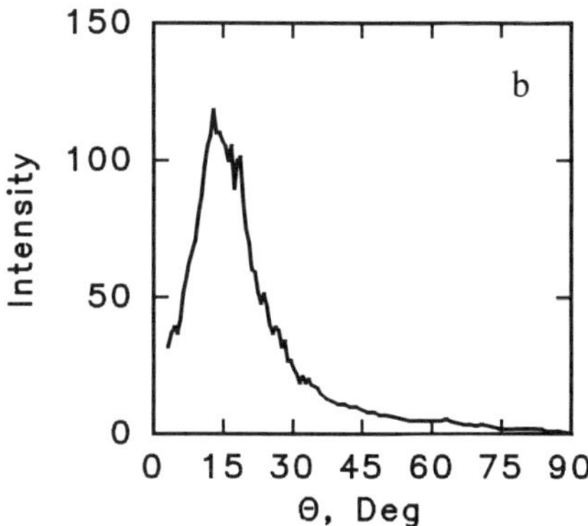

Figure 5 Angular distribution of the intensity of light scattered from the P(S/A) latex immobilized on the quartz slides.

Surface of quartz was modified with 2% APTS (a) and 6% APTS (b). Degree of coverage DC = 30.3 (a) and 37.8 (b)
Surface of quartz slides was covered essentially with the single layer of particles. Therefore, the multiple scattering did not have to be taken into account. However, because particles were arranged on the surface into the two-dimensional assemblies these arrangements modified scattering picture from single particles. The brief inspection of the SEM microphotographs indicates that latex assemblies vary significantly in shape and orientation. For the low degree of coverage the distribution of distances between the nearest assemblies is very broad and the interference of light scattered from the randomly distributed neighbor particles does not perturb significantly the distribution of light scatted from the single particles. In the case of the high degree of coverage the distribution of distances between the nearest assemblies, and the corresponding distribution of phase shifts of scattered light, are more narrow. Surface densely covered with the two-dimensional latex assemblies resembles surface with the irregular holes (areas not covered with particles) and the diffraction pattern of transmitted light should be characterized with maximum corresponding to the average distance between latex assemblies.

The distribution shown in Figure 5a is very close to the typical distribution of light scattered from single particles randomly distributed in suspension *(42)*, whereas for the high degree of coverage (cf Figure 5b) the scattering pattern displays well developed maximum at the scattering angle $\Theta = 16^{o}20'$. The rough approximation of the average distance between assemblies $d = \lambda/\sin(\Theta)$, calculated from the scattering picture shown in Figure 5b, is equal to 3.3 μm. For the same system the average distance determined from SEM microphotographs was equal to 3 ± 1 μm.

Two-Dimensional Assemblies of Fluorescent Derivatives of the P(S/A) Latex

Aldehyde groups on the surface of the P(S/A) latex particles can be used also for the immobilization of fluorescent labels. Labeling of P(S/A) latex with dansylhydrazine (DA), used in experiments with the two-dimensional assemblies, is illustrated in Scheme II.

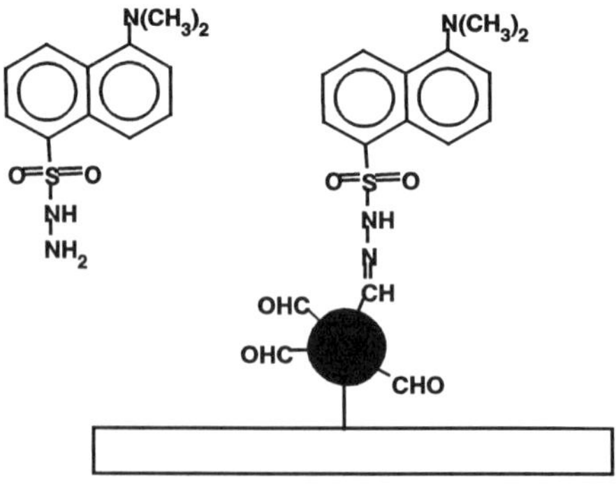

Scheme II.

It is worth to note that labeling, involving significant fraction of aldehyde groups on the surface of P(S/A) particles, did not decrease their ability to immobilize covalently protein macromolecules (B.Miksa, S.Slomkowski, J.Biomater.Sci.Polym.Ed., in press).

The two-dimensional assemblies of fluorescent particles were prepared by incubating quartz slides with immobilized P(S/A) particles, with ethanol solution of dansylhydrazine. Concentration of DA was equal $3.77 \cdot 10^{-2}$ mol/L. For the 1 cm^2 slides covered with particles and 30 mL of DA solution used for incubation the excess of DA, with respect to aldehyde groups on latex, was ca 10^7 fold. The immobilization of P(S/A) latex on quartz was carried out at room temperature during 15 h. Then, slides were rinsed with ethanol. Traces of not immobilized DA were removed by washing slides containing labeled latex with fresh portions of ethanol (ten times). During each step of the washing procedure slides were was kept in ethanol for 30 min. Finally, slides were dried and the steady state emission spectra of assemblies of latex particles labeled with dansyl groups were registered by the front face emission spectroscopy.

The emission spectra obtained for slides with various degree of coverage are shown in Figure 6.

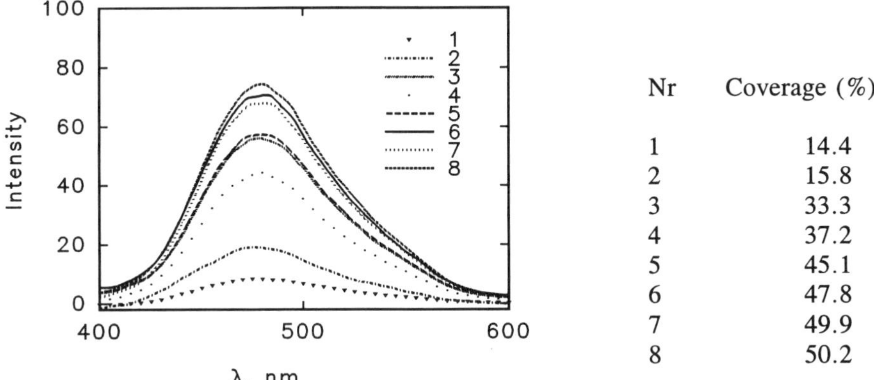

Nr	Coverage (%)
1	14.4
2	15.8
3	33.3
4	37.2
5	45.1
6	47.8
7	49.9
8	50.2

Figure 6 Front face emission spectra of two-dimensional assemblies of P(S/A) latex labelled with dansyl groups. Excitation at 345 nm.

The emission spectra obtained for the P(S/A) latex labeled wit dansyl and immobilized on the quartz slides (cf Figure 6) are similar to the spectra registered for this latex in suspension *(43)* and indicate that immobilized particles did not lose their ability to immobilize compounds with the hydrazine groups.

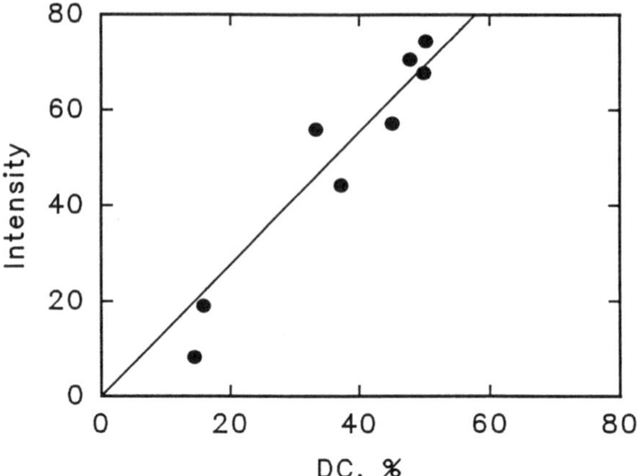

Figure 7 Dependence of the intensity of emission (at λ_{max} = 480 nm) on the degree of coverage (DC) for the P(S/A) latex assemblies labeled with dansylhydrazine.

It has been found (cf Figure 7) that the intensity of emission from assemblies of labeled particles is proportional to the degree of coverage. Thus, labeling with dansylhydrazine, after calibration for each type of the P(S/A) latex, could be used for determination of the degree of coverage of the surface of quartz slides.

Model Immunofluorescence Assays Involving Human Serum Albumin and/or anti-Human Serum Albumin Immobilized on Two-Dimensional Assemblies Bound to the Quartz Slides

Latex assemblies on quartz, suitable for manufacturing elements of biosensors for the diagnostic applications, should maintain the ability to immobilize proteins with the retention of their biologic activity. For model studies we have chosen human serum albumin (HSA) (Sigma, Cohn fraction V) and goat anti-HSA serum (Sigma) as the inexpensive and well characterized antigen-antibody system.

Two arrangements were investigated. In the test type I anti-HSA was immobilized on the two-dimensional latex assemblies on quartz slides and then we checked whether anti-HSA on these assemblies can be used for detection of particles with HSA on the surface. In this experiment we used the fluorescent P(S/A) latex, labeled with dansyl groups, with attached HSA (P(S/A)-DA-HSA). Experiment of type I is illustrated in Figure 8. In the test type II we used latex assemblies, with immobilized HSA, for the fluorescent sandwich test for the detection of anti-HSA. During this test latex assemblies with immobilized HSA were exposed to the analyzed solution of anti-HSA and subsequently to the suspension of the fluorescent latex P(S/A)-DA-HSA. According to the scheme shown in Figure 8, binding of P(S/A)-DA-HSA latex was expected to be possible only when anti-HSA was present in the analyzed solution.

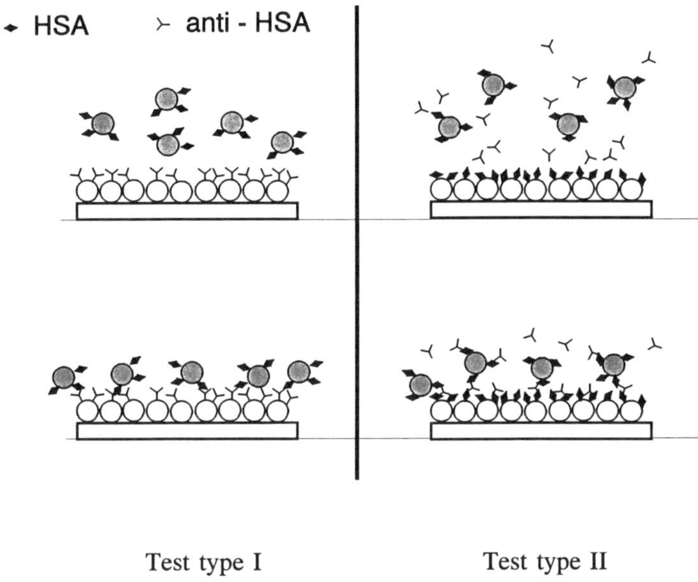

Test type I Test type II

Figure 8 Schemes of the model diagnostic tests.

For experiments we used quartz slides with the degree of coverage equal to 47.8%. In the test type I the slides were incubated with solution containing anti-HSA serum in PBS (27 000 fold dilution of the standard Sigma serum) for 18 h. Thereafter slides were rinsed with PBS and transferred to the suspension of P(S/A)-DA-HSA in PBS (concentration of latex 0.1 wt/vol %, surface concentration of immobilized HSA $8.8 \cdot 10^{-4}$ g/m^2). Incubation with P(S/A)-DA-HSA was carried out during 15 min. Finally, slides were washed with fresh portions of PBS and dried at room temperature. An example of the front face emission spectrum of the slide prepared according to the procedure described above is shown in Figure 9.

In the test type II the slides covered with P(S/A) latex were incubated with HSA ([HSA] = $3.0 \cdot 10^{-4}$ g/mL) for 4 h, rinsed 30 times with PBS, incubated with anti-HSA modelling analyte (standard anti-HSA diluted 27 000 times), and then with suspension of P(S/A)-DA-HSA (concentration of latex 0.1 wt/vol %, surface concentration of immobilized HSA $8.8 \cdot 10^{-4}$ g/m^2). Eventually, for slides rinsed with PBS and dried at room temperature the front face emission spectra were registered. An example is shown in Figure 9.

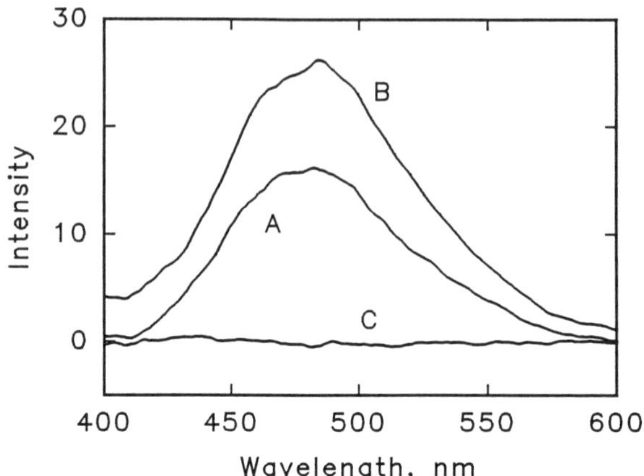

Figure 9 Emission spectra for latex assemblies: A – slide-P(S/A)-anti-HSA - P(S/A)-DA-HSA, B – slide-P(S/A)-HSA - anti-HSA - P(S/A)-DA-HSA, C – slide-P(S/A) - P(S/A)-DA-HSA.

From spectra in Figure 9 it follows that the nonspecific attachment of the P(S/A)-DA-HSA latex to the two-dimensional assemblies of the P(S/A) latex immobilized covalently on quartz is negligible. The emission spectrum, due to the dansyl labels, registered for the test type I indicate that antibodies of anti-HSA immobilized on the two-dimensional assemblies of P(S/A) latex on quartz retain their activity and are able to recognize HSA attached to the surface of the P(S/A) latex in suspension. The emission spectrum registered for the test type II indicates that epitopes of HSA immobilized on the P(S/A) particles are recognized by anti-HSA, in spite of their very high (27 000 times) dilution.

Conclusions

1. The poly(styrene/acrolein) latex is suitable for the formation and covalent immobilization of the two-dimensional latex assemblies on quartz.

2. The degree of coverage with the two-dimensional latex assemblies depends on the preparation of the surface of quartz and on the time of latex immobilization. Morphology these assemblies depends on the interparticle interactions.

3. Aldehyde groups on latex particles immobilized on quartz can be used for covalent binding of the fluorescent dansyl label.

4. HSA and anti-HSA, immobilized on the assemblies of P(S/A) latex, recognize their partners (anti-HSA and HSA respectively) in solution and/or immobilized on latex in suspension

All properties of the two-dimensional latex indicate suggest that the materials of this kind might be interesting for preparation of elements of biosensors with the fluorescent detection.

The work was supported by the M.Sklodowska-Curie Fund, Grant PAN/NIST-94-169

Literature Cited

1. Hall, E.A.H. In *Biosensors and Chemical Sensors, Optimizing Performance Through Polymeric Materials*; Edelman, P.G.; Wang, J., Eds; ACS Symp.Ser. 487; ACS: Washington, D.C., 1992; p.1.
2. Gabel, D.; Steinberg, I.Z.; Katchalsky, E. *Biochemistry* **1971**, *10*, 4661.
3. Soderquist, M.E.; Walton, A.G. *J.Colloid Interface Sci.* **1980**, *75*, 386.
4. Van Dulm, P.;.Norde, W. *J.Colloid Interface Sci.* **1983**, *91*, 248.
5. Norde, W.F.; MacRitchie, G.; Nowicka, G.; Lyklema, J. *J.Colloid Interface Sci.* **1986**, *112*, 447.
6. Andrade, J.D.; Hlady, V. *Adv.Polym.Sci.* **1986**, *79*, 1.
7. Lundström, I.; Ivarsson, B.; Jönsson, U.; Elvig H. In *Polymer Surfaces and Interfaces*; Feast, W.J.; Munro, H,S., Eds, Wiley: New York, N.Y., 1987, p. 201.
8. Sandwick, R.K.; Schray, K.J. *J.Colloid Interface Sci.* **1987**, *115*, 130.
9. Chen, J.P.; Kiaei, D.; Hoffman, A.S. *J.Biomater.Sci. Polymer Edn* **1993**, *5*, 167
10. Galisteo-González, F.; Martin-Rodrigiez, A.; Hidalgo-Alvarez, R. *Colloid Polym.Sci.* **1994**, *272*, 352
11. Tarcha, P.J.; Misun, D.; Finley, D.; Wong, M.; Donovan J.J. In *Polymer Latexes; Preparation, Characterization, and Applications*; Daniels, E.S.; Sudol, E.D.; El-Aasser, M., Eds; ACS Symp.Ser. 492; ACS: Washington, DC, 1991; p. 347.
12. Galanti, L.M.; Cornu, Ch.; Masson, P.L.; Robert, A.R.; Bechenau, D.; Lamy, M.E.; Cambiasso, C.L. *J.Virological Methods*, **1991**, *32*, 221.
13. Borque, L.; Rus, A.; Ruiz, R. *J.Clin.Lab.Anal.* **1991**, *5*, 175.
14. Lim, P.L.; Ko, K.H. *J.Immunol.Methods* **1990**, *135*, 9.
15. Millán, J.K.; Nustad, K.; Nørgaard-Pedersen, B. *Clin.Chem.* **1985**, *31*, 54.
16. Fanta, K.; Merétey, K.; Kovács, P.; Bánkuti, L.; Zsidai, J. *Ann.Immunol.hung.* **1985**, *25*, 319.
17. Rippey, J.H. *CRC Crit.Rev.Clin.Lab.Sci.* **1984**, *19*, 353.
18. Nustad, K.; Closs, O.; Ugelstad, J. In *Developments in Biological Standardization*; Barme, M.; Hennessen W., Eds; S.Karger, Basel, 1984, vol. 57, p. 321.
19. Nustad, K.; Monard-Hansen, H.P.; Paus, E.; Millán, J.L.; Nørgaard-Pedersen, B. and the DATECA study group In *Human Alkaline Phosphatases*; Stigbrand, T.; Fishman W.H., Eds, A.R.Liss, New York, N.Y., 1984, p.337.
20. Børmer, O. *Clin.Biochem.* **1982**, *15*, 128.
21. Hunter, W.M.; Budd, P.S. *J.Immunol.Methods*, **1981**, *45*, 255.
22. Ito, K.; Yoshida, H.; Ise, N. *ACS Polym.Prep.* **1992**, *33(1)*, 771.
23. Yoshino, S. *ACS Polym.Prep.* **1992**, *33(1)*, 773.

24. Okubo, T. In *Macro-ion Characterization, From Dilute Solutions to Complex Fluids*; Schmitz, K.S. Ed.; ACS Symp.Ser. 548; ACS, Washington, DC, 1994, 364.
25. Robinson, D.J.; Ernshaw, J.C. *Langmuir*, **1993**, *9*, 1436.
26. Ugelstad, J.; Berge, A.; Ellingsen, T.; Schmid, R.; Nilsen, T.-N.; Mørk, P.C.; Stenstad, P.; Hornes, E.; Olsvik, O. *Prog.Polym.Sci.* **1992**, *17*, 87.
27. Denkov, N.D.; Velev, O.D.; Kralchevsky, P.A.; Ivanov, I.B.; Yoshimur, H.; Nagayama, K. *Langmuir*, **1992**, *8*, 3183.
28. Meinders, J.M.; Busscher, H.J. *Colloid Polymer Sci.* **1994**, *272*, 478.
29. Sharma, M.M.; Chamoun, H.; Sita Rama Sarma, D.S.H.; Schechter, R.S. *J.Colloid Interface Sci.* **1992**, *149*, 121.
30. Meinders, J.M.; Noordmans, J.; Busscher, H.J. *J.Colloid Interface Sci.* **1992**, *152*, 265.
31. Adamczyk, Z.; Zembala, M.; Siwek, B.; Warszynski, P. *J.Colloid Interface Sci.* **1990**, *140*, 123.
32. Sjollema, J.; Busscher, H.J.; Weerkamp, A.H. *J.Microbiol.Methods*, **1989**, *9*, 73.
33. Sjollema, J.; Busscher, H.J.; Weerkamp, A.H. *J.Microbiol.Methods*, **1989**, *9*, 79.
34. Dabros, T. *Colloids Surf.* **1989**, *39*, 127.
35. Schaaf, P.; Talbot, J. *J.Chem.Phys.* **1989**, *91*, 4401.
36. Van de Ven, T.G.M. *Colloids Surf.* **1989**, *39*, 107.
37. Hubbe, M.A. *Colloids Surf.* **1985**, *16*, 227.
38. Basinska, T.; Slomkowski, S.; Delamar, M. *J.Bioact.Compat.Polym.* **1993**, *8*, 205.
39. Basinska, T.; Slomkowski, S. In *Uses of Immobilized Biological Compounds*, Guilbault, G.G.; Mascini M. Eds; Kluwer Academic Publishers, Dordrecht, Boston, London, 1993, p. 453.
40. Kowalczyk, D.; Slomkowski, S.; Wang, F.W. *J.Bioact.Compat.Polym.* **1994**, *9*, 282.
41. Weetal, H.H. *Biochim.Biophys.Acta*, **1970**, **212**, 1.
42. van de Hulst, H.C. *Light Scattering by Small Particles*; Wiley, New York, NY, Chapman & Hall, London, 1956, p 65 and 147.
43. Slomkowski, S.; Kowalczyk, D.; Basinska, T.; Wang, F.W. In *Macro-ion Characterization, From Dilute Solutions to Complex Fluids*, Schmitz, K.S. Ed; ACS Symp.Ser. 548; ACS, Washington, DC, 1994, p. 449.

RECEIVED December 6, 1995

Chapter 15

Molecular Recognition at a Monolayer Interface

2,4-Diaminopyrimidine–Succinimide Host–Guest Partners

Susan L. Dawson[1], James Elman[2], Douglas E. Margevich[2], William McKenna[2], David A. Tirrell[1,4], and Abraham Ulman[3]

[1]Department of Polymer Science and Engineering, University of Massachusetts, Amherst, MA 01003
[2]Eastman Kodak Company, Rochester, NY 14650
[3]Department of Chemistry, Polytechnic University, Brooklyn, NY 11201

Molecular recognition at a 2,4-diaminopyrimidine terminated monolayer **1** (host) by a succinimide derivative (guest) has resulted in the formation of bilayers. The bilayers were prepared either from solution or by transfer from the air-water interface to the monolayer substrate. The preorganized 2,4-diaminopyrimidine is oriented for recognition with a molecule of complementary hydrogen bonding ability, succinimide **2**. The resultant bilayer **3** is stabilized by the formation of three hydrogen bonds per host-guest pair. We present an example of a self-assembling process, wherein a relatively weak hydrogen bonding interaction (molecular recognition) leads to the formation of bilayers. The bilayer structures and the hydrogen bonding interactions were analyzed by external-reflection Fourier transform infrared spectroscopy (FTIR), ellipsometry, and X-ray photoelectron spectroscopy (XPS).

Molecular recognition, the selection and non-covalent binding of a guest molecule by a host (*1,2*), is ubiquitous in nature. Specific recognition processes, e.g. between antibody and antigen, enzyme and substrate (or inhibitor), or hormone and receptor, play an important role in mediating a variety of biological processes.

There is growing interest in applying the principles of biological recognition to the construction of new synthetic materials (*3, 4*). The issues that govern the interaction of large biological molecules do not always apply in a simple way to recognition events in smaller molecules (*5*); however, we can adopt two key elements -complementarity and specificity (*5, 6*)- to the design of molecular recognition sites in synthetic systems. Complementary molecules have geometries that allow binding through multiple, non-covalent intermolecular interactions; e.g., hydrogen bonding, van der Waals forces, and electrostatic interactions (*7*). The weak interaction energies associated with these forces result in modest enthalpic contributions to the Gibbs free energy of binding, and binding is easily prohibited by comparable, unfavorable entropic contributions (*8*). One approach to the minimization of the unfavorable entropic contribution of host-guest pairing is to preorganize the host (*9*) with rigid elements or through the formation of clefts and

[4]Corresponding author

0097–6156/96/0627–0187$15.00/0

cavities. For example, preorganization of cyanuric acid and melamine derivatives by tethering them to central hubs permits assembly of rosettes or linear tapes (*10*).

Although molecular recognition events have been investigated most thoroughly in solution, a recent approach has been to incorporate artificial receptors at monolayer interfaces. Studies at the air-water interface have involved chiral recognition (*11-13*), and amino acid or nucleic acid base complexation (*14*). Monolayers self-assembled on substrates have been utilized to probe interfacial recognition of a variety of host-guest partners; including dioctadecyldithiocarbamate and Cu^{2+}(*15*), thiolated β-cyclodextrin and ferrocene (*16*), a biotinylated polythiophene copolymer with streptavidin and a biotinylated photoactive protein (*17*), and 4-hydroxythiophenol with $Ru(NH_3)_6^{3+}$(*18*), with proposed technological applications ranging from chemical and biosensors to opto-electronic signal transducers.

The recognition of streptavidin by biotin-terminated monolayers has been extensively studied (*19-21*). The binding interaction of this host-guest pair (K_a= 10^{15} M^{-1}) (*21*) is so strong that it is essentially irreversible. In nature recognition is predominantly a reversible interaction and strong binding can result in inhibition of the process of interest (*22*). A weaker binding interaction can lead to reversible recognition and the construction of semi-permanent structures. The relatively weak hydrogen bonding interactions (K_a=10^2 to 10^4 M^{-1} in $CDCl_3$) (*23, 24*) of imides with 2,4-diamidoazines and 2,4-diaminoazines has provided a basis for the self-assembly of synthetic replicators (*25, 26*), supramolecular strands (*27, 28*), and supramolecular complexes (*10*). In aqueous systems the binding of nucleic acid bases to diaminotriazines has a comparable binding constant (10^2 M^{-1}) (*29*). Thus, binding energies such as those in the biotin-streptavidin system are not necessary to assemble supramolecules (*30, 31*) and nanostructures (*8*). In this report we present preliminary results on a self-assembly process, wherein relatively weak hydrogen bonding interactions lead to the formation of bilayers.

Results and Discussion

Bilayer Formation. We have engineered recognition sites into a self-assembled alkanethiolate monolayer (**1**) on a gold substrate. The preorganized 2,4-diaminopyrimidine terminal groups are oriented for recognition of molecules with complementary arrangements of acidic protons and electron pairs. This preorganization reduces the often substantial entropic barrier to recognition. Recognition of the monolayer **1** (host) by the succinimide derivative (**2**) (guest)

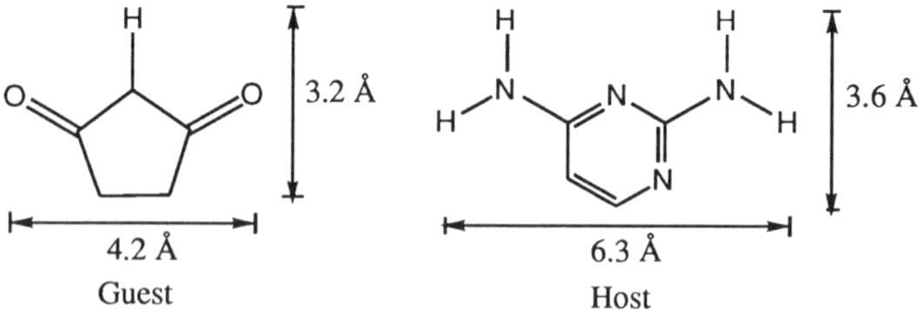

3.2 Å

4.2 Å

Guest

3.6 Å

6.3 Å

Host

results in the formation of a bilayer **3** (Figure 1), which is stabilized by the formation of three hydrogen bonds per host guest pair. The monolayer is easily prepared by solution self-assembly of the pyrimidine derivative on gold (*32, 33*). Ellipsometry measurements of film thickness are in good agreement with the calculated maximum thickness for an all-trans alkyl chain (22 Å).

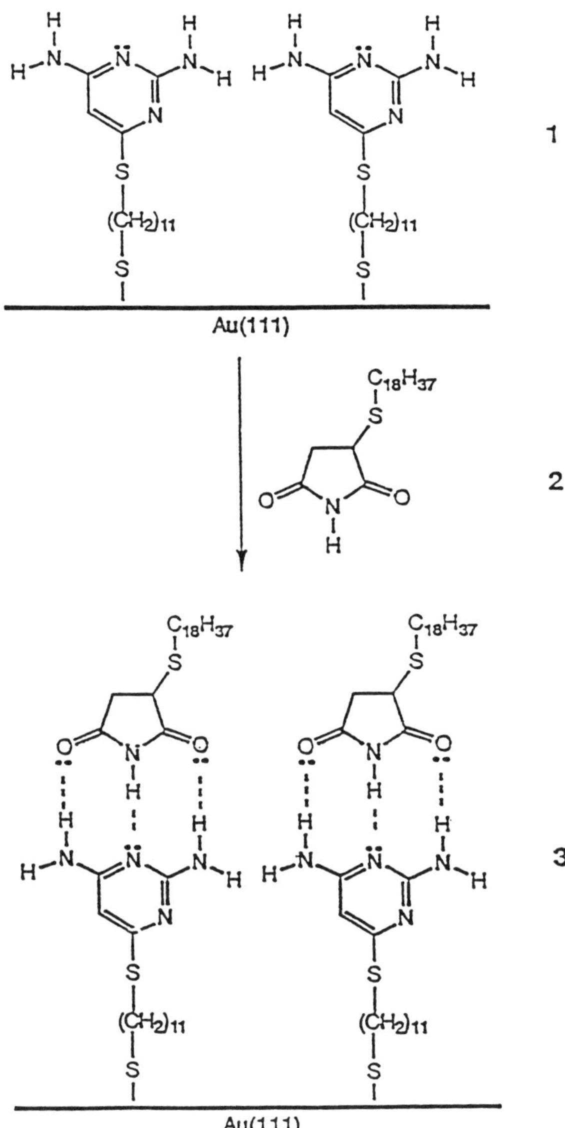

Figure 1. Formation of bilayer **3**. Recognition of the monolayer **1** (host) by the succinimide **2** (guest) results in the formation of a bilayer **3**, which is stabilized by the formation of three hydrogen bonds per host guest pair.

Langmuir-Blodgett Bilayer. A monolayer of **2** was first compressed at the air-water interface (Figure 2) and then transferred onto monolayer **1**. The resultant bilayer is comprised of host-guest molecules positioned at the interface. Figure 3A presents the external-reflection Fourier transform infrared (FTIR) spectrum for the monolayer **1**. The 2- and 4- amino (NH_2) scissoring vibrations appear at 1638 and 1665 cm^{-1}, respectively. The NH_2 scissoring vibration of the 2-amino group (2-aminopyrimidine) always appears at a higher frequency than that of the 4-amino group (4-aminopyrimidine) (*34, 35*). We assume the same assignment in 1,4-diaminopyrimidine. One of the four ring stretching modes, ν_8 (C_s, A'), appears at 1576 cm^{-1} (1563 cm^{-1} in KBr). The frequency of ν_{8a} for a number of substituted pyrimidines was reported by LaFaix and Lebas (*35*).

After transfer of **2**, there are some notable changes. First, there is a reduction in the amino scissoring intensity at 1665 cm^{-1} indicating a conformational change within the monolayer as the pyrimidine ring repositions itself to interact with **2**. [Notice that in a p-polarized IR spectrum, only the perpendicular components of the transition dipoles of the adsorbate are visualized (*36*).] Secondly, the symmetric and asymmetric carbonyl bands of the succinimide **2** appear at 1781 and 1722 cm^{-1} (*37*). Finally, the hydrogen bonding interaction raises the frequency of the ring stretching band at 1576 cm^{-1} by 9 cm^{-1}. A change in the ring stretching frequency with hydrogen bonding has been observed previously in diaminotriazine systems (*29*). Kurihara *et al.* report that Langmuir-Blodgett films of an amphiphilic

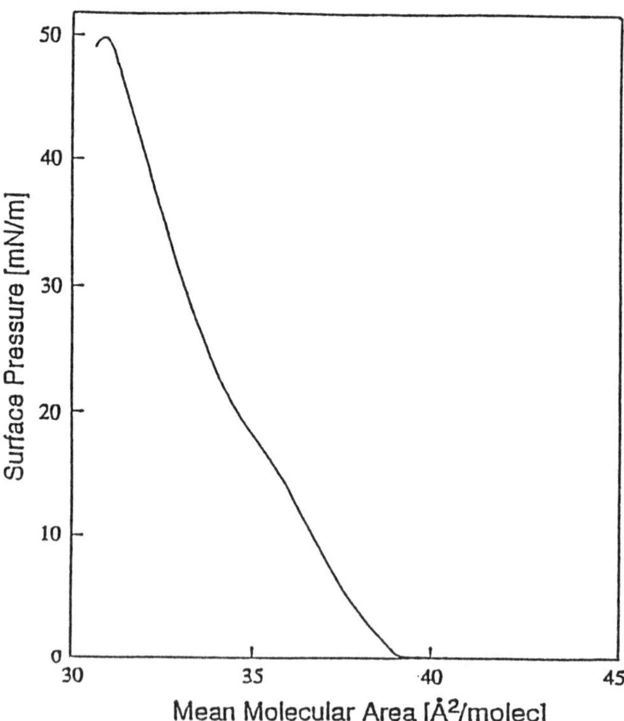

Figure 2. The π-A isotherm for **2**.

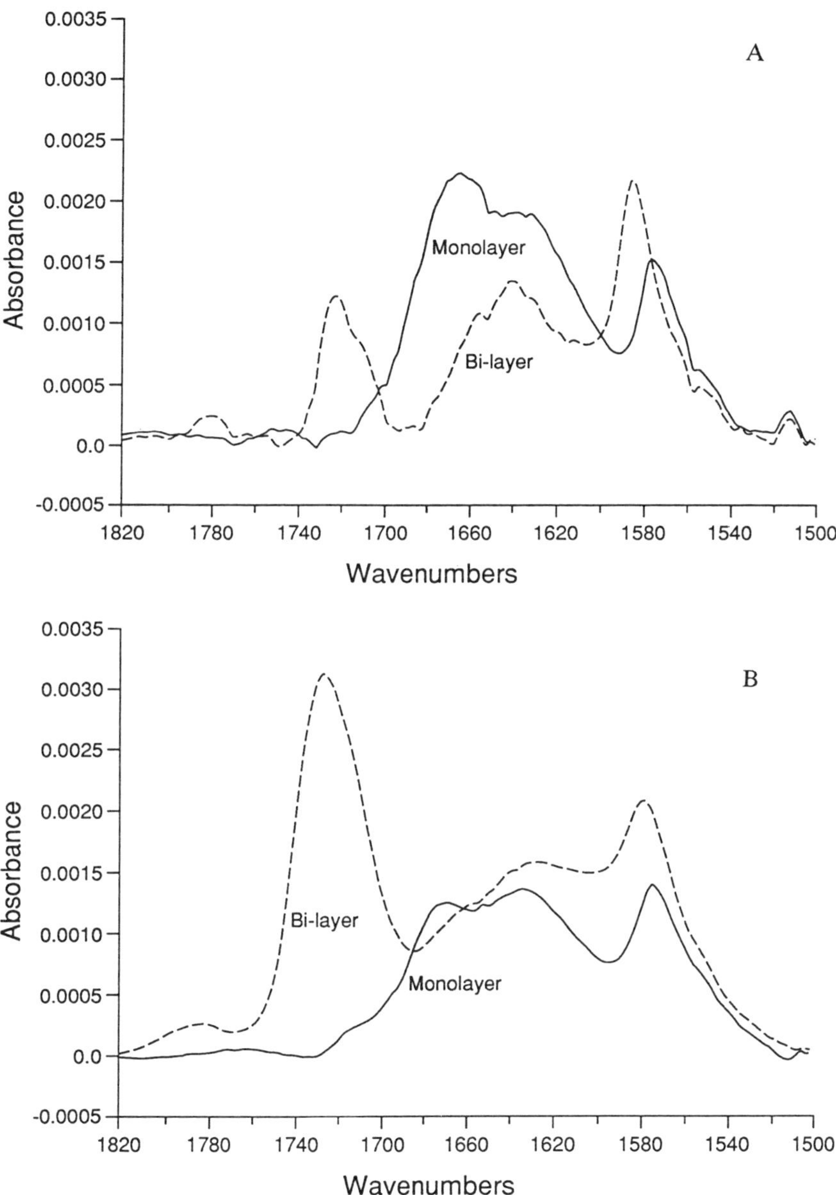

Figure 3. Infrared external reflection-absorption spectra with a p-polarized beam of the self-assembled monolayer on gold (**1**) and bilayers (**3**) in the mid frequency region. (**A**) The monolayer and the Langmuir-Blodgett bilayer. (**B**) The monolayer and the solution self-assembled bilayer.

diaminotriazine transferred from 0.01 M aqueous thymidine exhibit a C=N stretching vibration at 1557 cm^{-1}. Transfer from water shifts the vibration to 1571 cm^{-1}, consistent with the observations of Stidham and DiLella (*38*) and Takahashi *et al.* (*39*) on hydration of nitrogen heterocycles. In addition, DiLella and Stidham have shown in the Raman spectrum of pyridine that ν_8 shifts to higher frequency upon hydrogen bonding to water and Takahashi *et al.* observed a similar shift in the infrared spectrum.

Solution Self-Assembled Bilayer. In a second experiment we prepared a bilayer by solution self-assembly. Thus, a sample of monolayer **1** was immersed in a 1.0 mM toluene solution of **2** for 1.8 days, rinsed with solvent, and dried under nitrogen. The same trends that were observed previously are seen in Figure 3b. The carbonyl stretching bands at 1781 and 1726 cm^{-1} confirm the presence of **2**. Hydrogen bonding is evident by an increase of 4 cm^{-1} to 1578 cm^{-1} for the ν_8 band. Again we observe the effects of conformational changes in the amino scissoring bands.

High-Frequency Infrared Region. The high frequency infrared region characterizes the alkyl chains in the mono- and bilayers. There, the frequency of the asymmetric methylene stretching band, d^- ($\nu_{as}CH_2$), at 2921 cm^{-1} indicates that the monolayer is liquid-like (*40*). With the Langmuir-Blodgett transfer of **2** to the monolayer, however, there is a shift to a more solid-like frequency at 2919 cm^{-1} (Figure 4). The frequency in the KBr spectrum of pure **2** is 2919 cm^{-1}. The IR spectrum recorded one week later showed virtually no change, thus indicating the stability of the bilayer. Compression of **2** at the air-water interface should result in extended alkyl chains. Apparently, once compressed, the transferred **2** retains the extended structure. Notice that CH_2 groups from both molecules contribute to the observed IR spectrum; however, it is impossible to suggest conformational changes in monolayer **1** due to bilayer formation without specific deuteration. Ellipsometric measurements confirm that the alkyl chain is extended, in that there is a 25 Å increase in thickness upon adsorption of **2**.

In the self-assembled bilayer there is a shift in d^- to 2923 cm^{-1}; i.e., the bilayer is clearly liquid-like. This is not surprising, since the overlayer has not been closely packed prior to bilayer formation. An increase of only 16 Å in the film thickness is observed, suggesting that the succinimide overlayer is less dense than that prepared by Langmuir-Blodgett transfer, and that the alkyl chains are disordered. Although it may be inferred from the relative intensities of the asymmetric carbonyl bands in Figures 3A and 3B that there is more succinimide in the self-assembled bilayer, the relative intensities of the asymmetric methylene stretching vibrations in Figure 4 suggest the opposite, in agreement with ellipsometric results. Differences in packing and conformation in the overlayers probably account for the observed variation in the infrared intensities.

X-Ray Photoelectron Spectroscopy. The bilayers were examined by angle resolved X-ray photoelectron spectroscopy (XPS). In Table 1 the atomic concentrations for several electron take off angles (ETOA) are given. The concentration of gold is greater for the self-assembled sample, supporting a less densely covered surface. The one anomaly is the oxygen content for the solution self-assembled bilayer at 15° ETOA, which is likely a result of surface contamination. Succinimide overlayer calculations (14 Å) reveal that the more rigorous washing of the bilayer with solvent has removed more of **2** of this sample. The trends in elemental concentration as a function of ETOA are similar for the bilayers. Ellipsometric measurements of both bilayers before and after exposure to the high vacuum of the XPS showed no change in thickness and no apparent loss of material.

Table I. Atomic Percent versus Electron Take-off Angle

	Solution Self-assembled bilayer atomic percent				Langmuir-Blodgett bilayer atomic percent			
Element	15 °	25°	45°	75°	15°	25°	45°	75°
C	71.0	67.6	60.6	56.5	85.3	78.4	73.1	68.4
O	9.6	8.1	8.8	7.6	5.0	5.2	5.1	5.7
N	8.9	9.0	9.3	9.7	3.3	6.0	6.6	6.7
S	2.2	3.6	2.6	2.1	2.2	3.9	3.4	3.8
Au	8.4	11.7	18.8	24.2	4.2	6.6	11.8	15.4

Concentration Effects. The extent of bilayer formation depends on the concentration of the succinimide in the solution used in the self-assembly step. Thus, as the concentration of **2** is reduced from 1.0 to 0.5 to 0.25 mM, there is a corresponding reduction in coverage as indicated by a reduction in the intensity of the asymmetric carbonyl band at 1727 cm^{-1}, and the asymmetric methylene band at 2923 cm^{-1} (Figure 5). In all of these bilayer spectra there is an increase in the ring stretching frequency with hydrogen bonding. At concentrations below 1.0 mM, there are conformational changes due to the repositioning of the pyrimidine ring as described above. Intensity changes indicate that the 4-amino group is in a more nearly perpendicular orientation, and the 2-amino group and the ring are oriented parallel to the monolayer plane. One possible explanation for these conformational changes is that there is only one amino group and one carbonyl group involved in the recognition, similar to that found in base pairing mismatch (*41*). The breadth of the asymmetric carbonyl band indicates the possibility of several orientations.

Conclusion

The molecular recognition of components with complementary hydrogen bond donors and acceptors has created Langmuir-Blodgett and self-assembled bilayers stabilized by relatively weak hydrogen bonding interactions. This suggests that the prospects for building semi-permanent solid state structures with weak, non-covalent interactions by reversible recognition are good, with the logical extension being the incorporation of specific functionalities into guest-host partners to construct sensors and devices. Finally, beyond the specific host-guest system reported here, there is the prospect that the techniques described herein will be utilized for obtaining binding constants for host-guest systems (*42*).

Acknowledgments

S.L.D. would like to thank Eastman Kodak Company for the opportunity to conduct this work at their facility and acknowledges support from CUMIRP. We thank Professor Howard D. Stidham of the University of Massachusetts for a valuable discussion on pyrimidines, and D. J. Motyl of Eastman Kodak Company for his assistance in obtaining IR spectra.

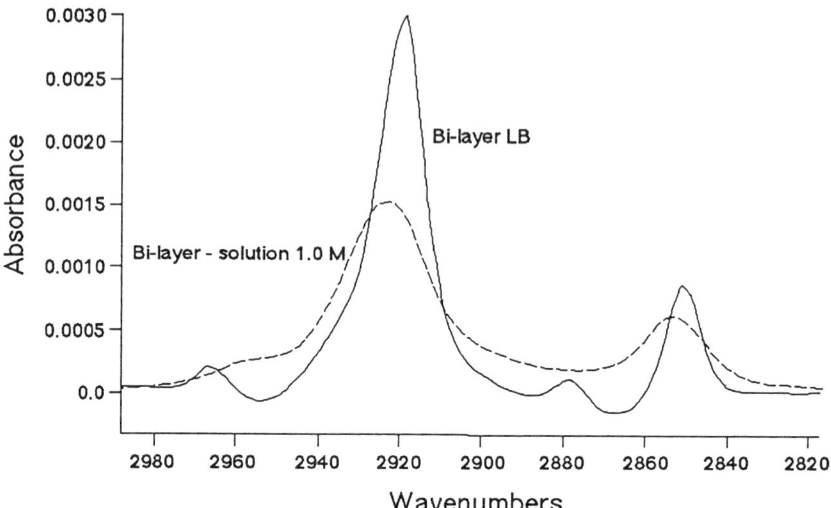

Figure 4. Infrared external reflection-absorption spectra with a p-polarized beam of the Langmuir-Blodgett and the solution self-assembled bilayers.

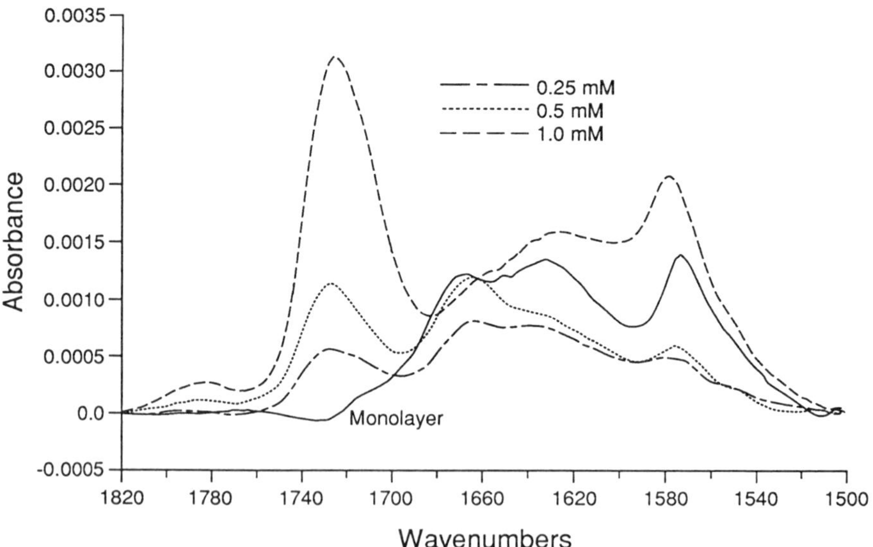

Figure 5. Infrared external reflection-absorption spectra with a p-polarized beam of the self-assembled monolayer on gold (**1**) and solution self-assembled bilayers (**3**) formed from 1.0 mM, 0.5 mM and 0.25 mM solutions.

Literature Cited

1. Lehn, J.-M. *Angew. Chem. Int. Ed. Engl.* **1988,** *27,* 90.
2. Ringsdorf, H.; Schlarb, B.; Venzmer, J. *Angew. Chem. Int. Ed. Engl.* **1988,** *27,* 113.
3. Lindsey, J. S. *New J. Chem.* **1991,** *15,* 153.
4. Schneider, H.-J. *Angew. Chem. Int. Ed. Engl.* **1991,** *30,* 1417.
5. Van Binst, G. *Design and Synthesis of Organic Molecules Based on Molecular Recognition*; Proceedings of the Solvay Conference on Chemistry; Springer-Verlag: New York, 1986; 2-4 and references within.
6. Chaiken, I.; Chiancone, E.; Fontana, A.; Neri, P. *Macromolecular Biorecognition Principles and Methods*; Humana Press: Clifton, New Jersey, 1987.
7. Alberts, B.; Bray, D.; Lewis, J.; Raff, M.; Roberts, K.; Watson, J. D. *Molecular Biology of the Cell,* Garland: New York, 1983; 91-98.
8. Whitesides, G. M.; Mathias, J. P.; Seto, C. T. *Science* **1991,** *254,* 1312.
9. Cram, D. J. *Angew. Chem. Int. Engl.* **1988,** *27,* 1009.
10. Whitesides, G. M.; Simanek, E. E.; Mathias, J. P.; Seto, C. T.; Chin, D. N.; Mammen, M.; Gordon, D. M. *Acc. Chem. Res.* **1995,** *28,* 37.
11. Bethel, D. *Advances in Physical Organic Chemistry*; Academic Press: New York, 1993, *28*; 45-138.
12. Quian, P.; Matsuda, M.; Miyashita, T. *J. Am. Chem. Soc.* **1993,** *115,* 5624.
13. Tsuruta, T.; Doyama, M.; Seno, M.; Imanishi, Y. eds *New Functional Materials*: Synthesis and Function Control of Biofunctionality Materials; Elvsevier: New York, 1993, **C**; 607-614.
14. Ikeura, Y.; Kurihara, K.; Kunitake, T. *J . Am. Chem. Soc.* **1991,** *113,* 7342.
15. Budach, W.; Ahuja, R. C.; Mobius, D.; Schrepp, W. *Thin Solid Films* **1992,** *210/211,* 434.
16. Rojas, M. T.; Koniger, R.; Stoddart, J. F.; Kaifer, A. E. *J. Am. Chem. Soc.* **1995,** *117,* 336.
17. Samuelson, L.A.; Kaplan, D. L.; Lim, J. O.; Kamath, M.; Marx, K. A.; Tripathy, S. K. *Thin Solid Films* **1994,** *242,* 55.
18. Chailapakul, O.; Crooks, R. M. *Langmuir* **1993,** *9,* 884.
19. Spinke, M.; Liley, M.; Schmitt, F.-J.; Gruder, H.-J.; Angemaier, L.; Knoll, W. *J. Chem. Phys.* **1993,** *99,* 7012.
20. Fujita, K.; Kimura, S.; Imanishi, Y. *J. Am. Chem. Soc.* **1994,** *116,* 2185.
21. Ahlers, M.; Muller, W.; Reichert, A.; Ringsdorf, H.; Venzmer, J. *Angew. Chem. Int. Ed. Engl.* **1990,** *29,* 1269.
22. Jencks, W. P. In *Design and Synthesis of Organic Molecules Based on Molecular Recognition;* Van Binst, G., Ed.; XVIIIth Solvay Conference on Chemistry; Springer-Verlag: New York, 1983; 59.
23. Motesharei, K.; Myles, D. C.; *J. Am.. Chem.. Soc.* **1994,** *116,* 7413.
24. Hamilton, A. D.; van Engen, D. *J. Am. Chem. Soc.* **1987,** 109, 5035.
25. Park, T. K.; Schroeder, J.; Rebek, J., Jr. *J. Am. Chem. Soc.* **1991,** 113, 5125.
26. Feng, Q.; Park, T. K.; Rebek, J., Jr. *Science* **1992,** 256, 1179.
27. Lehn, J.-M. *Pure & Appl.Chem.* **1994,** *66,* 1961.
28. Ahuja, R.; Caruso, P.-L.; Mobius, D.; Wolfgang, P.; Ringsdorf, H.; Wildburg, G. *Angew. Chem. Int. Ed. Engl.* **1993,** *32,* 1033.
29. Kurihara, K.; Ohto, K.; Honda, Y.; Kunitake, T. *J. Am. Chem. Soc.* **1991,** *113,* 5077.
30. Roberts, S. M., Ed.; *Biomolecular Recognition: Chemical and Biological Problems II;* Second International Chemical and Biochemical Problems in Molecular Recognition; Royal Society of Chemistry: Cambridge, 1992.
31. Lehn, J.-M. *Angew. Chem. Int. Ed. Engl.* **1990,** 29, 1304.

32. Ulman, A. *An Introduction to Ultrathin Organic Films: From Langmuir-Blodgett to Self-Assembly;* Academic Press, Inc.: New York, 1991.
33. Evans, S.; Urankar, E.; Ulman, A.; Ferris, N. *J. Am. Chem. Soc.* **1991**, 113, 4121.
34. Thompson, W. K. *J. Chem. Soc.* **1962**, 617.
35. LaFaix, A. J.; Lebas, J. M. *Spectrochim. Acta.* **1970**, *26A*, 1243.
36. Greenler, R. J. *J. Chem. Phys.* **1966**, *44* , 310.
37. Uno, T.; Machida, K. *Bull. Chem. Soc.* (Japan) **1962**, 35, 276.
38. Stidham, H. D.; DiLella, D. P. *J. Raman Spectroscopy* **1979**, *8*, 180.
39. Takahashi, H.; Mamola, K.; Plyler, E. K. *J. Mol. Spectroscopy* **1966**, *21*, 217.
40. Porter, M. D.; Bright, T. B.; Allara, D. L.; Chidsey, C. E. D. *J. Am. Chem. Soc.* **1987**, *109,* 3559.
41. Kennard, O.; Hunter, W. *Angew. Chem. Int. Ed. Engl.* **1991**, *30,* 1254.
42. We thank one of the referees for suggesting this possibility.

RECEIVED July 6, 1995

Chapter 16

Conjugation of DNA with Functional Vinyl Polymers by Using Vinyl Derivatives of Psoralen as a Linker

Mizuo Maeda, Daisuke Umeno, Chitoshi Nishimura, and Makoto Takagi

Department of Chemical Science and Technology, Faculty of Engineering, Kyushu University, Fukuoka 812–81, Japan

We describe a vinyl monomer having a psoralen moiety, which can form a photoadduct with double-helical DNA. This monomer (**2**) which is bound covalently to DNA gives the double-helical DNA a polymerizable function. The covalently bound DNA with vinyl groups can be copolymerized with a comonomer to give rise to a DNA-vinyl polymer conjugate. In contrast to our previously reported psoralen-containing monomer (**1**) which has a relatively short spacer from ethylenediamine, the present monomer provides a highly efficient conjugation. Therefore, the psoralen-containing monomer **2** should be a useful tool for anchoring double-helical DNA on polymeric materials when applying the DNA as an affinity ligand for bioseparation and bioanalysis.

Immobilization of biological macromolecules may give rise to new possibilities in bioseparation and biosensing. A molecular recognition function employed in such techniques as immunoassays and biosensors has been provided, in most cases, by proteins including enzymes and antibodies. On the other hand, DNA is also an important host molecule in biological affinity reactions. For example, polycyclic aromatic compounds, some of which are mutagens and/or carcinogens, may bind to DNA by inserting themselves between the base pairs of the double helix (intercalation). Metal ions can bind to DNA with some selectivity with the DNA double helix preferring magnesium ions to other alkali and alkaline earth metal ions. Some proteins can recognize a certain sequence of nucleic bases and bind to that site on DNA double strands. Thus double-helical DNA immobilized on a substrate would be a useful tool for fractionating or detecting protein molecules.

In fact, hybrid materials containing double-helical DNA have been utilized for separation and detection of DNA-binding substances. Affinity purification of a sequence-specific DNA-binding protein was attained by using Sepharose resin with a covalently coupled double-helical DNA which has a recognition site for the protein (*1*). The interaction of some antibiotics with DNA was studied by DNA-cellulose column chromatography (*2*). A high-performance affinity chromatography method was

0097–6156/96/0627–0197$15.00/0

developed to study the DNA-drug interactions by using a DNA-carrying silica support (*3*) and a DNA-immobilized gold electrode to monitor electrochemically a DNA-binding drug molecule (*4*).

In addition to these DNA-anchoring immobile supports, soluble conjugates are also of interest. A macromolecular complex between DNA and poly(N-isopropylacrylamide) (polyNIPAAM) was exploited for the thermally-induced separation of DNA-binding substances (*5*). The complex was found to acquire a thermoresponsive nature; the DNA was precipitated from aqueous solution when the temperature was raised above 31 ℃, due to the coil-to-globule transition of polyNIPAAM chains associated with the DNA. A preliminary experiment showed that a restriction endonuclease, *Hin*d III, was also co-precipitated with the thermoresponsive DNA upon transition. On the other hand, DNA was reported to be solubilized in organic media as the polyion complex with cationic lipid molecules (*6*). In order for the highly-sensitive detection of DNA, polyamide-oligonucleotide conjugate was prepared as DNA probes which carry one or more non-radioactive markers (*7*).

Recently, we reported a facile approach to a soluble conjugate from DNA and a vinyl polymer, in which the two components are covalently linked to each other (*8*). The method relied on a vinyl monomer having a psoralen moiety, which can form a photoadduct with DNA double-strands. This monomer (**1**) was proved to be bound covalently to DNA, endowing a polymerizable function to it (see Scheme 1). The vinyl-derivative of DNA can be copolymerized with a comonomer such as acrylamide to give DNA-vinyl polymer conjugates. However, the monomer **1** did not seem a very good linker since the polymerizability of vinyl groups which are introduced to DNA was relatively low. A longer and more hydrophilic spacer chain between psoralen and the vinyl group was suggested to improve the conjugation efficiency.

We describe in this chapter an improved linker monomer (**2**) for the preparation of the DNA-vinyl polymer conjugate. The efficient conjugation was attained by the use of N,N-bis(3-aminopropyl)methylamine as a spacer between the psoralen moiety and the vinyl group. The psoralen-containing monomer is expected to be a useful tool for immobilizing or anchoring double-helical DNA on polymeric materials with the intention of separating and monitoring DNA-binding substances.

Experimental

Materials. Acrylamide (AAm) was obtained from Wako Pure Chemicals (electrophoresis grade) and used without purification. N-Isopropylacrylamide (NIPAAM) was obtained from Tokyo Kasei Kogyo and recrystallized from a mixture of benzene and hexane. pBR322 DNA was purchased from Takara Shuzo. λ Phage DNA and *Hin*d III were purchased from Nippon Gene. Trioxalen was obtained from Aldrich.

Psoralen-Containing Monomers (1 and 2). Monomer **1** was prepared according to the method described in our previous paper (*8*): 4'-[[N-(2-aminoethyl)amino]methyl]-4,5',8-trimethylpsoralen, which was synthesized according to a method described by Lee, *et al.* (*9*), was reacted with N-methacryloyl succinimide

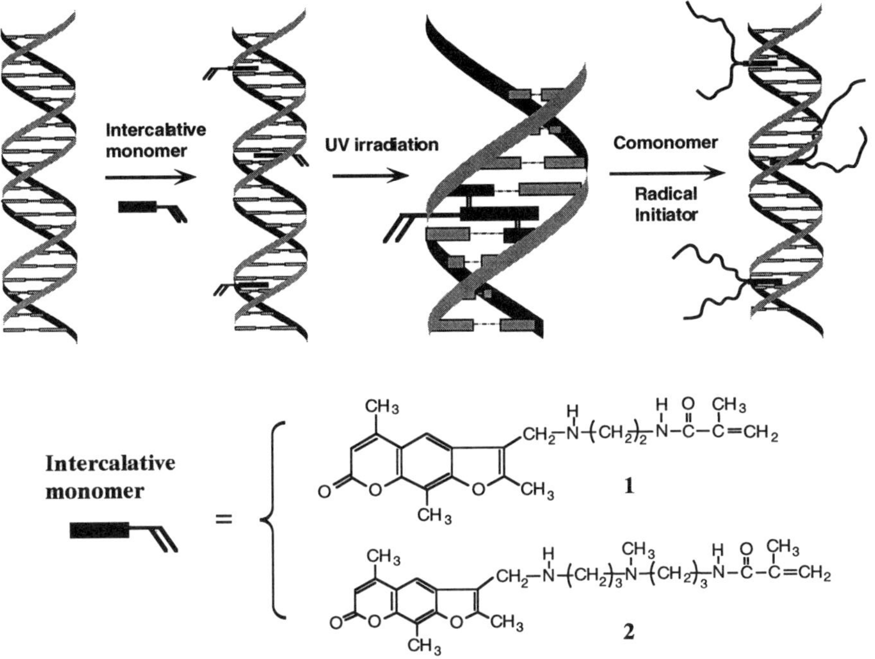

Scheme 1. Light-induced conjugation of double-helical DNA with the vinyl-derivatives of psoralen.

(Tokyo Kasei Kogyo). Since **1** was poorly soluble in water, an ethanol solution of **1** (10 mM) was diluted with TE buffer (10 mM Tris-HCl, 1 mM EDTA, pH 7.9) to prepare a stock solution (100 μ M) containing 1 vol % of ethanol.

Monomer **2** was prepared in a similar way but with N,N-bis(3-aminopropyl) methylamine as a spacer between the psoralen moiety and the vinyl group. The synthetic route to the monomer **2** is summarized in Scheme 2. Although the positive charges in the monomer **2** had been expected to provide water solubility to the relatively insoluble psoralen part, we had to use an ethanol solution of **2** (as well as in the case of the monomer **1**) and dilute it to the TE stock solution. A PE (8 mM Na_2HPO_4, 2 mM NaH_2PO_4, 1mM EDTA, pH 7.4) solution of **2** was prepared similarly.

The DNA-binding abilities of **1** and **2** were compared to each other by measuring the melting curves of sonicated calf thymus DNA (CT DNA) when one of the monomers (5 μ M) was added to the DNA solution (CT DNA, 0.1 mM in nucleotide; 10 mM MES (2-morpholinoethanesulfonate), 1 mM NaCl, pH 5.7): absorbance vs. temperature profiles were measured at 260 nm on a Hitachi U-3210 Spectrophotometer equipped with a SPR-10 Temperature Controller. The melting temperature (Tm) is defined as the temperature where the fraction of strands in the double helix is 0.5.

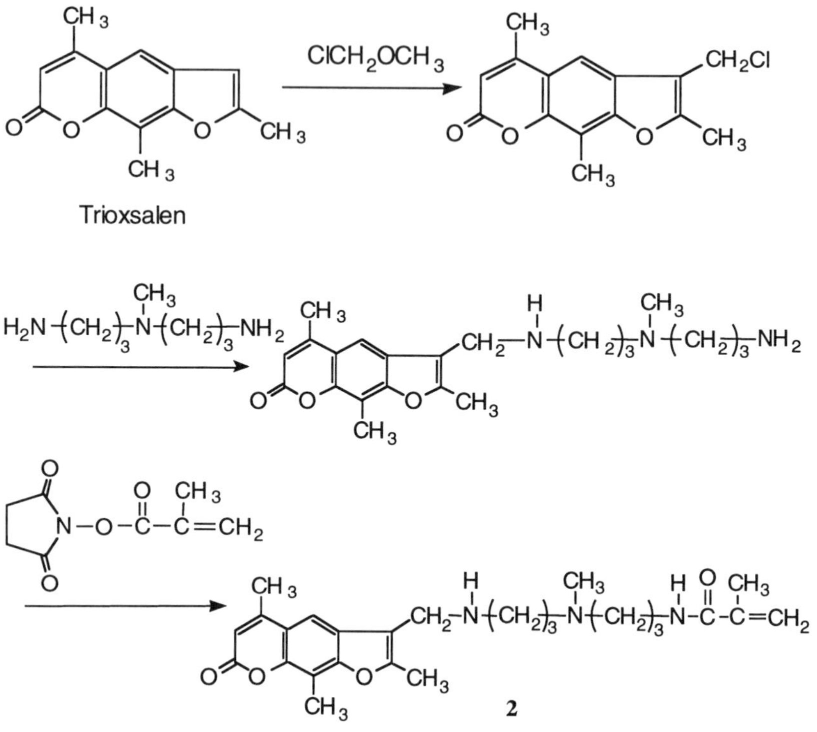
CH3
ClCH2OCH3
CH2Cl
Trioxsalen
H2N-(CH2)3-N(CH3)-(CH2)3-NH2
CH2-NH-(CH2)3-N(CH3)-(CH2)3-NH2
N-O-C(=O)-C(CH3)=CH2
CH2-NH-(CH2)3-N(CH3)-(CH2)3-NH-C(=O)-C(CH3)=CH2

2

Scheme 2. Synthetic route to the psorlen monomer **2**.

Photo-Reaction with DNA. Plasmid pBR322 DNA was linearized with *Hin*d III, and the DNA was purified by phenol extraction, followed by ethanol precipitation and resuspension in TE buffer at 100 μ g/L. To 2 μ L of the DNA solution in a micro test tube (Eppendorf, 1.5 mL) was added a TE solution of **2** from the stock solution. Then the total volume was made up to 12 μ L by adding TE buffer. The final concentration of the DNA was 51 μ M (in nucleotide). The concentration of **2** was varied in the range of 1.7 μ M and 83 μ M. Each solution was irradiated (ca. 60 mW/cm^2) on an ice bath with a 500-W ultra-high pressure Hg lamp equipped with a high-pass filter (Toshiba, UV-31) for 10 min.

To the reaction mixture was added 3 μ L of aqueous NaOH (1.2 M) in order to denature the DNA which was then analyzed by gel electrophoresis. For the analysis, the mixture was combined with 3 μ L of gel-loading solution consisting of glycerin and water (7:3, v/v), the resulting mixture was loaded on a 1 %-agarose gel, and the gel electrophoresis was performed at 7 V/cm for 1 h in TAE buffer (80 mM Tris-acetate, 1 mM EDTA, pH 7.6). After electrophoresis, the DNA in the gel was stained with ethidium bromide.

Polymerization. Before polymerization, a photo-reaction was carried out as follows: to a micro test tube, a TE solution of λ DNA (150 μg/mL; 5 μL) and PE solution of **2** (10 μM; 1-10 μL) were charged. With the addition of PE buffer, the total volume was made to be 20 μL; the concentrations of DNA and **2** were 115 μM and 0.5-5 μM, respectively. Then the mixture was irradiated by UV light in a similar manner to that described above. The reaction mixture was extracted twice by chloroform-isoamyl alcohol (3-methylbutanol) (24:1, v/v) in order to remove **2** which was not covalently bound to the DNA.

After the photo-reaction, a PE solution of AAm (0.35 M; 40 μL), a PE solution of N,N,N',N'-tetramethylethylenediamine (TEMED) and aqueous ammonium peroxodisulfate were succesively introduced to the mixture under nitrogen atmosphere. The total volume of the reaction mixture was 0.1 mL (TE:PE = 1:19, v/v) with the addition of PE buffer. For polymerization, the concentration of the respective components were as follows: DNA, 22.5 μM (in nucleotide); AAm, 140 mM; ammonium peroxodisulfate, 0.3 mg/mL; and TEMED, 10 mg/mL. Polymerization was carried out at 24 ℃ for 1 h.

After polymerization, the mixture was analyzed by gel electrophoresis: 10 μL portion of each of the sample solutions was combined with 8 μL of gel-loading solution consisting of glycerin and water (1:1, v/v). The mixture was loaded on a 0.5 %-agarose gel, and the gel electrophoresis was performed at 7 V/cm for 1 h in TBE buffer (89 mM Tris-borate, 2.5 mM EDTA, pH 8.0). After electrophoresis, DNA in the gel was stained with ethidium bromide.

Modification with PolyNIPAAM. The modification of λ DNA with polyNIPAAM was carried out similarly, except that gel electrophoresis for the analysis was performed at 10 ℃ since precipitaion of polyNIPAAM and its conjugates takes place from aqueous glycerin (25 %) at ambient temperature (*5*). Another difference in the case of NIPAAM was that the total volume of the polymerization mixture was adjusted to 50 μL but not to 100 μL: thus, the concentration of each component was twice as much as that for AAm.

The post-polymerization mixture was centrifuged (15000rpm, 30 min) at 35 ℃ in order to precipitate the DNA-polyNIPAAM conjugates. After centrifugation, the supernatant was collected and subjected to gel electrophoresis. The DNA band due to unprecipitated DNA was evaluated by scanning densitometry in order to estimate a precipitation efficiency.

Precipitation Separation of Ethidium. By using the temperature-responsive DNA, *i.e.*, DNA-polyNIPAAM conjugates, ethidium, which is a well-known DNA-binding dye, was separated from the solution. First the reaction mixture just after the modification of λ DNA with polyNIPAAM was combined with TE solution of ethidium bromide (Caution!; a strong carcinogen). The resulting mixture containing the conjugate (45 μM as DNA in nucleotide) and ethidium (1.25 μM) in 1 mM Tris-EDTA was left at 25 ℃ for 15 min. Then the mixture was centrifuged (15000 rpm, 5 min) at 40 ℃. The supernatant was collected and was evaluated for the concentration of ethidium using a Hitachi 650-60 Fluorescence Spectrophotometer (λex=546 nm,

λ em=595 nm). Since the fluorescence intensity of ethidium is strongly dependent on the concentration of co-existing DNA, the measurement was made in the presence of a large excess of CT DNA (2 mM in nucleotide) in order to avoid a possible influence by slightly-contaminating DNA in the supernatant.

Results and Discussion

Binding affinity with DNA. The psoralen-containing monomer (**2**) consists of a psoralen moiety, connected to polymerizable vinyl group by a positively charged linker. The positive charges confer good water solubility to the relatively insoluble psoralen part, and aid in binding to negatively charged nucleic acids. As to the water solubility, however, we had to use an ethanol solution of **2** when preparing a stock solution.

The DNA-binding affinity of the monomer was estimated by its ability to stabilize double-helical structure of DNA. The effect of the monomers **1** and **2** on the DNA-melting temperature (Tm) evaluated from absorbance (260 nm) vs. temperature profiles was summarized in Table 1. The increase in Tm (ΔTm) due to the presence of **2** was found to be 10 ℃, which is twice larger than that for **1**. This suggests that the monomer **2** is bound to DNA more efficiently than the monomer **1**.

Table 1. Effect of psoralen monomers on the DNA-melting temperature (Tm)[a]

additive	Tm/℃	ΔTm/℃
none	46.7	-
1	51.6	4.9
2	56.7	10

[a] Tm is defined as the temperature where the fraction of strands in double helix is 0.5. A psoralen monomer (5 μM) was added to a solution of sonicated calf thymus DNA (0.1 mM in nucleotide; 10 mM MES, 1 mM NaCl, pH 5.7): absorbance vs. temperature profiles were measured at 260 nm.

Photo-induced covalent binding with DNA. Psoralens are known to intercalate into DNA double strands in the dark and form covalent bonds at their 3,4 and 4',5' double bonds with pyrimidines upon near-UV irradiation (*10*). If both sites of the psoralen are reacted, the result is an interstrand DNA crosslink; both strands are connected to each other. The ability of **2** to form DNA crosslinks was tested according to the literature (*11*), by reacting linear double-stranded plasmid DNA with **2** and near-UV light.

The photo-reacted DNA was alkali denatured to be the single-stranded form and was loaded onto a nondenaturing agarose gel. Crosslinked DNA immediately renatures in the neutral gel buffer (pH 7.6) and runs as the double-stranded form (see Scheme 3), while non-crosslinked DNA remains single-stranded and runs with greater

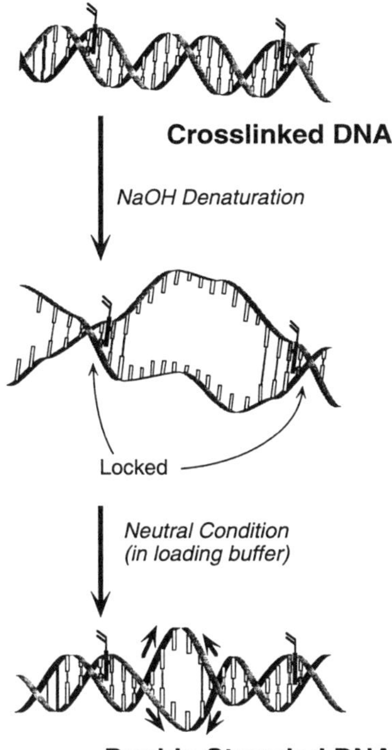

Scheme 3. Renaturing of the psoralen-crosslinked DNA.

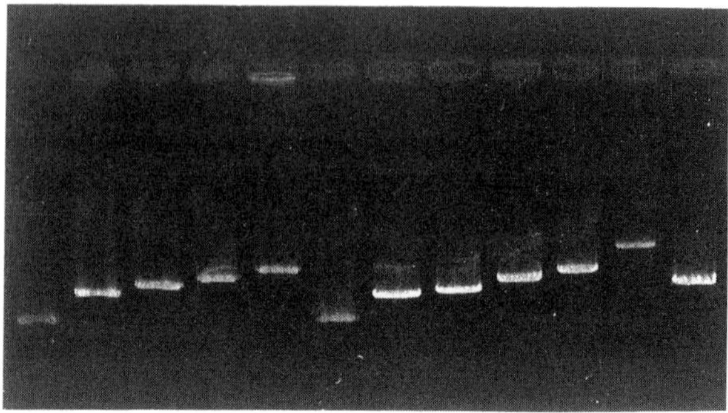

Figure 1. DNA crosslinking by the psoralen-containing monomer (**2**). Linear pBR322 DNA (51 μM) was irradiated by UV light in the presence of increasing amount of **2**. Samples from lane 1 to lane 6 were alkali denatured. All the samples were loaded on a nondenaturing gel from 1 % agarose. Lanes 1 and 7, [**2**] = 0; lanes 2 and 8, [**2**] = 1.7 μM; lanes 3 and 9, [**2**] = 8.3 μM; lanes 4 and 10, [**2**] = 16.7 μM; lanes 5 and 11, [**2**] = 83 μM; lane 6 and 12, [**2**] = 83 μM, but no UV irradiation.

mobility. As seen in Figure 1, addition of **2** resulted in crosslinking after UV irradiation (lanes 2-5), while even the highest concentration of **2** produced no crosslinking in the absence of light (lane 6). Thus the monomer **2** was proved to be bound covalently to double-helical DNA through a photochemical reaction when irradiated by UV light.

Shown in Figure 1 is that increasing the amount of **2** (1.7 - 83 μM) resulted in an appreciable retardation of DNA migration when UV irradiated (lanes 2-5); the degree of the retardation nicely corresponds to those seen for nondenatured DNA (lanes 8-11). The retardation should be ascribed to **2**-dependent elongation of the plasmid DNA.

Modification of DNA with vinyl polymers. λ DNA which had been reacted photochemically with **2** was co-incubated in the polymerization system of acrylamide (AAm). The gel electrophoresis of the product showed retarded migration as well as broadening of the DNA band, as seen in Figure 2, lanes 3-9. The degree of retardation increased with increasing concentration of **2** in the photo-reaction mixture. In contrast, the light-induced binding of **2** with λ DNA did not affect the migration significantly in the concentration range examined here (0.2 - 5 μM) at the photo-reaction. Since the mobility in gel electrophoresis is primarily a function of size and charge, the retarded migration was ascribed to the increase in size or the 'fattening' of the migrating species brought about by the modification of DNA with nonionic polyAAm chains.

On the other hand, lane **2** for the control in which the DNA was not derivatized with **2** showed no retardation after the polymerization; the migration profile of the

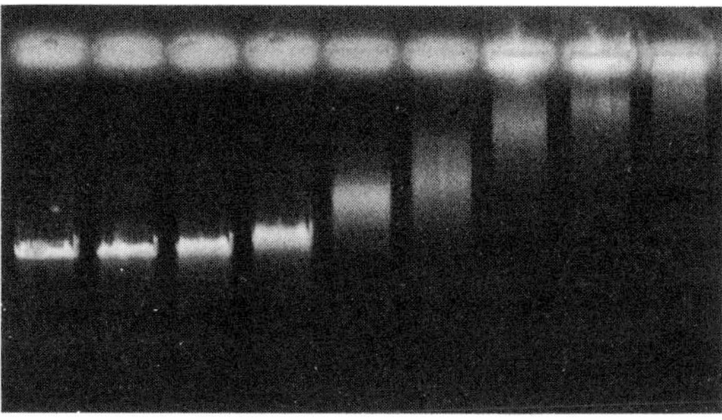

Figure 2. Gel electrophoresis of DNA-polyAAm conjugates. λ DNA (115 μM) was irradiated by UV light in the presence of **2**. After removing unbound **2**, samples were co-incubated in the polymerization mixture of AAm, and then loaded on 1 % agarose gel. Lane 1, DNA alone; lane 2, [**2**] = 0 (control); lane 3, [**2**] = 0.2 μM; lane 4, [**2**] = 0.5 μM; lane 5, [**2**] = 1 μM; lane 6, [**2**] = 2 μM; lane 7, [**2**] = 3 μM; lane 8, [**2**] = 4 μM; lane 9, [**2**] = 5 μM. The concentrations of **2** ([**2**]) represent those at the antecedent photochemical reaction.

DNA was the same as that for native DNA (lane 1). This indicated that the control sample was a simple mixture of DNA and homo-polymer of AAm; direct conjugation of vinyl polymers on DNA does not take place. Thus the vinyl-derivatized DNA incubated in the polymerization system of AAm was concluded to conjugate with polyAAm by means of **2** residues, which had been previously introduced in the DNA and then took part in copolymerization with AAm.

Modification of DNA with poly(N-isopropylacrylamide) was also possible. λ DNA which had been reacted photochemically with **2** was co-incubated in the polymerization system of NIPAAM. The gel electrophoresis of the product showed migration behavior very similar to those in Figure 2.

Thermally-induced precipitation of DNA-polyNIPAAM conjugates. The modification of DNA with polyNIPAAM should make the conjugate temperature-responsive since the modifying chains from polyNIPAAM have a property to precipitate from aqueous media around 31 ℃ (*5*). For the present **2**-dependent conjugate, the temperature-responsiveness was evaluated. The reaction mixture after polymerization was centrifuged at 40 ℃ so that the precipitation of DNA-polyNIPAAM conjugate whould take place. The supernatant was analyzed by gel electrophoresis to determine the amount of unprecipitated DNA. The result is shown in Figure 3 in terms of the degree of precipitation with respect to DNA originally used. Thus more than 80 % of the DNA was found to acquire the temperature-responsive property when as little as 0.5 μM of **2** was used for 115 μM of DNA at the photo-reaction.

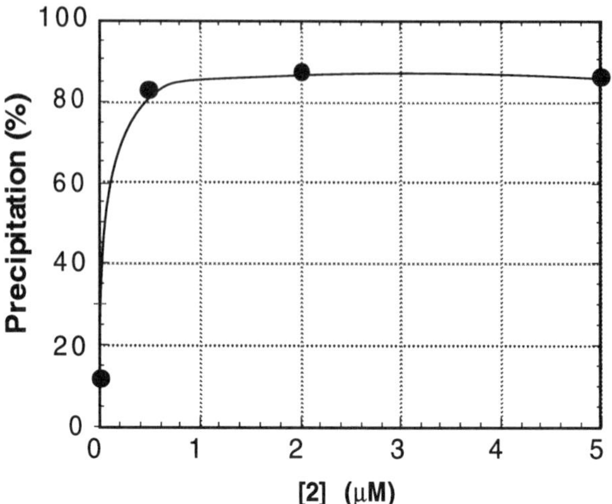

Figure 3. Amount of λ DNA which acquired the temperature-responsive property upon conjugation with polyNIPAAM; amount of thermally-precipitated DNA expressed in percent to the DNA originally added is plotted against the concentration of **2** at the antecedent photochemical reaction.

We wish to emphasize the fact that the efficiency of themally-induced precipitation of the present conjugates was strikingly improved as compared with those values in our previous studies, where the monomer **1** was used (*8*). In these studies only 45-58 % of DNA was found to acquire the temperature-responsive property when 5-50 μ M of **1** was used at the photochemical reaction. The improved efficiency may be ascribed to the positively charged, longer and more hydrophilic spacer chain between psoralen and the vinyl group of **2**.

The DNA-polyNIPAAM conjugate presented here should separate DNA-binding substances upon the thermally-induced affinity precipitation. As a preliminary experiment, the separation of a well-know DNA-binding dye, ethidium, was examined. The result is shown in Figure 4. More than 80 % of ethidium in the solution was removed by the thermal transition of the conjugate, while only 20 % was incorporated into the precipitate from homopolymer of NIPAAM as can be seen at [**2**]=0 in Figure 4.

In conclusion, we describe a vinyl monomer (**2**) having a psoralen moiety, which can form a photoadduct with double-helical DNA. This monomer was designed to be bond covalently to the DNA, endowing it with a polymerizable function (Scheme 1, c). The resulting DNA, furnished with vinyl groups, was copolymerized with a comonomer to give rise to a DNA-vinyl polymer conjugate (Scheme 1, d). In contrast

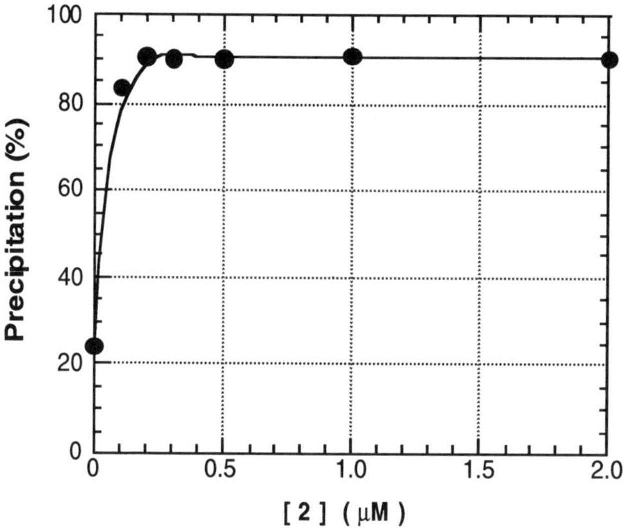

Figure 4. Separation of ethidium from an aqueous solution by using the thermally-induced precipitaion of DNA-polyNIPAAM conjugates. [**2**] stands for the concentration of **2** at the photo-reaction prior to the polymerization.

to our previously-reported psoralen-containing monomer (**1**) which has a relatively short spacer from ethylenediamine (*8*), the present monomer provides a highly efficient conjugation. Therefore, the psoralen-containing monomer **2** should be a useful tool for anchoring double-helical DNA onto polymeric materials when applying the DNA as an affinity ligand for bioseparation and bioanalysis. In addition, conjugation of DNA with vinyl polymers which carry reporter groups such as biotin would be a promising way of highly sensitive detection of DNA, since the present method should allow a large number of labels on DNA.

Acknowledgments

The authors are grateful to Dr. Toshihiro Ihara of Kyushu University for his discussion. This work was supported in part by Nakatani Electronic Measuring Technology Association of Japan. Financial support by a Grant-in-Aid for Scientific Research from Ministry of Education, Science and Culture of Japan is also acknowledged.

Literature Cited

1. Kadonaga, J.T.; Tjian, R. *Proc. Natl. Acad. Sci. USA*, **1986**, 83, 5889.

2. Inagaki, A.; Kageyama, M. *J. Biochem.*, **1970**, 68, 187.
3. Kashiwagi, S.; Ohmori, K.; Maeda, M.; Takagi, M. *Anal. Sci.*, **1992**, 261.
4. Maeda, M.; Mitsuhashi, Y.; Nakano, K.; Takagi, M. *Anal. Sci.*, **1992**, 83.
5. Maeda, M.; Nishimura, C.; Inenaga, A.; Takagi, M. *Reactive Polym.*, **1993**, 21, 27.
6. Ijiro, K.; Okahata, Y. *J. Chem. Soc., Chem. Commun.*, **1992**, 1339.
7. Haralambidis, J.; Duncan, L.; Angus, K.; Tregear, W. *Nucleic Acids Res.*, **1980**, 18, 493.
8. Maeda, M.; Nishimura, C.; Umeno, D.; Takagi, M. *Bioconjugate Chem.*, **1994**, 5,527.
9. Lee, B.L.; Murakami, A.; Blakes, K.R.; Lin, S.-B.; Miller, P.S. *Biochemistry*, **1980**, 27, 3197.
10. Cimino, D.G.; Gamper, H.B.; Isaacs, S.T.; Hearst, J.E. *Ann. Rev. Biochem.*, **1985**, 54, 1151.
11. Saffran, W.A.; Welsh, J.T.; Knobler, R.M.; Gasparro, F.P.; Cantor, C.R.; Edelson, R.L. *Nucleic Acids Res.*, **1988**, 16, 7221.

RECEIVED December 30, 1995

Chapter 17

pH-Sensitive Swelling and Drug-Release Properties of Chitosan–Poly(ethylene oxide) Semi-interpenetrating Polymer Network

Vijay R. Patel and Mansoor M. Amiji

Department of Pharmaceutical Sciences, Northeastern University, Boston, MA 02115

Chitosan-poly(ethylene oxide) (PEO) semi-interpenetrating polymer network (semi-IPN) was developed for pH-sensitive drug delivery into the stomach. In simulated gastric fluid (SGF, pH 1.2), chitosan-PEO semi-IPN had a swelling ratio of 18.4 after 6 h. In contrast, the swelling ratio in simulated intestinal fluid (SIF, pH 7.2) was only 1.80. The presence of pepsin and pancreatin in SGF and SIF, respectively, did not effect the swelling behavior. The release of riboflavin from chitosan-PEO hydrogel was also found to be dependent on the pH of the medium. More than 50% of the drug was released from chitosan-PEO semi-IPN after 6 h in SGF. In SIF, on the other hand, only 29% was released after 6 h. The presence of enzymes did decrease the release of drug from the hydrogels by complexing with the cationic chitosan matrix. The results clearly show that chitosan-PEO semi-IPN, containing high molecular weight PEO, meets the requirements of pH responsive drug delivery system, and could be useful for site-specific delivery into the stomach.

Hydrogels are defined as three-dimensional crosslinked polymeric material with the ability to absorb significant amount of water (1,2). Due to the high water content, hydrogels are known to possess excellent biocompatibility. They have been used in various biomedical applications including implants (3), contact lenses (4), and controlled-release drug delivery systems (5). Recently, hydrogels that can undergo a reversible phase transition in response to environmental stimuli such as changes in pH and temperature are being actively considered as physiological-responsive drug delivery systems (6,7).

On the basis of the chemical structure of the monomer, hydrogels with pendant acidic or basic functional groups can swell and collapse in response to changes in the pH of the medium (8). pH-sensitive swelling hydrogels, when used for oral drug delivery, offer beneficial properties of "triggered" drug release in the specific region of the gastro-intestinal (GI) tract. Cationic hydrogels, for instance, would swell and release the drug in the low pH environment of the stomach. Localized antibiotic delivery into the stomach would be highly beneficial for the treatment of *Helicobacter pylori* infection in peptic ulcer disease (9).

0097–6156/96/0627–0209$15.00/0

Chitosan, a linear polymer of β (1→4)-linked 2-amino-2-deoxy-D-glucopyranose, is a natural cationic polysaccharide derived from N-deacetylation of chitin (10,11). Chitin is isolated from the exoskeleton of crustaceans such as crabs, krill, and shrimps. Chitin and chitosan, therefore, are abundant biopolymers obtained from a renewable marine resource. The D-glucosamine residue of chitosan has an approximate pK_a of 6.4 (12). In the low pH medium, crosslinked chitosan hydrogels would swell extensively due to the positive charges on the network. Chitosan hydrogels offer additional benefits of being non-toxic and biodegradable (10).

In this chapter, we describe the preliminary results obtained from chitosan-poly(ethylene oxide) (PEO) semi-interpenetrating network (semi-IPN) as a pH-sensitive drug delivery system. Dynamic swelling and drug release properties of chitosan-PEO semi-IPN were evaluated in simulated gastric fluid (SGF, pH 1.2) and simulated intestinal fluid (SIF, pH 7.2).

MATERIALS AND METHODS

Materials. Chitosan (SeaCure®-240) with a degree of deacetylation of >85% was obtained from Pronova Biopolymers (Raymond, WA). PEO of molecular weight ranging from 10,000 to 1,000,000 daltons was either purchased from Sigma Chemical Company (St. Louis, MO) or obtained from Union Carbide (Danbury, CT). The crosslinking agent, glyoxal, was purchased from Aldrich Chemicals (Milwaukee, WI). Porcine pepsin (480 units/mg), porcine pancreatin, and riboflavin were all purchased from Sigma. Deionized distilled water (NANOpure® II, Barnsted/Thermolyne, Dubuque, IO) was used exclusively to prepare all aqueous solutions. All other chemicals were of analytical grade or better.

Synthesis of Hydrogels. Chitosan was dissolved in 0.1 M acetic acid to prepare a 2.0% (w/v) solution. Chitosan solution was filtered through glass wool and was allowed to degas overnight. PEO, dissolved in 0.1 M acetic acid, was added to chitosan solution to prepare a blend with PEO in a weight ratio ranging from 10% to 30%. Increasing the concentration of PEO above 30% resulted in gels with poor mechanical properties. Glyoxal was added to 50 ml of stirred solution of chitosan-PEO blend to give a final concentration 8.0 mg/ml. The feed mixture was allowed to gel for 24 h at room temperature. Chitosan-PEO semi-IPN was cut into round discs (~20 mm diameter) and neutralized with 0.1 M sodium hydroxide solution, followed by extensive washing with deionized distilled water. The hydrogel was allowed to dry to a constant weight at room temperature. Control chitosan hydrogel was prepared according to the same procedure without the addition of PEO.

Swelling Studies. Chitosan and chitosan-PEO hydrogels were allowed to swell in 50 ml of SGF (pH 1.2) or SIF (pH 7.2) at 37°C. SGF and SIF were prepared according to the procedure described in the United States Pharmacopeia (13). The concentration of pepsin and pancreatin in SGF and SIF were 0.1 mg/ml and 10 mg/ml, respectively. In order to examine the effect of enzymes on the swelling behavior, in some experiments, pepsin and pancreatin were omitted from SGF and SIF, respectively. At different time points, the gels were removed from the swelling medium and were blotted to remove excess moisture on a piece of Kimwipe® tissue (Kimberly-Clark, Roswell, GA). The swelling ratio (Q) was calculated according to the following expression:

$$Q = W_s/W_d \qquad (1)$$

where W_s is the weight of the swollen hydrogel and W_d is the weight of the dry hydrogel.

In the sequential swelling studies, chitosan and chitosan-PEO hydrogels were first swollen in pepsin-free SGF for up to 6 h, followed by swelling in pancreatin-free SIF for an additional 6 h. The results represent mean ± S.D. from four independent experiments.

Degradation Studies. The degradation of chitosan and chitosan-PEO hydrogels was examined by measuring the formation of glucosamine in SGF and SIF, with and without the enzymes. One-hundred μl sample of the swelling medium was removed at different time intervals and assayed for glucosamine using the *o*-toluidine assay (14). The green-colored product was quantitated by measuring the absorbance at 630 nm with a Shimadzu UV-160U (Columbia, MD) spectrophotometer. The concentration of glucosamine in the swelling medium was calculated from appropriate calibration curves. The results represent mean ± S.D. from four independent experiments.

Drug Loading and Release Studies. Riboflavin, a model drug, was mixed with chitosan and chitosan-PEO solutions in 0.1 M acetic acid to give a final concentration of 1.0 mg/ml. Drug-loaded hydrogel was prepared by crosslinking with glyoxal as described above. Riboflavin-containing chitosan and chitosan-PEO hydrogels were placed in 50 ml of SGF or SIF, with and without the enzymes, at 37°C. At specified time intervals, 1.0 ml of the medium was removed and assayed for the released drug at 445 nm. The release kinetics of riboflavin from chitosan and chitosan-PEO hydrogels was analyzed using the following equation developed by Peppas (15,16):

$$M_t/M_\infty = kt^n \tag{2}$$

where M_t is the amount of drug released at time t, M_∞ is the total amount of drug released, k is a release constant, and n is the release exponent. The value of n determines the relationship between the release rate and time. The results represent mean ± S.D. from four independent experiments.

RESULTS AND DISCUSSION

pH-Sensitive Swelling of Chitosan and Chitosan-PEO Hydrogels. In order to determine the optimum molecular weight and amount of PEO necessary for pH-sensitive swelling behavior, chitosan-PEO semi-IPN containing different molecular weights of PEO and different weight ratios of chitosan to PEO were swollen in pepsin-free SGF at 37°C. At initial time points, the swelling ratios of chitosan-PEO (weight ratio 80:20) semi-IPN with increasing molecular weight of PEO up to 1,000,000 daltons were not very different as shown in Figure 1. As the duration of swelling continued, however, the swelling ratios of the semi-IPN increased with increasing molecular weight of incorporated PEO. After 6 h, the swelling ratio of control chitosan hydrogels in SGF was 5.30. The swelling ratio of chitosan-PEO semi-IPN after 6 h, on the other hand, was 11.9 for PEO 100,000 daltons and 14.3 for PEO 1,000,000 daltons. Addition of high molecular weight PEO contributed significantly to the initial swelling of the hydrogels. High molecular weight PEO is expected to act as an osmotic agent in facilitating the initial uptake of water into the hydrogels.

Figure 2 shows the swelling behavior of chitosan-PEO semi-IPN containing PEO of molecular weight 1,000,000 daltons in different weight ratios of chitosan to PEO. After 6 h, the swelling ratio increased from 7.74 in chitosan-PEO semi-IPN containing chitosan and PEO in a weight ratio of 90:10 to 29.8 when the amount of PEO in the semi-IPN was increased to 70:30. Although increasing the amount of PEO did increase the swelling ratio, addition of PEO above 30% also compromised

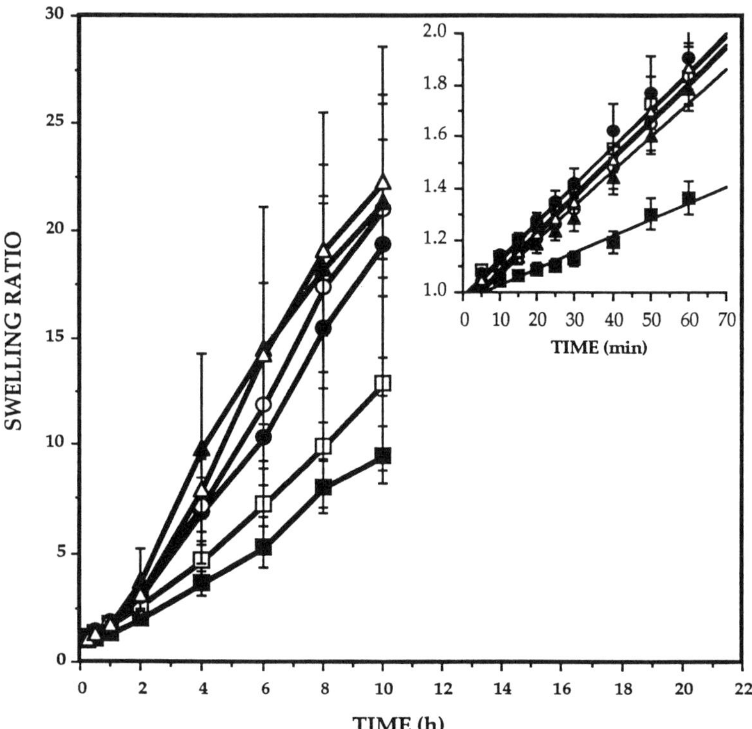

Figure 1. Swelling kinetics of chitosan and chitosan-PEO hydrogels, containing PEO of different molecular weights, in pepsin-free simulated gastric fluid (pH 1.2) at 37°C. The inset shows swelling behavior at the lower time points. The symbols represent chitosan (■) and chitosan-PEO (weight ratio of 80:20) hydrogels prepared with PEO of molecular weight 10,000 (□), 20,000 (●), 100,000 (○), 600,000 (▲), and 1,000,000 (Δ) daltons.

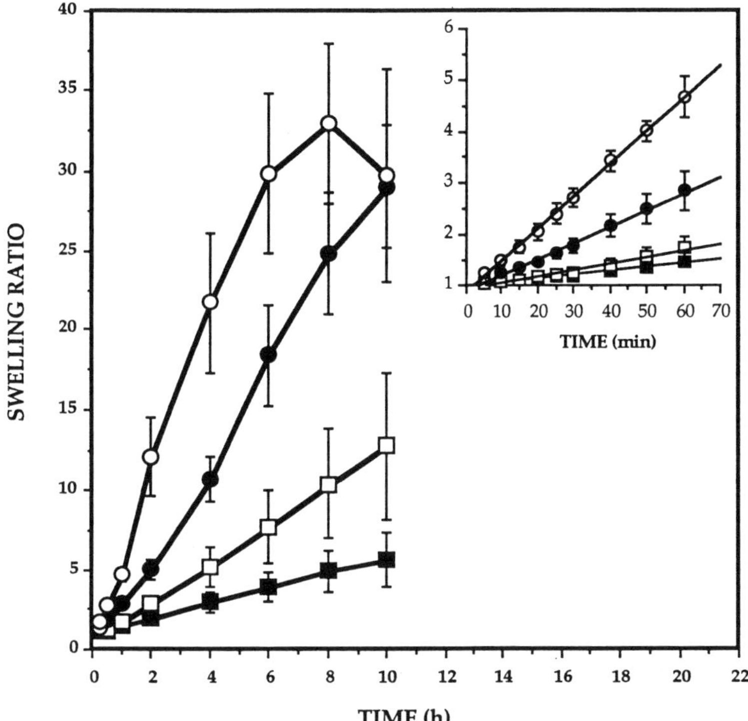

Figure 2. Swelling kinetics of chitosan and chitosan-PEO hydrogels, containing different amounts of PEO (Mol. wt. 1,000,000 daltons), in pepsin-free simulated gastric fluid (pH 1.2) at 37°C. The inset shows swelling behavior at the lower time points. The symbols represent chitosan (■), and chitosan-PEO hydrogels containing chitosan and PEO in a weight ratio of 90:10 (□), 80:20 (●), and 70:30 (○).

the mechanical properties of the hydrogels. On the basis of the above results, chitosan-PEO semi-IPN containing PEO of molecular weight 1,000,000 daltons in a weight ratio of 80:20 (chitosan to PEO) was used for further evaluation.

pH-sensitive swelling behavior of chitosan and chitosan-PEO hydrogels is shown in Figure 3. The hydrogels were swollen in enzyme-free SGF and SIF for various time intervals. At initial time points, it is important to note that chitosan-PEO semi-IPN was swollen to about the same extent in SGF and SIF (Figure 3, inset). It seems that the addition of high molecular weight PEO influenced the uptake of water regardless of the pH of the medium. At later time points, however, the influence of pH on the swelling became apparent. After 6 h, for instance, the swelling ratio of chitosan-PEO semi-IPN was 18.4 in SGF and only 1.80 in SIF. Ionization of glucosamine residues by the hydronium ions of SGF led to the electrostatic repulsion between like charges in adjacent polymer chains in chitosan-PEO hydrogels. This effect is important for the extensive swelling of chitosan-PEO hydrogels in SGF. Chitosan hydrogels, on the other hand, did not swell extensively in SGF or SIF. Intermolecular association between chitosan and PEO in the semi-IPN may have also decreased the crystallinity of chitosan matrix resulting in higher swelling of chitosan-PEO semi-IPN as compared to chitosan hydrogels.

De-Yao et al. (17,18) have examined the pH-sensitive swelling behavior of chitosan-poly(proplyene glycol) (PPG, Mol. wt. 3,300 daltons) semi-IPN. The maximum swelling was observed at pH 3.19 and the minimum swelling at pH 13.0. In addition, using infra-red analysis, the authors noted that the NH_3^+ to NH_2 transition was reversible when the hydrogels were transferred from an acidic medium to a basic medium (18). Being a relatively hydrophobic polymer, low molecular weight PPG, would have less effect on the swelling of the hydrogels as compared to high molecular weight PEO.

Chitosan, when positively charged, is known to form ionic complexes with various proteins in aqueous solution (19). It is, therefore, important to characterize the effect of pepsin and pancreatin on the swelling of chitosan and chitosan-PEO hydrogels in SGF and SIF, respectively. The results, presented in Table I, show that the swelling ratio of chitosan-PEO semi-IPN in pepsin-free SGF was 29.1 and remained at around 25.0 after the addition of pepsin. Overall, the results clearly suggest that there was no difference between the swelling ratios of the hydrogels in the absence and presence of the enzymes. Lack of enzyme effect on the swelling behavior can be attributed to the rapid influx of hydronium ions into the hydrogels even in the presence of the enzymes.

Chitosan and chitosan-PEO hydrogels were swollen sequentially in SGF for 6 h, followed by swelling in SIF for an additional 6 h. The swelling ratio of chitosan-PEO semi-IPN, as shown in Figure 4, increased in SGF with increasing duration of incubation. After 6 h, the swelling ratio of chitosan and chitosan-PEO hydrogels were 11.1 and 17.8, respectively. When the hydrogels were transferred to SIF, they continued to swell even in the high pH medium. After 2 h in SIF, the swelling ratios of chitosan and chitosan-PEO hydrogels increased to 17.9 and 30.4, respectively. It is interesting to note that the exchange of hydronium ions from the swollen hydrogel matrix is a gradual process taking up to 2 h. At later time points in SIF, the swelling ratio decreased as the hydronium ions concentration inside the matrix had decreased substantially.

Degradation of Chitosan and Chitosan-PEO Hydrogels. Recently, a study by Pantaleone and Yalpani (20) suggests that some uncommon enzymes including amylase, protease, and lipase may be effective in degrading chitosan. The effect of pepsin and pancreatin in SGF and SIF, respectively, on degradation of the hydrogels was examined by measuring the amount of D-glucosamine formed. Although glucosamine was detected in SGF and SIF, as shown in Table II, the degradation could not be accounted for by the presence of enzymes, since the same amount of

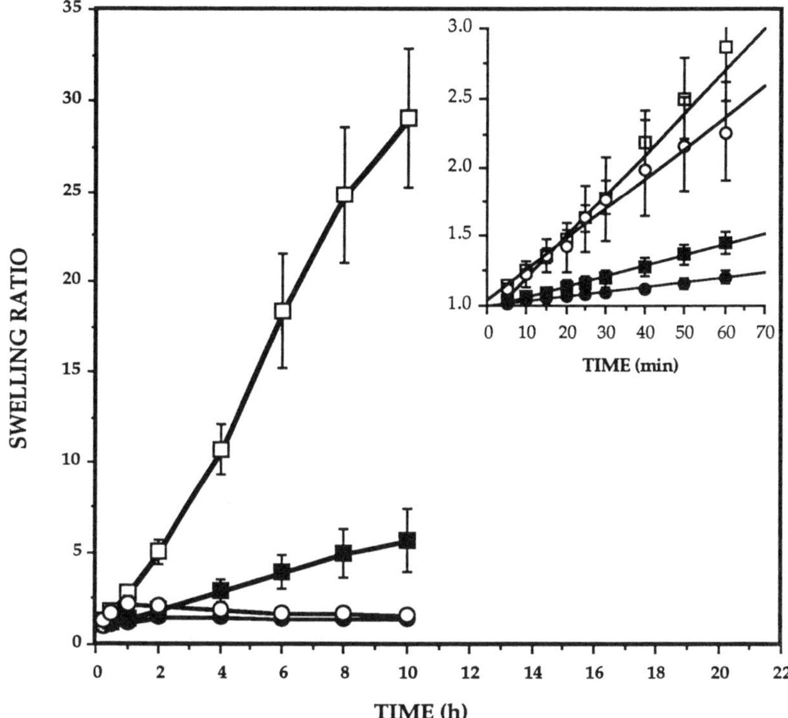

Figure 3. Swelling kinetics of chitosan and chitosan-PEO hydrogels in pepsin-free simulated gastric fluid (SGF, pH 1.2) and pancreatin-free simulated intestinal fluid (SIF, pH 7.2) at 37°C. The inset shows swelling behavior at the lower time points. The symbols represent chitosan (■) and chitosan-PEO (□) in SGF and chitosan (●) and chitosan-PEO (○) in SIF. Chitosan-PEO (weight ratio of 80:20) semi-IPN was prepared with PEO of molecular weight 1,000,000 daltons.

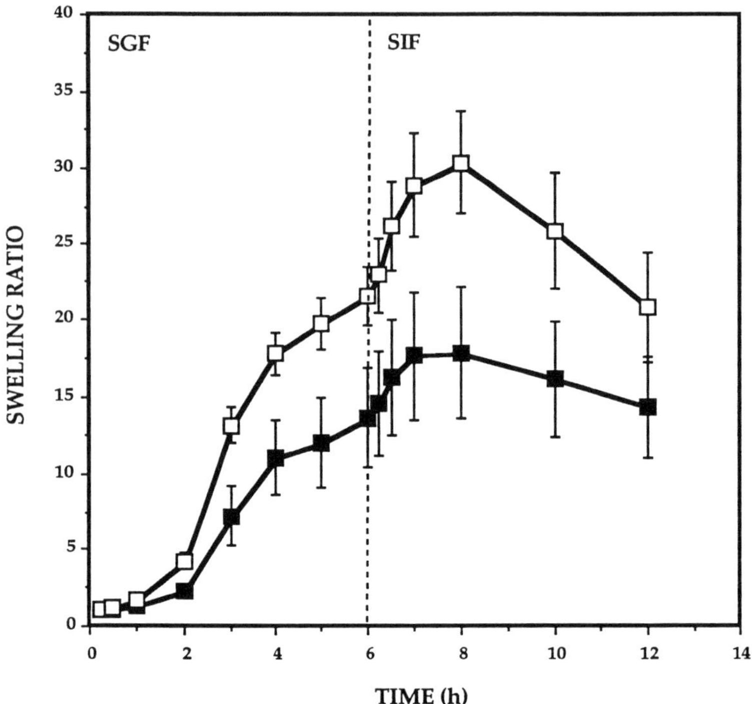

Figure 4. Sequential swelling of chitosan and chitosan-PEO hydrogels in pepsin-free simulated gastric fluid (pH 1.2) and pancreatin-free simulated intestinal fluid (pH 7.2) at 37°C. The symbols represent chitosan (■) and chitosan-PEO (□) hydrogels. Chitosan-PEO (weight ratio of 80:20) semi-IPN was prepared with PEO of molecular weight 1,000,000 daltons.

glucosamine was formed even in the absence of enzymes. For chitosan-PEO semi-IPN in SGF, for instance, 1.61 μM and 1.68 μM of glucosamine was formed in the absence and presence of pepsin, respectively. Furthermore, there was no structural alteration in the hydrogels to indicate possible degradation. Chitosan oligomers leaching out of the hydrogels could have contributed to the glucosamine formation.

TABLE I. Effect of Enzymes on the Swelling Behavior of Chitosan and Chitosan-PEO Hydrogels in SGF and SIF[a]

Hydrogel Type	Swelling Medium	Swelling Ratio
Chitosan	SGF - Without Pepsin	5.67 ± 1.72
	SGF - With Pepsin	7.82 ± 0.77
Chitosan-PEO[b]	SGF - Without Pepsin	29.08 ± 3.78
	SGF - With Pepsin	24.98 ± 3.85
Chitosan	SIF - Without Pancreatin	1.34 ± 0.03
	SIF - With Pancreatin	1.32 ± 0.02
Chitosan-PEO	SIF - Without Pancreatin	1.86 ± 0.03
	SIF - With Pancreatin	1.63 ± 0.03

[a] Chitosan and chitosan-PEO hydrogels were swollen in simulated gastric fluid (SGF) and simulated intestinal fluid (SIF) in the absence and presence of pepsin and pancreatin, respectively, for 10 h at room temperature.

[b] Chitosan-PEO semi-IPN was prepared with chitosan and PEO (Mol. wt. 1,000,000 daltons) in a weight ratio 80:20. The crosslinker concentration was 8.0 mg/ml.

TABLE II. Degradation of Chitosan and Chitosan-PEO Hydrogels in SGF and SIF[a]

Hydrogel Type	Swelling Medium	Glucosamine Concentration (μM)
Chitosan	SGF - Without Pepsin	1.28 ± 0.61
	SGF - With Pepsin	1.00 ± 0.11
Chitosan-PEO	SGF - Without Pepsin	1.61 ± 0.00
	SGF - With Pepsin	1.68 ± 0.39
Chitosan	SIF - Without Pancreatin	0.78 ± 0.11
	SIF - With Pancreatin	0.68 ± 0.22
Chitosan-PEO	SIF - Without Pancreatin	1.22 ± 0.61
	SIF - With Pancreatin	1.50 ± 1.06

[a] Chitosan and chitosan-PEO hydrogels were incubated in simulated gastric fluid (SGF) and simulated intestinal fluid (SIF) in the absence and presence of pepsin and pancreatin, respectively, for 10 h at room temperature.

Riboflavin Release from Chitosan and Chitosan-PEO Hydrogels. Figure 5 shows the riboflavin release kinetics from chitosan and chitosan-PEO hydrogel. The release exponent (n) for each case was calculated to be approximately 0.5, suggesting that drug release occurs mainly by the Fickian diffusion mechanism, regardless of the pH of the medium or the type of hydrogel (15). After 6 h, more than 50.4% of the drug was released from chitosan-PEO semi-IPN in pepsin-free SGF. In contrast, only 28.8% of the drug was released from chitosan-PEO semi-IPN in pancreatin-free SIF. Drug release from chitosan hydrogels was significantly lower than from chitosan-PEO semi-IPN. Only 30.2% and 6.65% of riboflavin was released after 6 h in SGF and SIF, respectively. Riboflavin release from chitosan and chitosan-PEO hydrogels decreased by 30% to 40% in SGF and SIF containing pepsin and pancreatin, respectively, as shown in Table III. In SGF, for instance, drug released from chitosan-PEO semi-IPN decreased from 64.2% in pepsin-free medium to 48.4% in pepsin-containing medium. Adsorbed enzymes on the surface of the hydrogel matrix could have decreased the release of riboflavin by increasing the number of diffusional layers or by blocking the pores.

TABLE III. Effect of Enzymes on the Release of Riboflavin from Chitosan and Chitosan-PEO Hydrogels in SGF and SIF[a]

Hydrogel Type	Swelling Medium	Riboflavin Released (%)
Chitosan	SGF - Without Pepsin	40.9 ± 4.73
	SGF - With Pepsin	27.2 ± 5.15
Chitosan-PEO	SGF - Without Pepsin	64.2 ± 10.5
	SGF - With Pepsin	48.4 ± 6.36
Chitosan	SIF - Without Pancreatin	16.4 ± 0.47
	SIF - With Pancreatin	9.5 ± 0.24
Chitosan-PEO	SIF - Without Pancreatin	33.5 ± 1.58
	SIF - With Pancreatin	27.2 ± 2.79

[a] Riboflavin-loaded chitosan and chitosan-PEO hydrogels were incubated in simulated gastric fluid (SGF) and simulated intestinal fluid (SIF) in the absence and presence of pepsin and pancreatin, respectively, for 10 h at room temperature.

CONCLUSIONS

Chitosan-PEO semi-IPN was developed as pH-sensitive hydrogel for swelling and drug release in the low pH environment. As the molecular weight and the amount of incorporated PEO increased, the swelling ratio of the hydrogel increased also. Chitosan-PEO semi-IPN had a ten times higher swelling ratio in SGF than in SIF. In the low pH medium, chitosan-PEO semi-IPN had swollen extensively due to the ionization of the glucosamine residues. In addition, when compared to chitosan hydrogels, chitosan-PEO semi-IPN had a significantly higher swelling ratio in SGF. More than 50% of riboflavin was released after 6 h from chitosan-PEO semi-IPN in SGF. Clearly, chitosan-PEO semi-IPN meets the requirements for pH dependent release, and could be useful for localized drug delivery in the stomach.

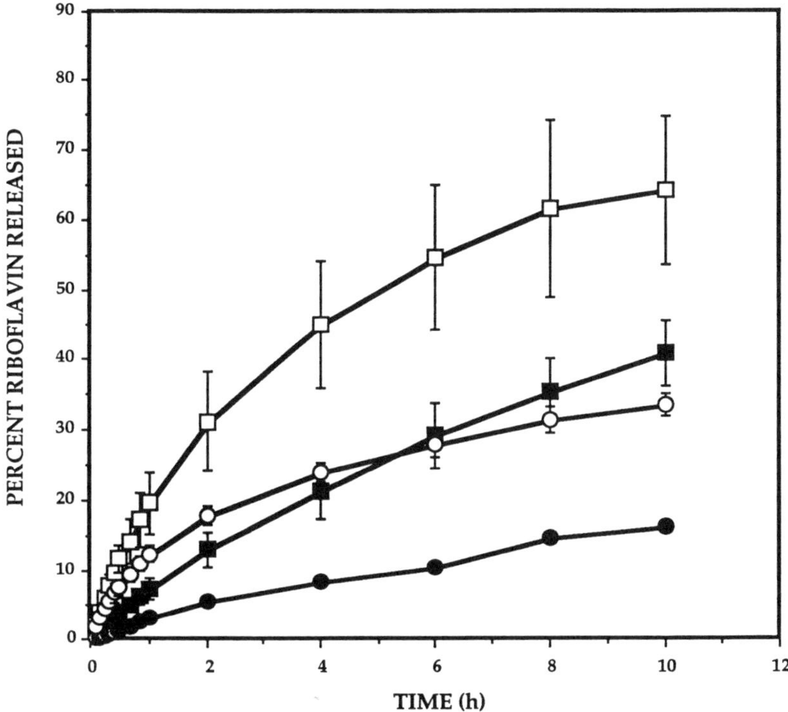

Figure 5. Riboflavin release from chitosan and chitosan-PEO hydrogels as a function of time in pepsin-free simulated gastric fluid (SGF, pH 1.2) and pancreatin-free simulated intestinal fluid (SIF, pH 7.2) at 37°C. The symbols represent chitosan (■) and chitosan-PEO (□) in SGF and chitosan (●) and chitosan-PEO (○) in SIF. Chitosan-PEO (weight ratio of 80:20) semi-IPN was prepared with PEO of molecular weight 1,000,000 daltons.

ACKNOWLEDGMENTS

This study was supported in part by the Biomedical Research Support Grant SO7-RR-05830-12. We thank Pronova Biopolymers and Union Carbide for supplying chitosan and PEO, respectively.

LITERATURE CITED

1. O. Wichterle and D. Lim. *Nature* **185:** 117 (1960).
2. N.A. Peppas. (ed). *Hydrogels in Medicine and Pharmacy: Volume I. Fundamentals.* CRC Press, Boca Raton, FL. 1986.
3. B.D. Ratner and A.S. Hoffman. In J.D. Andrade (ed.) *Hydrogels for Medical and Related Applications.* ACS Symposium Series, Volume 31. American Chemical Society, Washington, DC. 1976, pp 1.
4. D.G. Pedley, P.J. Skelly, and B.J. Tighe. *Br. Polym. J.* **12:** 99 (1980).
5. N.B. Graham and M.E. McNeill. *Biomaterials* **5:** 27 (1984).
6. T. Tanaka. In R.S. Harland and R.K. Prud'homme (eds.) *Polyelectrolyte Gels.* ACS Symposium Series, Volume 480. American Chemical Society, Washington, DC. 1992, pp 1.
7. J. Heller. *J. Controlled Release* **8:** 111 (1988).
8. H. Brønsted and J. Kopecek. In R.S. Harland and R.K. Prud'homme (eds.) *Polyelectrolyte Gels.* ACS Symposium Series, Volume 480. American Chemical Society, Washington, DC. 1992, pp 285.
9. A. Ateshkadi, N.P. Lam, and C.A. Johnson. *Clin. Pharmacy* **12:** 34 (1993).
10. S. Hirano, H. Seino, Y. Akiyama, and I. Nonaka. In C.G. Gebelein and R.L. Dunn (eds.) *Progress in Biomedical Polymers.* Plenum Press, New York, NY 1990, pp 283.
11. T. Chandy and C.P. Sharma. *Biomat., Art. Cells, Art. Org.* **18:** 1 (1990).
12. M.W. Anthonsen, K.M. Vårum, O. Smidsrød. *Carbohydrate Polym.* **22:** 193 (1993).
13. *The United States Pharmacopeia. Volume XXII.* The United States Pharmacopeial Convention, Inc. Rockville, MD. 1990 pp 1786.
14. N.W. Tietz (ed.) *Textbook of Clinical Chemistry.* W.B. Saunders Company, Philadelphia, PA. 1986 pp 793.
15. N.A. Peppas and R.W. Korsmeyer. In N.A. Peppas. (ed). *Hydrogels in Medicine and Pharmacy: Volume III. Properties and Applications.* CRC Press, Boca Raton, FL. 1986 pp 109.
16. L. Brannon-Peppas and N.A. Peppas. *J. Controlled Release* **8:** 267 (1989).
17. K. De-Yao, T. Peng, M.F.A. Goosen, J.M. Min, and Y.Y. He. *J. Appl. Polym. Sci.* **48:** 343 (1993).
18. K. De-Yao, T. Peng, H.B. Feng, Y.Y. He. *J. Polym. Sci.: Part A: Polym. Chem.* **32:** 1213 (1994).
19. Q. Li, E.T. Dunn, E.W. Grandmaison, and M.F.A. Goosen. *J. Bioact. Compat. Polym.* **7:** 370 (1992).
20. D. Pantaleone and M. Yalpani. In M. Yalpani (ed.) *Carbohydrates and Carbohydrate Polymers: Analysis, Biotechnology, Modifications, Antiviral, Biomedical, and Other Applications.* ATL Press, Inc. Mount Prospect, IL. 1993 pp 44.

RECEIVED January 3, 1996

Chapter 18

Microbes in Polymer Chemistry

V. Crescenzi and M. Dentini

Department of Chemistry, University La Sapienza, P. le A. Moro 5, 00185 Rome, Italy

A few examples of the impact that conventional and advanced microbial technologies already have and may, eventually, have on polymer chemistry are outlined. From microbial biomass in addition to enzymes (not considered in this review), a variety of polymeric materials of potential commercial importance can be obtained including carbohydrate polymers, aliphatic polyesters and polyaminoacids. In addition, a number of monomeric compounds of interest in polymer chemistry for the synthesis of industrially important macromolecules (e.g. polyamides and polyesters) can be obtained *via* fermentation of wild, mutated and/or genetically engineered microbial strains. A better exploitation of natural, renewable resources, e.g. microbial biomass, for the production of a larger number of bifunctional species amenable to polymerization will also help increase the ecological friendliness of raw materials industries.

Microorganisms have long been utilized to produce, in some cases at the industrial level, a variety of organic compounds, in particular high added-value biologically active species. This has promoted the development of conventional and advanced biotechnologies.

In addition, the many possibilities offered to the "polymer community" by a wise exploitation of microbial biomass is not yet fully recognized. Indeed, selected microbial cultures are, or may become, valuable sources of important classes of polymers and of a number of monomers or reactive intermediates.

Microbial polymers include, as is well known, enzymes, polysaccharides, polyesters, and polyaminoacids: some of these natural macromolecules are presently considered as irreplaceable commodities or specialties for a number of applications.

Monomers most conveniently derived from microbial fermentation include L-lactic acid and acrylonitrile, but the list is likely to widen

0097–6156/96/0627–0221$15.00/0

considerably in the near future: this should be significant for the ecologically oriented chemical industry.
There follows a few specific examples of the impact of microbial technologies on polymer chemistry. For brevity, the presentation is limited to two cases (with which the authors have direct working experience), namely : I). - industrial microbial polysaccharides ; II). - exocellular polyaminoacids.
In addition, the case of monomers from microbial biomass - a topic not yet reviewed, to the authors' knowledge - is presented in a schematic way including some actual, stimulating examples.

I) - **Microbial Polysaccharides.**

- Some comments on "where we are".

A large number of exocellular and capsular polysaccharides are known and new examples are being discovered and described in the scientific literature at a steadily increasing pace.

Limiting attention to examples of actual interest for the polymer industry, a list of polysaccharides already commercialized and obtainable exclusively - with the exception of hyaluronic acid and cellulose - from microbial fermentations is given in Table I. Most prominent application of each species is briefly indicated in the foot-note to Table I. Detailed discussions on this topic can be found in recent reviews (1-3): suffice here to mention that for such end uses, the polysaccharides listed in Table I are considered rather unique and, often, irreplaceable. They include practically all desirable properties for technological biopolymers.

A number of studies on the solution and gelling properties of these microbial polysaccharides and their derivatives have been carried out in the last two decades in our laboratory. In particular, polysaccharides studied include: xanthan (4-6), the "gellan family" polysaccharides (7-10), succinoglycan (11-12), and scleroglucan (13-14).

One important feature shared by these biopolymers, which attracts particular attention, is their ability to adopt ordered stable conformations even in very dilute aqueous solutions. These conformations, with the notable exception of succinoglycan, are double (the "gellan family" polysaccharides) or triple stranded (scleroglucan).

The main aim of our work, like that of many other researchers, has been to contribute to the understanding of how chemical structure determines the adoption of ordered conformation, and how the latter controls the formation of three dimensional networks (physical gels).

Table I. Industrial Microbial Polysaccharides

biopolymer	main microbial source	typical applications
Polyelectrolytes		
hyaluronan	*Streptococcus* spp.	a
xanthan	*Xanthomonas campestris*	c, d, e
gellan	*Aureomonas elodea*	d, f
welan	*Alcaligenes* spp.	i, g
succinoglycan	*Pseudomonas* spp.	i, g
Non-ionic		
cellulose	*Acetobacter xylinum*	a
curdlan	*Agrobacter.* spp.	f, d
pullulan	*Aureobasidium pullulans*	h
dextran	*Leuconostoc* spp.	a, b
scleroglucan	*Sclerotium* spp.	i, g

a = pharmaceuticals ; b = chemical gels ; c = oil recovery ; d = food additives ; e = textiles ; f = physical gels ; g = drilling fluids; h = films ; i = rheology modifier.

Clearly, a detailed knowledge of the characteristic conformational ordering of stereoregular microbial polysaccharide chains is fundamental to obtaining a deeper insight of the molecular basis of their novel macroscopic solution properties, especially rheological and gelation. This, in turn, permits a better industrial exploitation of these biopolymers.

In addition, in view of the biocompatibility of these polysaccharides we have also prepared novel derivatives with the purpose of 1) studying the influence of chemical functionalization on the stability of ordered conformations and hence gelling ability (e.g. benzyl esters of gellan of different degrees of esterification (15-16); 2) preparing and characterizing novel biomaterials - including soluble derivatives and cross-linked hydrogels - potentially suitable for sustained drug release formulations (17). In the latter context, we recently succeeded in preparing a variety of chemical hydrogels built up by gellan chains engaged at random through their carboxylate functions in ester formation with simple difunctional reagents (18).

The materials obtained exhibit interesting swelling properties when in contact with aqueous salt media, as shown in Fig. 1. They appear to have potential as controlled release agents for selected cationic drugs, as shown by the preliminary, in vitro results outlined in Fig. 2.

While continued basic research on this topic has obvious relevance, it could be claimed that applied research is no longer necessary. We consider that this is not so provided that some practical guidelines are observed.

In this context, we wish to submit the following schematic considerations which, for a comprehensive overview of the topic, should be considered in conjunction with other reviews (2,3).

a) - In some cases, the performance/cost ratio of microbial polysaccharides appears questionable for certain end uses. An approach to reducing polysaccharides production costs and, at the same time, alleviating specific pollution problems consists of the "target oriented" search, identification, and use of strains - good producers of polysaccharides - living on waste materials.

In this regard at least two recent, interesting examples can be mentioned dealing with the production of polysaccharides - displaying good rheological properties - by different bacterial strains fed either with whey permeate (19) or with maple sap (20).

If such bacterial strains can also be proved "beyond all doubts" from the biocompatibility standpoint - as, for instance, is obviously the case for *Lactobacteriaceae* (21) - the polysaccharides produced should

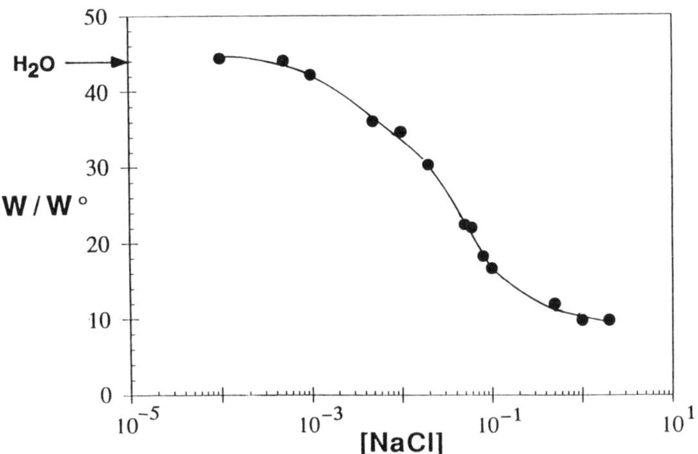

Figure 1. Swelling equilibria of a gellan cross-linked hydrogel at different ionic strengths (NaCl, 25°C). The sample was prepared following the procedure given elsewhere (6) for the synthesis of simple, linear alkyl and arylalkyl esters of gellan but using α,ω-dibromo octane as reticulating agent. Estimated degree of cross-linking: 0.5.

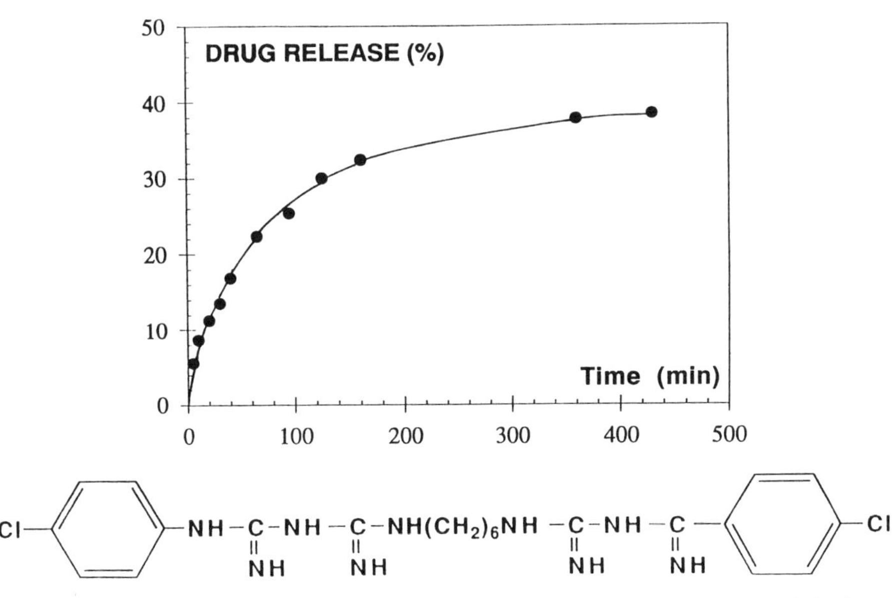

Figure 2. Time course of the release of Chlorohexidine from the cross-linked gellan hydrogel of Fig. 1.
(t=0 conditions: 4mg of drug/70mg of hydrogel; 0.1 M NaCl, 25°C).

find easier approval for food uses by relevant authorities (thus cutting R&D costs and gaining in commercial value).

b)- A further development of what today may be called "microbial polysaccharide science and technology" - an interdisciplinary field encompassing expertises such as microorganism genetics and physiology, carbohydrate chemistry and biochemistry, and polymer chemistry - will certainly result from a more extensive use of advanced biotechnologies.

This may allow the enzymatic synthesis in cell free systems of valuable biopolymers (e.g. very pure, very high molecular weight hyaluronic acid (22)) and the production of tailor-made microbial polysaccharides for specific biomedical or food applications where parameters such as production costs/yields may become only marginally relevant.

c) - The understanding of the fine-tuned relationships between structure and properties - in the solid state, in solution, and in the gel state - of the often structurally complex microbial polysaccharides is still being developed.

This understanding is essential for a better exploitation of polysaccharides already known, for the selection among polysaccharides being discovered of those better suited for industrial R&D, and for an intelligent guide to the synthesis of novel polysaccharide-based derivatives.

The latter include chemically functionalized polymeric materials with potential in biomedicine (e.g. as pro-drugs) and highly hydrophylic chemical gels target oriented primarely to controlled drug-delivery formulations.

II) **Microbial Polyaminoacids.**

- Biopolymers "in search of applications" (23).

In the early twenties, a milestone in understanding the structure and the biological significance of carbohydrate polymers from pathogenic bacteria was represented by the discovery that serological, type specific antigens of bacteria of the genus *Streptococcus* are polysaccharides (24). This opened the way to important progress in the use of such polysaccharides as immunogens, a subject of actual studies on the formulation of human vaccines.

In 1925, Lemoigne, a microbiologist at the Pasteur Institute in Paris, first described poly(3-hydroxybutyrate) as an endocellular polyester obtainable from bacterial cultures (25).

These were indeed "roaring" years for microbial polymers. In fact, between 1921 and 1937 high molecular weight polyglutamates of microbial origin were also identified, isolated and partially characterized (26).

Ironically, a fate common to the three types of biopolymers mentioned above - with the possible exception of polysaccharides - is that, after their discovery, many years had to elapse before they fully deserved in polymer chemistry. Indeed, in the case of polyglutamates (more in general of microbial polyaminoacids) one can still speak of "biopolymers in search of application" (23).

At present, polyglutamates (which consist of glutamic acid units linked between the α-amino and the γ-carboxylic acid functionalities and hence are named poly-γ-glutamates (γ-PGA)) of different stereoregularity (i.e. D- and L-glutamate content) and molecular weight (normally exceeding 10^5) are produced exocellularly in very good yields from species of the genus *Bacillus*.

Recently, a revival in interest in these biopolymers has lead to a number of publications dealing in particular with the biosynthesis, the resulting chains stereochemistry, and with the production of derivatives with potential as biomaterials (23, 27).

In a companion paper (28), we present original data on the physico-chemical characterization of γ-D-PGA (approximately 96 % D) in dilute aqueous solution. In our laboratory, work is in progress also on the synthesis of novel γ-PGA based hydrogels.

In the context of this review, it should be underlined that γ-PGAs can be easily obtained by fermentation of selected *Bacillus licheniformis* strains with yields of tens of g/L. They are water soluble polyelectrolytes capable of interacting specifically with divalent counterions (without precipitation) and tervalent counterions (with gelation/precipitation). They exhibit novel conformation-dependent solution properties (28), and lend themselves to derivatization leading to organic solvent soluble alkyl and arylalkyl polyesters (23, 27) for example, and to cross-linked, highly water swellable hydrogels (18).

Finally, it is worth mentioning that from the culture broths of *Streptomyces albulus*, low molecular weight ε-poly-L-lysine can be extracted (29). This unusual polyaminoacid, whose properties have been studied so far in a preliminary fashion, but sufficient to characterize ε-poly-L-lysine chains as conformationally adaptable and as efficient sequestrant of dye molecules and divalent ions (30), deserves more attention from polymer chemists.

III) Examples of Monomers from Microbial Sources.

- An ecological future for the polymer industry.

An incomplete list of monomeric compounds of importance for the synthesis of a variety of macromolecules and which are produced, or may be produced in a probably near future, using microbial biomass is given in Table II. Some of these monomeric compounds are now discussed.

In polymer chemistry, **lactic acid** is an important feedstock for the production of biodegradable lactide polymers which are especially attractive for biomedical applications. A number of publications have recently appeared on the fermentative production of L-lactic acid in which attention is focused on yields and product recovery. For example, an innovative system for lactic acid recovery by ion-exchange resins has been proposed (31): the final yield obtained using *Lactobacillus casei* subsp. *casei* (glucose as main nutrient) is quite high and the resulting monomer (exclusively the L form) of purity adequate (> 99 %) for its use in the synthesis of polylactides.

Acrylamide can be produced industrially in a convenient fashion by selective hydration of acrylonitrile using as catalyst *Rhodococcus rhodocrous* J1 nitrile hydratase. According to a recent report (19), after fermentation for 72 h at around 20 °C with a cell yield of 28 g/L, the acrylamide productivity is higher than 7000 g/g cells. The final concentration of acrylamide is 40 % maintaining the acrylonitrile concentration at 6 %. Nitto Chemical Industry Co. of Japan claims to produce acrylamide at kiloton scale using *R. rhodochrous* J1 fermentation (32).

Concerning **1,3-propanediol**, a fermentative production from glycerol using *Clostridium butyricum* , up to a scale of 2 m^3, has been reported (33). Selected strains of *C. butyricum* are able to convert up to 110 g/L of glycerol to 56 g/L of 1,3-propanediol in 29 h. These results are quite interesting in the context of the present review for the following two main reasons.

Firstly, glycerol is available in large amounts at low price mainly as a by-product of the detergent industry. Secondly, 1,3-propanediol is a monomer suitable for the synthesis of a variety of polyesters, in particular for light-stable and/or biocompatible plastics.

In addition, it should be pointed out that conventionally produced 1,3-propanediol has not gained economic and technological importance mainly because its synthesis is carried out by hydrolysis of acrolein and subsequent catalytic hydrogenation with relatively low yields (about 43 % (21)).

2,3-Butanediol is an interesting chemical which can find a

Table II. Industrially relevant monomers which are produced or may be produced from microbial sources

monomer (substrate)	**microorg.** (recent example)	**stage**
L(+)-lactic acid (carbohydr.)	*Lactobacillus casei.*	industrial (70% of total)
acrylamide (acrylonitrile)	*Rhodococcus rhodocrous* J1	industrial
1,3-propanediol (glycerol)	*Clostridium butyricum*	pilot
2,3-butanediol (carbohydr.)	*Enterobacter aerogenes*	research
fumaric acid (carbohydr.)	*Rhizopus arrhizus*	research
2,3 and 3,4-cis dihydrodiols (benzene & deriv.)	*Pseudomonas spp*	research
adipic acid (glucose)	engineered *E. coli*	research

number of applications, in addition to polymer synthesis. Recent reports suggest that for this diol also microbial production (e.g. using *Enterobacter aerogenes* strains) may soon become competitive with chemical synthesis (34).

The development of a really competitive microbial production of **fumaric acid** has proved more difficult. Fumaric acid is conventionally prepared by acid isomerization of maleic acid, a bulk chemical commodity. Nevertheless, it is of interest at least to mention that this dicarboxylic monomer can be obtained from carefully controlled fermentation of *Rhizopus arrhizus* at a final concentration of 47 g/L (5 days, 47 % yield (35)).

Passing to the "**cis-diol**" type compounds, it should be recalled that the biosynthesis of 5,6-dihydroxycyclohexa-1,3-diene by means of the bacterial oxidation of benzene using genetically manipulated *Pseudomonas putida* strains has been reported in 1970 (36). Simple ester derivatives of this compound are readily polymerized, using radical initiators, to give high molecular weight polymers which are soluble in common organic solvents. Moreover, such polymers undergo smooth conversion, on heating at 140-240 °C, into polyphenylene a polymer of interest for its thermal stability and electrical conductivity (when doped), but obtainable only in oligomeric form *via* conventional chemical routes.

P. putida strains are an interesting example of the intelligent search for microorganisms living on waste materials (see also Section I). In this case, environmentally dangerous chemical wastes are used, which can produce monomeric compounds not otherwise obtainable in comparable yields and steric purity.

More recently, it has been reported (37) that it is now possible to obtain 2,3- and 3,4-cis-dihydrodiols of either enantiomeric form in a predictable manner by the action of *P. putida* and *E. coli* engineered strains on carefully selected substrates (i.e. monosubstituted benzenes) and hydrogenolysis of ensuing metabolites. Such dihydrodiols should prove useful chiral syntons in organic synthesis as well as in polymer synthesis.

Of similar keen interest are recent reports (38) indicating that lactones and lactames related to carbohydrates can be obtained by the functionalization of 1-chloro-2,3-dihydroxycyclohexa-4,6-diene biocatalytically generated by fermentation of cholorobenzene with *P. putida* 39D.

This section concludes with what we consider a real "up-beat note" by making reference to a recent paper by Draths and Frost (39). These authors have addressed with success on a laboratory scale the problem of the biocatalytic conversion of D-glucose into cis,cismuconic acid (using heavily genetically engineered *E. coli* strains) and subsequent catalytic hydrogenation to afford **adipic acid**.

The possibility of obtaining in the future adipic acid, primarely used in production of nylon-6,6 and whose annual global demand is of the order of billion Kg, following the route schematically outlined above (thus, eventually, reducing at least partially the use of conventional, highly polluting chemical processes) is very exciting and probably one of the most convincing examples of the impact microbial technologies may have on polymer chemistry and technology.

Acknowledgements.

The A.A. wish to acknowledge the financial support of the Italian National Research Council, Progetto Finalizzato Chimica II, Rome.

References

1. Sutherland, I. W.; *Biotechnology of Microbial Polysaccharides*, Cambridge Studies on Biotecnology - 9, Cambridge University Press, **1990**.
2. Crescenzi, V.;*Trends in Polymer Science* **1994**, *2*, 104.
3. Crescenzi, V.; *Biotechnol. Progress*, in press.
4. Dentini, M.; Crescenzi, V.; Blasi, D.; *Int. J. Biol. Macromol.* **1984**,*6 (2)* , 93.
5. Coviello, T.; Kajiwara, K.; Burchard,W.; Dentini, M.; Crescenzi, V.; *Macromolecules* **1986,** *19*, 2826.
6. Coviello, T.; Burchard, W.; Dentini, M.; Crescenzi, V.; *Macromolecules* **1987**, *20*, 1102 .
7. Crescenzi,V.; Dentini, M.; Dea, I.C.M.; *Carbohydr. Res.* **1987**,160, 283 .
8. M. Dentini; T. Coviello; W. Burchard; V. Crescenzi; Macromolecules **1988**, *21*, 3312.
9. Crescenzi,V.; Dentini, M.; Coviello, T.; in: *Novel Biodegradable Microbial Polymers* . E. A. Dawes Ed., NATO ASI Series E: Applied Sciences, Vol. 186, Kluwer Academic Publs., Dordrecht, Netherlands, **1990**, pp 277-284 .
10. Crescenzi, V.; Dentini, M.; Coviello, T.; in: *Frontiers in Carbohydrate Research 2.*, R.Chandrasekaram, Ed., Elsevier, New York, **1991**, pp 100-114 .
11. Dentini, M.; Crescenzi, V.; Fidanza, M.; Coviello, T.; *Macromolecules* , **1989**, *22*, 954.
12. Fidanza, M.; Crescenzi,V.; Dentini, M.; Del Vecchio, P.; *Int. J. Biol. Macromol.* **1989**, *11*, 372.
13. Crescenzi, V.; Gamini, A.; Rizzo, R.; Meille, S.V.; *Carbohydrate Polymers* **1988**, *9*, 169.
14. Coviello,T.; Dentini, M.; Crescenzi, V.; Vincenti, A.; *Carbohydrate Polymers* **1995**, *26*, 5..
15. Crescenzi, V.; Dentini, M.; Segatori, M.; Tiblandi, C.; Callegaro, L.; Benedetti, L.; *Carbohydrate Research* **1992**, *231*, 73.
16. Crescenzi, V.; Dentini, M.; Maschio, S.; Segatori, M.; *Makromol. Chem., Makromol. Symp.*,**1993**, *76*, 95 .

17. Sanzgiri, Y.D.; Maschio, S.; Crescenzi, V.; Callegaro, L.; Topp, E.M.; Stella, V.J.; *J. Controlled Release* **1993**, *26*, 195.
18. Crescenzi, V.; Dentini, M.; in preparation.
19. Flatt, J.H.; Hardin, S.H.; Gonzalez, J. M.; Dogger, D. E.; Lightfoot, E. N.; Cameron, D. C.; *Biotechnol. Progress* **1992**, *8*, 327.
20. Morin, A.; Moresoli, C.; Rodrigue, N.; Dumont, J.; Racine, M.; Poitras E.; *Enzyme Microbiol. Technol.* **1993**, *15*, 500.
21. Gruter, M.; Leeflang, B. R.; Kuiper, J.; Kamerling J. P.; Vliegenthart, F. G.; *Carbohydr. Res.* **1993**, *239*, 209.
22. Dougherty, B.A.; van de Rijn, I.; *J. Biol. Chem.* **1993**, *268*, 7118.
23. Gross, R. A.; Birrer, G. A.; Cromwick, A.-M.; Giannos, S. A.; McCarthy, S. P.; *Biotechnological Polymers: Medical, Pharmaceutical, and Industrial Applications*, Gebelein, C. G., ed., Technomic Publ. Co., **1993**, Part III, pp 200-214.
24. Heidelberger, M.; *J. Am. Chem. Soc.* **1993**, *115*, 2558.
25. Lemoigne, M.; *Ann. Inst. Pasteur (Paris)*, **1925**, *39*, 144.
26. references in: Nitecki, D. E.; Goodman, J. W.; *Chemistry and Biochemistry of Aminoacids, Peptides and Proteins*, Weinstein, B., ed., Marcel Dekker, **1971**, Vol. I, pp 87-126.
27. - a) - Giannos, S. A.; Shah, D.; Gross, R. A.; Kaplan, D. L.; Mayer, J. M.; in: *Novel Biodegradable Microbial Polymers*., Dawes, E. A., ed., Series E, Applied Sciences, Vol. 186, Kluwer Academic Publ.,**1990**, pp 457-460.
 - b) - Borbély, M.; Nagasaki, Y.; Borbély, J.; Fan, K.; Bhogie, A.; Sevolan, M.; *Polymer Bulletin*, **1994**, 32, 127.
28. Crescenzi, V.; D'Alagni, M.; Dentini, M.; Segre, A.-L.; Ferretti, J. A.; *Biorelated Polymers*; Am. Chem. Soc. Symposium Series, Am. Chem. Soc., Washington, D.C. in press.
29. Shima, S.; Fukuhara, Y.; Sakai, H.; *Agric. Biol. Chem.* **1982**, *46*, 1917.
30. Takagishi, T.; Kuroki, N.; Shima, S.; Sakai, H.; Yamamoto, H.; Nakazawa, A.; *J. Polymer Sci., Polymer Letters Ed.* **1986**, *23*, 1917.
31. Vaccari, G.; Gonzalez-Vara y R., A.; Campi, A. L.; Dossi, E.; Brigidi, P.; Matteuzzi, D.; *Appl. Microbiol. Biotechnol.* **1993**, *40*, 23.
32. Nagasawa, T. Shimizu, H., Yamada, H., *Appl. Microbiol. Biotechnol.*, **1993**, *40*, 189.
33. Gunzel, B.; Yonsel, S.; Deckwer, W.-D.; *Appl. Microbiol. Biotechnol.*. **1991**, *36*, 289.
34. Zeng, A.-P; Biebl, H.; Deckwer, W.-D.; *Appl. Microbiol. Biotechnol.* **1990**, *33*, 264.
35. Moresi, M.; *J. Chem. Technol. Biotechnol.* **1992**, *54*, 283.
36. - a) - Gibson, D. T.; Cardini, G. E.; Maseles, F. C.; Kalio, R. E.; *Biochemistry* **1970**, *9*, 1631.
 - b) - Ballard, D. G. H.; Courtis, A.; Shirley, I. M.; Taylor, S. C.; *J. Chem. Soc., Chem. Comm.* **1983**, 954.
37. Boyd, D. R.; Sharma, N. D.; Barr, S. A.; Dalton, H.; Chima, J.; Whited, G.; Semayer, R.; *J. Am. Chem. Soc.* **1994**, *116*, 1147.
38. Hudlicki, T.; Rouden, J.; Luna, H.; Allen, S.; *J. Am. Chem. Soc.* **1994**, *116*, 5099.
39. Draths, K. M.; Frost, J. W.; *J. Am. Chem. Soc.* **1994**, *116*, 399.

RECEIVED January 3, 1996

Chapter 19

Aqueous Solution Properties of Bacterial Poly-γ-D-glutamate

V. Crescenzi[1], M. D'Alagni[2], M. Dentini[1], and B. Mattei[1]

[1]Department of Chemistry, University La Sapienza, P. le A. Moro 5, 00185 Rome, Italy

[2]Centro di Studio per la Chimica dei Recettori e delle Molecole Biologicamente Attive, CNR, 00168 Rome, Italy

A preliminary physico-chemical characterization of a bacterial poly-γ-glutamate sample (96% D-glutamic acid content), γ-D-PGA, in dilute aqueous solutions has been carried out by means of potentiometric, viscosimetric, infrared and chiroptical spectroscopic experiments.
The biopolymer exhibits properties strikingly dependent on a number of parameters, mainly: polymer concentration, pH, ionic strength, and nature of added salt. In dilute solutions (polymer concentration around 0.1% w/V) and for pH > 7, γ-D-PGA chains assume elongated, stiff conformations while upon protonation (pH < 3) globular states would prevail. Addition of divalent counterions (Ca(II)) also leads to compact γ-D-PGA conformations.

Poly-γ-glutamates (γ-PGA) of different stereoregularity and molecular weight can be produced in very good yields mainly from species of the genus *Bacillus* (1). These natural polyamides, which consist of glutamic acid units linked between the γ-carboxylic and the α-amino functionalities (Fig.1), have been known since the pioneering fundamental studies on macromolecules of natural origin, particularly those from microbial sources, carried out in the early twenties (2). Recently, a revival in interest in these peculiar biopolymers has resulted in a number of publications dealing with their biosynthesis, the resulting stereochemistry of the chains, as well as with the preparation of derivatives with potential as novel biomaterials (3-4). However, relatively little unambiguous data is available on the conformation dependent solution properties of these poly-γ-glutamates, for which the

0097–6156/96/0627–0233$15.00/0

relationships between structure and physico-chemical properties are still a matter of controversy (5-6). Thus, the potentially large number of applications that these water soluble, biocompatible polymeric materials may exhibit remain essentially unexploited. In our laboratories a detailed physico-chemical characterization of series of poly-γ-glutamate samples all produced by *Bacillus licheniformis* strains but differing in D/L glutamic acid content is in progress. Results obtained so far on the properties of γ-D-PGA (96% D-glutamic acid content, from *B. licheniformis* ATCC 9945-a) in dilute aqueous solution are herein reported and discussed.

Experimental

Polymer samples have been obtained (*Bacillus Licheniformis* ATCC 9945-a) and purified following procedures reported in the literature (1). The sample used in the present investigation (sodium salt) had a D-glutamic acid content of 96% .

Optical activity measurements were performed with a Perkin-Elmer 241 polarimeter using a 10 cm path-length, the temperature was controlled by means of a Lauda circulating-water bath .

Viscosity measurements for the γ-D-PGA samples were performed at 25°C using a Shott-Geraete automatic viscosimeter equipped with a water thermostat. A range of ionic strengths (or pH) were investigated; these strengths (or pH) were controlled by varying the level of NaCl (or $HClO_4$) added .

Circular dichroism (CD) spectra were recorded at 25°C using a Jasco J500-A CD apparatus (calibrated with androsteron) and quartz cells of 1.0, 0.5, and 0.1 cm optical paths. During the experiments the apparatus was fluxed with high purity nitrogen.

Infrared spectra of γ-D-PGA solutions in D_2O were recorded at room temperature using a BIO-RAD, FTS-40 A spectrometer and BaF_2 cells of 0.2 mm optical path.

The calcium salt form of γ-D-PGA was prepared by addition of an excess of $Ca(ClO_4)_2$ to a solution of the sodium salt form of the biopolymer, followed by extensive dialysis against distilled water and final freeze-drying of the product.

Results and Discussion

Data to be illustrated in what follows have been collected using γ-D-PGA concentrations less than 0.3% w/V. At higher polymer concentrations the solutions in water are exceedingly viscous and, in addition, weak gel formation and/or precipitation may be observed particularly at low pH values.

The physico-chemical properties of γ-D-PGA under acidic and basic conditions differ significantly. Data pertaining to Na^+-γ-D-PGA in water at neutral or basic pH are considered first. Potentiometric measurements of the "activity coefficient" , γ^+ , of the sodium counterions were carried out at 25°C as a function of polyelectrolyte concentration (Fig. 2). Variation of the latter results in little change in the value of γ^+ (from about 0.42 to 0.49). Such behavior is typically found for relatively highly charged polyelectrolytes. The observed values of γ^+ are close to those predicted from Manning's equation (7):

$$\ln \gamma^+ = - 0.5 - \ln x$$

in which the linear charge density parameter, x , is equal to 0.716/b where b (nm) is the distance projected on the chain axis between neighboring fixed charges (the carboxylate groups) . For a fully extended γ-PGA chain, b is approximately 0.6 nm, and thus x = 1.19 and $\gamma^+ = 0.5$. These results suggest that the γ-PGA chains <u>in water</u> assume a rather expanded conformation .

Addition of NaCl to aqueous Na^+-γ-PGA does quite naturally promote a reduction in average chain dimensions, as shown in a concise manner by the set of intrinsic viscosity data reported in Fig. 3 as a function of the inverse square root of the ionic strength I. These data exibit a regular trend and permit estimation of the parameter B, defined as (8-9) :

$$B = S / ([h]_{0.1})^{1.3}$$

(The slope S of the linear plot of Fig. 3 is normalized by the intrinsic viscosity measured for I=0.1). The results lead to a value of B=0.19.

Comparison of various B values for different natural and synthetic polyelectrolytes, from the flexible polyacrylate to very rigid DNA and Xanthan (11), suggests that the γ-D-PGA chains might be considered of "intermediate stiffness". Similarly, the change in Na^+-γ-D-PGA optical activity with increasing ionic strength (NaCl, 25°C, Fig. 4) although quite large shows a regular trend and would reflect the gradually more compact conformation assumed by γ-D-PGA chains in response to the screening effect exerted by the additional Na^+ ions (NaCl) on intrachain electrostatic repulsions .

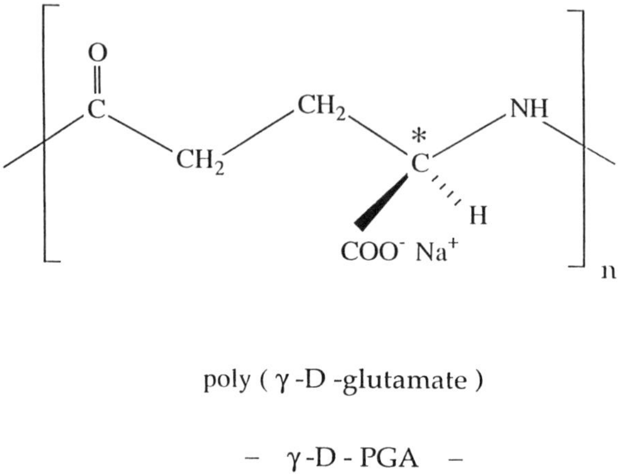

poly (γ -D -glutamate)

– γ-D - PGA –

Figure 1. Repeating unit of poly-γ-glutamate (γ-D-PGA).

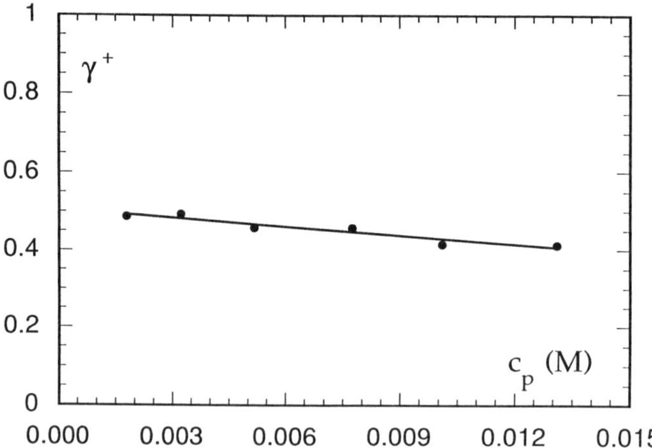

Figure 2. Potentiometric activity coefficient, γ^+, of the sodium counterions at 25°C as a function of polyelectrolytes (poly-γ-D-glutamate) concentration. Polymer concentration, c_p, in moles repeat unit/L .

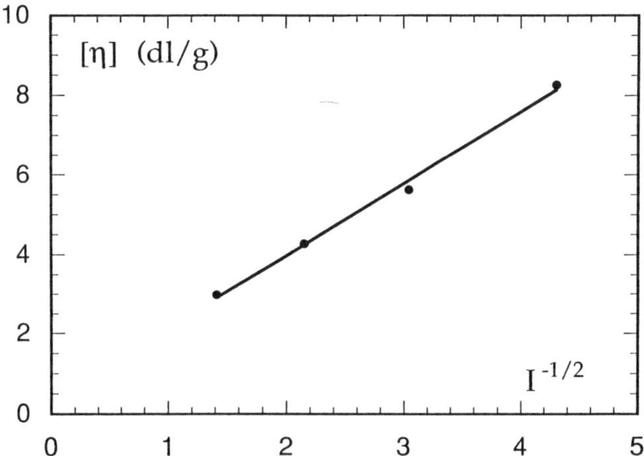

Figure 3. Dependence of poly-γ-D-glutamate intrinsic viscosity (25°C) on ionic strength (NaCl, moles/L) .

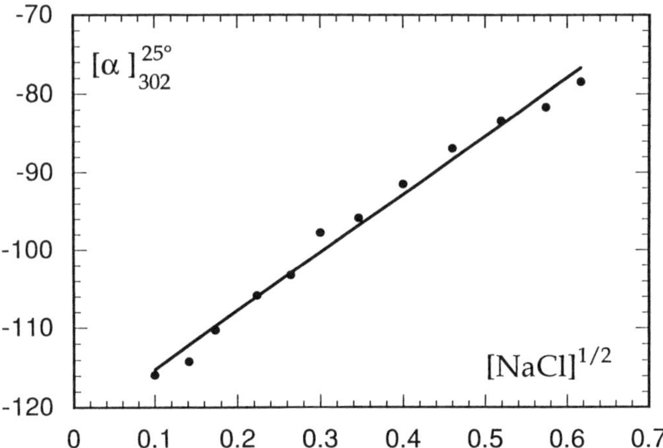

Figure 4. Dependence of poly-γ-D-glutamate optical activity on ionic strength (NaCl, moles/L). Polymer concentration 0.1% w/v (6.6 mmoles/L).

The behaviour of aqueous γ-D-PGA under acidic conditions suggests significantly different conformational behaviour from that observed for the polypeptide under basic conditions. The circular dichroism (CD) spectra of γ-D-PGA at two limiting pH values in water are reported in Fig. 5,a) while in Fig. 5,b) the ellipticity values recorded at 207 and 215 nm (from a number of additional spectra) are plotted as a function of pH. The change in chiroptical properties of γ-D-PGA with protonation is marked and features a trend which suggests a conformational change in the polypeptide chains with a midpoint around pH 5.

Interestingly, upon protonation the viscosity of γ-D-PGA aqueous solutions begins to drop dramatically around pH 5 (Fig. 6). A nearly obvious interpretation of such a marked drop is that fully protonated g-PGA chains assume a globular structure which is quite compact. At high pH where the side chains are negatively charged, the polyions would assume a more extended structure.

Very compact conformations would be assumed by γ-D-PGA chains also in the presence of added Ca(II) salts (neutral pH) as clearly demostrated by the viscosity data reported in Fig. 7.

In fact from Fig. 7 it is seen that the relative viscosity (25°C) of a 0.08 % (w/V) solution of γ - D-PGA in water is pratically reduced to unity at a $Ca(ClO_4)_2$ concentration as low as 30 mM.

In Fig. 8 the CD spectrum is reported of the Ca(II) salt form of γ-D-PGA (neutral pH) is reported: the spectral features are quite close to those exibited by the sodium salt of γ-D-PGA at high pH.

In view of the different conformational states populated by γ-D-PGA chains in the two salt forms and pH conditions, a possible deduction is that the CD characteristics of γ-D-PGA are mainly dependent on the ionization state of the carboxylate chromophores.

Similar considerations might apply also to IR spectral data given in Fig. 9 showing a marked shift in the characteristic amide band frequencies with changing pH. In fact, also in this case the spectra of the sodium salt form and of the calcium salt form of γ-D-PGA result almost superimposable.

In the context, it is interesting to mention that production of γ-D-PGA by fermentation is normally carried out at relatively high ionic strengths (about 0.15 M) and with added divalent cations concentration (Ca(II) and Mn(II) of the order of 20mM. In these conditions, according to our data, the viscosity of the fermentation broths should be quite low: this is a factor that contributes to making it possible to reach γ-D-PGA yields as high as 50-60 g/L, a production practically

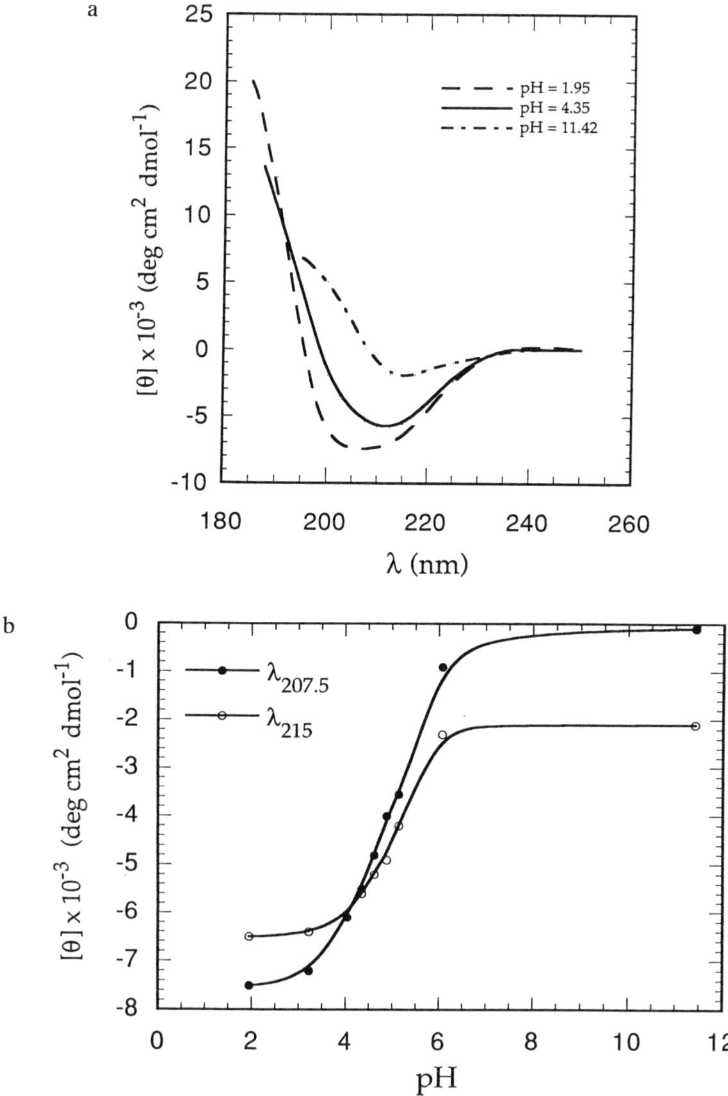

Figure 5. a) Representative circular dichroism spectra of aqueous poly-γ-D-glutamate at different pH values. b) pH dependence of aqueous poly-γ-D-glutamate ellipticity values measurements at 207.5 and at 215 nm. Polymer concentration 0.87 mmoles/L.

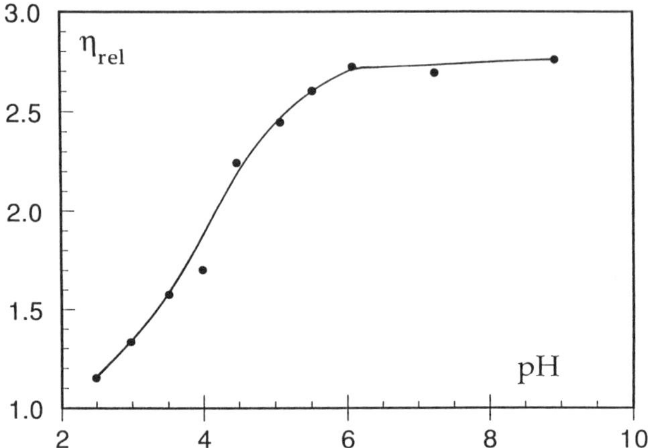

Figure 6. Relative viscosity of poly-γ-D-glutamate in NaCl 0.01 aqueous solutions as a function pH (25°C). Polymer concentration 0.1% (w/V) (6.6 mmoles/L).

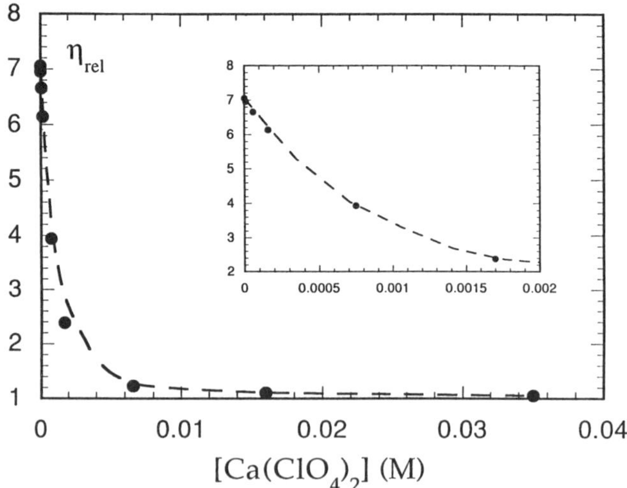

Figure 7. Relative viscosity of Na^+-γ-D-PGA aqueous solutions at different $Ca(ClO_4)_2$ concentration (25°C). Polymer concentration 0.08% (w/V) (6.2 mmoles/L). In the insert the magnified first part of the curve.

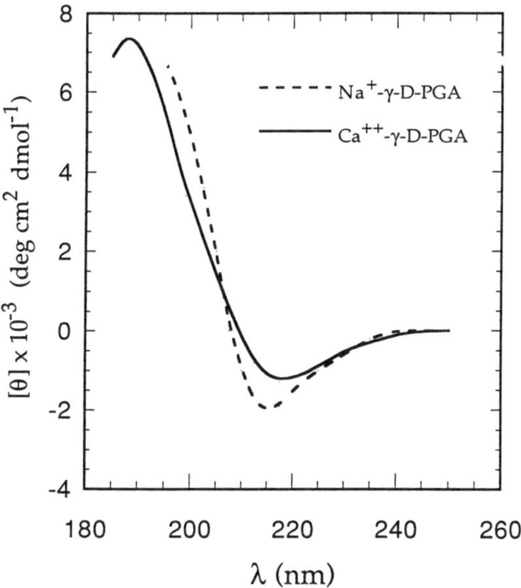

Figure 8. CD spectrum of Ca^{++}-γ-D-PGA (at neutral pH) in comparison with that of Na^+-γ-D-PGA at pH=11.42. Polymer concentration 0.87 mmoles/L.

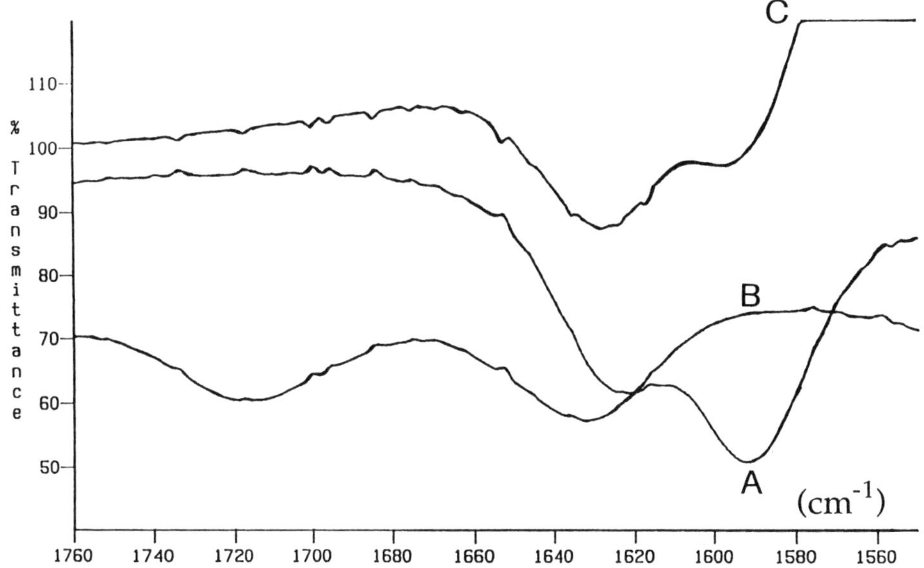

Figure 9. Infrared spectra (IR) of Na^+-γ-D-PGA/D_2O solutions at A) pD=11.2 and B) pD=1.5. C) IR of Ca^{++}-γ-D-PGA/D_2O solution at neutral pD. Polymer concentration 17 mmoles/L.

impossible in the case of other microbial biopolymers characterized by stiff chains, of average dimensions poorly sensitive to added salt, like many polysaccharides.

In conclusion, γ-D-PGA chains in dilute aqueous solution exhibit basically two different, limiting shapes, i.e. highly stretched and tightly globular, which interconvert rather sharply in response to minor changes in pH or added salt concentration (Ca(II) salts) within critical intervals.

This stimulates additional research aimed at a better understanding of the nature of such limiting conformations and at the exploitation of the associated sharp changes in chains average dimensions in, for example, the formulation of pH/Ca(II) sensitive hydrogels. Studies are in progress in our laboratories along these lines.

Acknowledgements

This work has been carried out with financial support of the Italian Ministry for University and Scientific and Technological Research (MURST, Rome), "60%-Ateneo" funds.

The selection and fermentation of highly mucoid cellular culture of *B. Licheniformis* (ATCC 9945-a) were carried out in the laboratory of Prof. C. Palleschi (Department of Cellular Biology & Development, University "La Sapienza", Rome), whose expert collaboration is sincerely appreciated.

We are grateful also to Dr. B. Sambuco for a number of spectroscopic measurements.

References

1. Gross, R. A.; Birrer, G. A.; Cromwick, A. M.; Giannos, S.A.; Mc Carthy, S.P.; *Biotecnologycal Polymers: Medical, Pharmaceutical and Industrial Applications* , Gebelein, C.G., Ed., Technomic Publ. Co.,1993, Part III, pp 200-214.
2. Lemogne, M.: *Ann. Inst. Pasteur* (Paris) **1925**, *39*, 144.
3. Giannos, S.A.; Shah, D.; Gross, R. A.; Kaplan, D. L.; Mayer, J. M.; *Novel Biodegradable Microbial Polymers*, Dawes, E. A., Ed., Series E, Applied Sciences, Vol. 186, Kluver Academic Publishers, Netherlands, 1990, pp 457-460.
4. Borbely, M.; Nagasaki, Y.; Borbely, J.; Fan, K.; Bhogle, A.,;Sevolan, M.; *Polymer Bull.* **1994**, *32*, 127.
5. Malborough, D. I.; *Biopolymers* **1973**, *12*, 1083.
6. Balasubramian, D., Kalita, C.C., Kovacs, J. *Biopolymers* **1973**, *12*, 1089.
7. Manning, G. S.: *Ann. Rev. Phys. Chem.* **1972**, *23*, 117.
8. Fixmann, M.; *J. Chem. Phys.* **1964**, *41*, 3772.
9. Smidsrod, O.; Haug, A.; *Biopolym.* **1971**,*10,* 1213.

RECEIVED January 3, 1996

Chapter 20

Bioactive Polymeric Dental Materials Based on Amorphous Calcium Phosphate

Effect of Coupling Agents

J. M. Antonucci[1], D. Skrtic[2,3], and E. D. Eanes[2]

[1]Polymers Division, National Institute of Standards and Technology, Gaithersburg, MD 20899
[2]Bone Research Branch, National Institute of Dental Research, National Institutes of Health, Bethesda, MD 20892
[3]Department of Chemistry, Ruder Boskovic Institute, 41000 Zagreb, Croatia

The effectiveness of acrylic-based visible light curable composites with amorphous calcium phosphate (ACP) as the filler phase to release Ca^{2+} and phosphate (PO_4) ions in aqueous media was enhanced with the incorporation of zirconyl dimethacrylate (ZrM) or 3-methacryloxypropyltrimethoxysilane (MPTMS) as coupling agents. Relatively hydrophobic resin-based composites formulated with these coupling agents were found to more rapidly release these ions over a longer period to establish solution Ca^{2+} and PO_4 concentrations much higher than those obtained from similar composites without these coupling agents. The increased effectiveness of the former composites was comparable to that observed with more hydrophilic composites formulated with 2-hydroxyethyl methacrylate. In addition, composite strength was not compromised by the incorporation of MPTMS or ZrM. These results suggest that coupling agents make ACP-filled composites even more effective as bioactive dental materials for use in clinical applications where preventing demineralization or promoting remineralization of tooth structures is desirable.

Because of their excellent biocompatibility with both soft and hard tissue, crystalline calcium orthophosphates, especially hydroxyapatite (HAP) and those capable of converting to HAP are finding increasing use as prophylactic, prosthetic, adhesive and restorative materials in dental and other biomedical applications (*1-5*). Polymeric calcium phosphate composites and cements have also been developed (*6-11*) in an effort to enhance the toughness, strength and handling properties of these materials.

By contrast, amorphous calcium phosphate (ACP), which has an approximate compositional formula of $Ca_3(PO_4)_2 \cdot 3H_2O$ (*12*) and is an important intermediate in

HAP, $Ca_{10}(PO_4)_6(OH)_2$, formation (*12-15*), has received far less attention as a biomaterial. The two principal reasons for this are ACP's high solubility in aqueous media and its rapid conversion to HAP. These very properties, however, suggest its use as a bioactive filler in polymeric materials, e.g. composites, sealants, adhesives, that can preserve teeth against demineralization and possibly promote remineralization of defective tooth structure (Skrtic,D.; Eanes,E.D.; Antonucci, J.M. Chapter 25 in *Industrial biotechnological polymers*; Gebelein,C.G., Carraher,C.E.Jr., Eds.; Technomic Publishing Co., Inc.: Lancaster, PA, 1995, in press). In this study, several visible light curable composites based on acrylic monomer systems containing coupling agents and a stabilized ACP were prepared, polymerized, and evaluated for their ability to release calcium and phosphate ions in the presence of aqueous environments in a sustained fashion and at levels adequate to favor HAP formation.

Materials and Methods

Resin Formulations. The resins used in this study were formulated by mixing the following commercial monomers in the proportions indicated in Table I: 2,2-bis[p-(2'-hydroxy-3'-methacryloxypropoxy)phenyl]propane (BisGMA), triethylene glycol dimethacrylate (TEGDMA), 2-hydroxyethyl methacrylate (HEMA), 3-methacryloxypropyltrimethyloxy silane (MPTMS), and zirconyl dimethacrylate (ZrM). Camphorquinone (CQ) and ethyl-4-N,N-dimethylaminobenzoate (4EDMAB) were used as the photooxidant and photoreductant components, respectively, of the visible light photoinitiator system (*8*). Figure 1 shows the chemical structures of the monomers, coupling agents and components of the photoinitiator system.

Table I. Composite disk resin formulations

Resin	Resin component (wt%)						
system	BisGMA	TEGDMA	ZrM	MPTMS	HEMA	CQ	4EDMAB
RS#1	49.5	49.5	-	-	-	0.2	0.8
RS#2	49.1	49.1	0.8	-	-	0.2	0.8
RS#3	44.5	44.5	-	10.0*	-	0.2	0.8
RS#4	35.5	35.5	-	-	28.0	0.2	0.8

* In some experiments, the MPTMS was used to silanize the ACP filler before mixing with RS#1.

ACP Preparation. Pyrophosphate ($P_2O_7^{4-}$)-stabilized ACP was prepared by rapidly adding, with stirring, a 800 mmol/L $Ca(NO_3)_2$ solution to an equal volume of a 525 mmol/L Na_2HPO_4 solution containing 11 mmol/L $Na_4P_2O_7$. The phosphate solution was

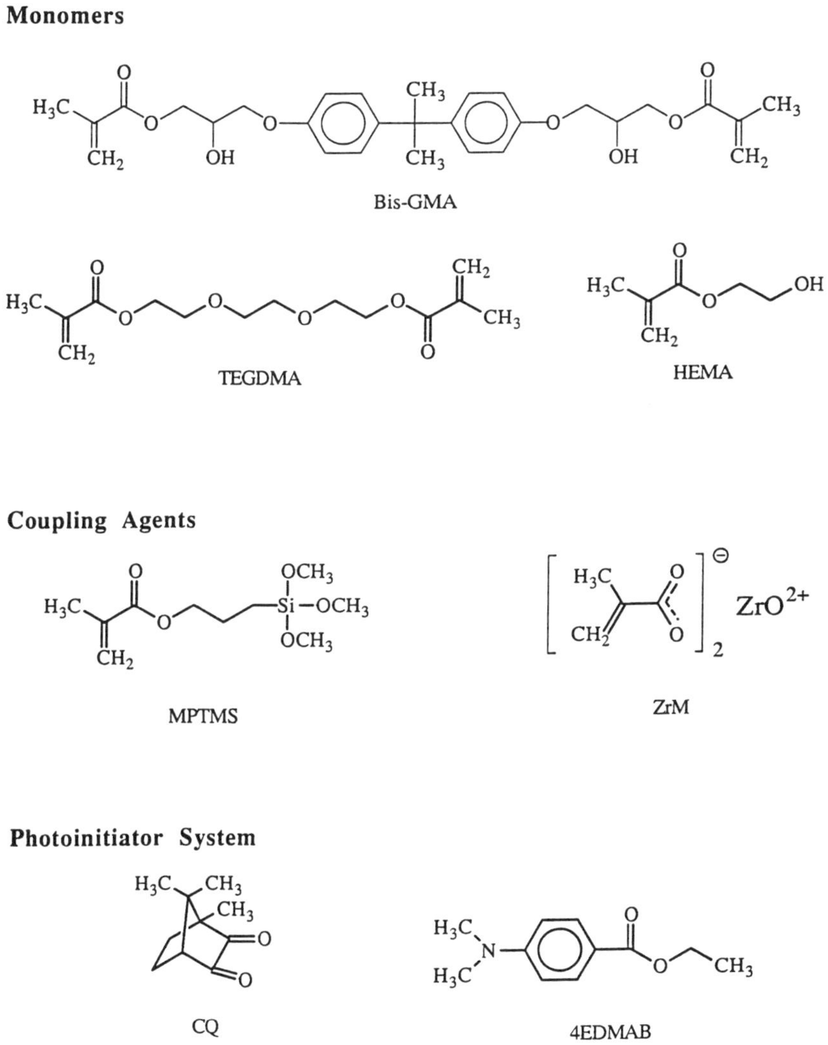

Monomers

Bis-GMA

TEGDMA

HEMA

Coupling Agents

MPTMS

ZrM

Photoinitiator System

CQ

4EDMAB

Fig. 1. Chemical structures of monomers, coupling agents and photoinitiator system used in formulation of ACP composites.

brought to pH 12.5 with 1 mol/L NaOH before mixing with the Ca^{2+} solution. The preparation was carried out at 22°C in a closed system under nitrogen in order to minimize CO_2 adsorption. After filtration of the fine precipitate, the $P_2O_7^{4-}$-ACP was washed with ammoniated water (pH 10.5) and lyophilized. The amorphous state of the lyophilized product was verified by X-ray diffraction (XRD).

For some experiments, the $P_2O_7^{4-}$-ACP was presilanized with MPTMS as follows: MPTMS (10 wt% based on ACP) was mixed into a slurry of the ACP powder in cyclohexane containing 3 wt% n-propylamine (based on ACP). The flask containing the silane-ACP mixture was attached to a rotary evaporator and the solvent, n-propylamine and volatile byproducts (H_2O, CH_3OH) were removed by heating (23-100°C) under moderate vacuum (2.7 kPa). The silanized powder was then heated at 100°C for 30 minutes, cooled and then washed with cyclohexane to remove residual MPTMS and unattached products. Finally, the silanized ACP powder was redried under vacuum at room temperature and characterized as still amorphous by IR spectroscopy and XRD analysis. In other experiments the ACP powder was silane treated by addition of MPTMS to Bis-GMA HEGDMA (1/1), a process referred to as *in situ* silanization (RS#3, Table I). The application of the zirconate coupling agent ZrM to ACP was carried out in a similar fashion by its addition to BisGMA/TEGDMA to give RS#2 (Table I).

Composite Preparation. Composite pastes were prepared by hand spatulation using 40 wt% ACP as the particulate filler and 60 wt% of each of the monomer systems shown in Table I as the resin component. After mixing, the homogenized pastes were deaerated by placing them under a moderate vacuum (ca. 2.7 kPa) for 24 hours. Disk specimens were prepared by inserting the paste into circular teflon molds (approximately 16 mm in diameter and 2.0 mm in height), covering the open ends with mylar lined glass slides, and irradiating each side for 120 s with a visible light source (Prismatics Lite, Caulk/Dentsply, Milford, DE). The specimens were postcured in their mold assemblies at 37°C for 18 h and, after removal from their molds, the composite disks were weighed and examined by XRD. Disk specimens containing the presilanized ACP were prepared only with resin system RS#1.

Composite Immersion Experiments. Each composite disk was immersed in 100 ml of a continuously stirred, 10 mmol/L HEPES-buffered (pH 7.4) 125 mmol/L sodium chloride solution (240 mOsm) at 37°C. The release of calcium (Ca^{2+}) and total phosphate ($H_2PO_4^-$, HPO_4^{2-}, PO_4^{3-}) ions (designated hereafter as PO_4) from the disks was kinetically monitored for up to 350 hours using atomic absorption spectroscopy (Ca) and UV/VIS spectrophotometry (PO_4) measurements of filtered aliquots taken at regular time intervals. Upon completion of the immersion tests, the disks were removed, air dried, reweighed, and characterized by XRD.

Thermodynamic (ΔG^*) Calculations. The thermodynamic stability (ΔG^*) of the immersion solutions containing the maximum concentrations of Ca^{2+} and PO_4 ions released from the disks was calculated with respect to stoichiometric HAP using the Gibbs free-energy expression:

$$\Delta G^{*} = -2.303(RT/n)\ln(IAP/K_{sp})$$

where IAP is the ion activity product for HAP i.e.$(Ca^{2+})^{10}(PO_4^{3-})^6(OH^-)^2$, K_{sp} is the corresponding thermodynamic solubility product, R is the ideal gas constant, T is the absolute temperature, and n is the number of ions in the IAP (n = 18). The solution chemical equilibrium software program EQUIL (MicroMath Scientific Software, Salt Lake City, UT) was used to calculate ΔG^{*} using the program provided value of 117.1 for the K_{sp} of HAP.

Composite Strength Testing. The mechanical strength of composites prepared from untreated ACP and resin systems RS#1 and RS#3 (Table I) and from presilanized ACP and resin system RS#1 was measured in tension under both biaxial flexure (*16,17*), and diametral (*18*) compression conditions before and after immersion in aqueous media.

Results

All the ACP/resin composites examined in this study steadily released Ca^{2+} and PO_4 ions upon immersion in buffered pH 7.4 saline (Figures 2 and 3, respectively) until solution accumulations exceeded the saturation threshold for HAP (Table II). However,

Table II. Thermodynamic stability (ΔG^{*}) of solutions containing maximum concentrations of Ca^{2+} and PO_4 ions released from ACP composites prepared with different coupling agents

Coupling agent[a]	N[b]	-logIAP[c]	ΔG^{*}kJ/mole)
none	3	104.6 ± 1.8	- 3.9 ± 0.9[d]
ZrM	3	100.8 ± 0.9	- 5.2 ± 0.3
MPTMS "presilanized"	6	99.7 ± 0.6	- 5.5 ± 0.2
MPTMS "in situ"	3	98.7 ± 1.2	- 5.8 ± 0.4

[a] Coupling agent added to RS#1 except (presilanization) where MPTMS was used to silanize ACP directly.
[b] Number of experiments.
[c] IAP = ion activity product for HAP.
[d] Negative ΔG^{*} values indicate solutions were supersaturated with respect to HAP.

composites that contained the coupling agents ZrM and MPTMS more rapidly released these ions over a longer period to establish solution Ca^{2+} and PO_4 concentrations that were much higher than those attained by composites formulated with only the relatively

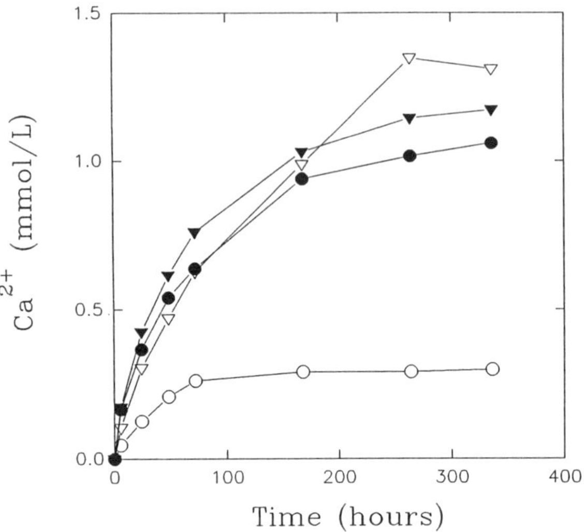

Fig. 2. Effect of ZrM and MPTMS on the concentration of Ca^{2+} released into buffered saline solution over time by the dissolution of $P_2O_7^{4-}$-stabilized ACP fillers in BisGMA:TEGDMA (1/1) (RS#1) composites. ○ RS#1; ● RS#2; ∇ RS#3; ▼ RS#1 only but with the ACP filler presilanized with MPTMS.

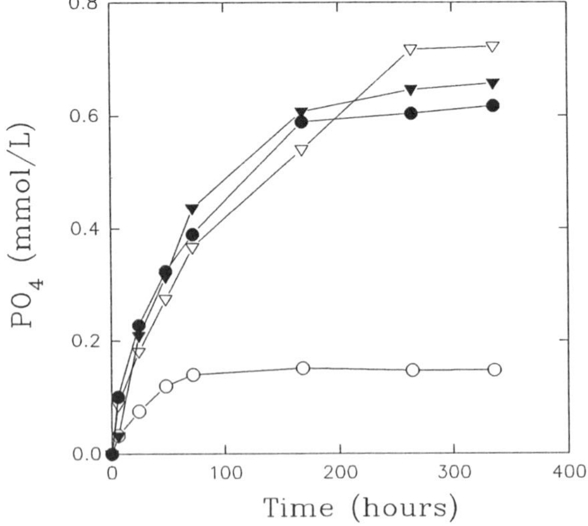

Fig. 3. PO_4 release from the same systems as described in Fig.2.

hydrophobic BisGMA/TEGDMA resin system (RS#1) and untreated ACP (Figs. 2,3). The enhanced ion release effected with the use of these coupling agents was comparable to that observed when ACP/RS#1 composites were modified by the addition of the hydrophilic monomer HEMA to the relatively hydrophobic base resin Bis-GMA/TEGDMA (1/1) (Figures 4,5).

The data in Table III show that the high ion concentration levels were achieved with the dissolution of only a relatively small (< 7%) amount of the ACP filler. XRD analysis (Figure 6) revealed, however, that after immersion of the composite a portion of the ACP had also converted to HAP within the composite. Comparison of the relative intensities of the HAP diffraction profiles of d and e in Fig.6 suggest that more ACP converted internally to HAP when ACP was presilanized than when it was mixed with MPTMS-containing resins.

Table III. Dissolution of ACP filler after long term immersion[a]

Type of composite	N	Maximum ion concn. (mM) Ca^{2+}	PO_4	Amount of filler dissolved (mg)	% of initial ACP amount
ACP + RS#1	3	0.297	0.147	2.6	1.3
ACP + RS#2	3	1.061	0.617	10.1	5.1
ACP + RS#1 (presilanized)	6	1.175	0.657	11.1	6.2
ACP + RS#3 ("in situ" silanized)	3	1.313	0.723	12.1	6.7

[a] The amount of filler dissolved upon soaking composite specimens in 100 ml buffered NaCl solution (pH 7.4 and ionic strength 0.129 mol/L) for 336 hours, was calculated from the corresponding Ca^{2+} and PO_4 levels obtained in solution using the formula $Ca_3(PO_4)_2$.
Results given are the average values of several kinetic runs for each system (number of runs given by N). All data "normalized" to initial total disk weight of 500 mg, which corresponds to 200 mg ACP in RS#1 and RS#2 composites, and 180 mg ACP in silane-containing composites.

The results of the biaxial flexure and diametral compression tests (Table IV) showed that the strength of the composites was not adversely affected either by immersion or by silanization.

Discussion

Previously we showed that the stabilization of ACP with $P_2O_7^{4-}$ was necessary in order for composites containing this bioactive filler material to be able to release calcium and phosphate ions at high levels in a sustained manner and, thus, have the potential to

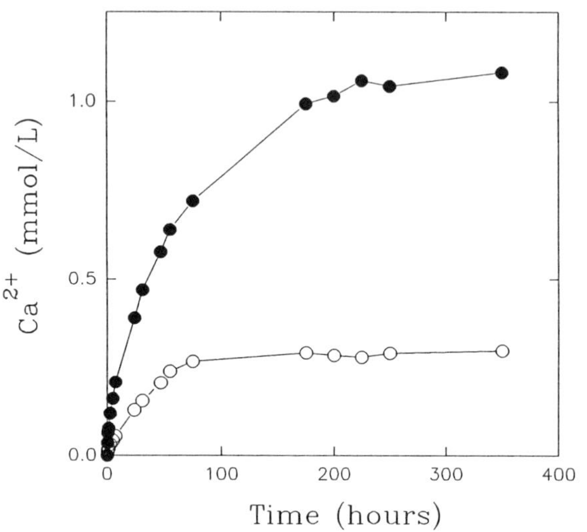

Fig. 4. Effect of HEMA on the concentration of Ca^{2+} released from $P_2O_7^{4-}$-ACP/RS#1 composites into buffered saline solutions. ○ RS#1; ● RS#4.

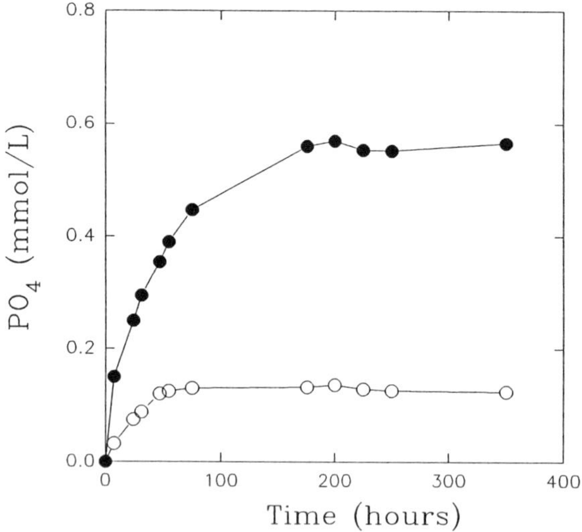

Fig. 5. PO_4 release from the same systems as described in Fig. 4.

Table IV. Effect of silanization on biaxial tensile strength (BTS) and diametral tensile strength (DTS) of ACP-filled composite disk specimens

ACP type	DTS (MPa) before immersion	DTS (MPa) after immersion	BTS (MPa) before immersion	BTS (MPa) after immersion
unsilanized[a]	15 ± 2(3)	15 ± 2(3)	54 ± 9(8)	62 ± 12(12)
presilanized[a]	15 ± 3(4)	15 ± 2(6)	47 ± 2(2)	68 ± 13(4)
"in situ" silanized[b]	19 ± 3(3)	18 ± 2(3)	63 ± 4(2)	64 ± 25(3)

( ) Number of samples tested.
[a] ACP mixed with RS#1 (Table I).
[b] Unsilanized ACP mixed with RS#3 (Table I).

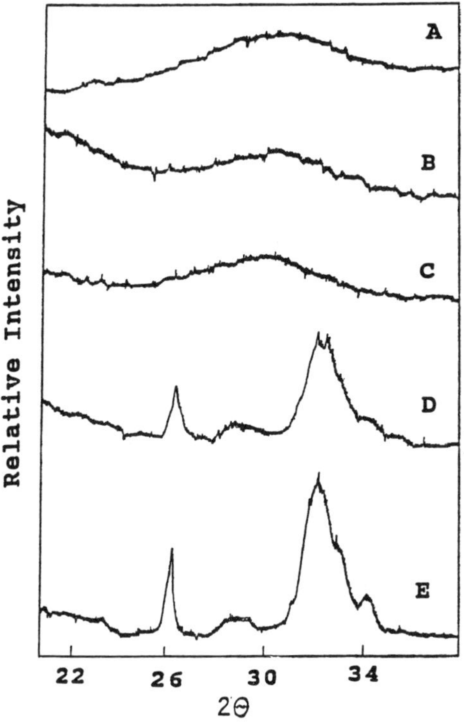

Fig. 6. X-ray diffraction patterns of (A) $P_2O_7^{4-}$-stabilized ACP filler material, (B) disk prepared from 40% $P_2O_7^{4-}$-ACP and 60% RS#3, (C) disk prepared from 36% ACP presilanized with MPTMS and 64% RS#1, (D) disk (B) after 350 hours immersion in buffered saline, (E) disk (C) after same immersion treatment.

promote remineralization of defective tooth structures such as those damaged by caries (Skrtic,D.; Eanes,E.D.; Antonucci, J.M. In *Industrial biotechnological polymers*, **1995**, in press). Without this stabilizer, internal conversion of ACP to HAP took place too rapidly for ions from the dissolving ACP to be released from the composite at a useful rate, and in high enough concentrations for significant mineralization to occur.

The present study shows that the inclusion of coupling agents such as ZrM and MPTMS in the formulation of stabilized-ACP composites further enhanced their ability to release mineral ions into external milieus. In particular, the finding that solutions in contact with ACP composites containing these surface modifying agents became more supersaturated with respect to HAP than those without these additives indicated that the corresponding concentrations of free Ca^{2+} and PO_4 ions within these composites were also elevated. The establishment of such higher internal ion levels would, in turn, enhance the effectiveness of composites in remineralizing damaged teeth by promoting a more rapid flux of ions across the composite-tooth surface interface and into the demineralized zones.

The mechanism by which the coupling agents increased free ion concentrations within the composite is not clear. One factor may be an increase in the area of contact between ACP particulates and resin because of the ability of these surface active agents to effect greater homogeneous dispersion of the ACP within the composite. It is difficult to see, however, how this greater intimacy between filler particles and surrounding polymer would lead directly to higher ion concentrations unless the diffusion of water to active ACP sites is also enhanced. Although direct evidence is lacking, comparison with the results from HEMA enriched composite suggests possibly that the coupling agents ZrM and MPTMS increased internal ion saturation by allowing the accommodation of more matrix water and/or better accessibility of the ACP to the water already entrained.

Only a small portion of the initial ACP filler was consumed during release of Ca^{2+} and PO_4 ions into the immersion solution by dissolution from the various composites (Table III). Instead, during immersion, a sizeable fraction of the ACP in the disks converted to HAP (Figure 6), as expressed by idealized equation (*12-15*):

$$10Ca_3(PO_4)_2 \cdot 3H_2O \xrightarrow{aq} 3Ca_{10}(PO_4)_6(OH)_2 + 2HPO_4^{2-} + 4H^+ + 24H_2O$$

The interesting feature of this conversion is that, despite the buildup of HAP within the composite, Ca^{2+} and PO_4 concentrations in the immersion solution steadily rose to high levels of supersaturation. In an earlier study (Skrtic,D.; Eanes,E.D.; Antonucci,J.M. In *Industrial biotechnological polymers*, **1995**, in press) it was found that the immersion solution will remain stable at these high ion levels unless seeded with exogenous HAP crystals, whereupon rapid depletion of solution Ca^{2+} and PO_4 by precipitation occurs. That a similar destabilization did not take place when HAP formed within the disks indicates isolation by intervening polymeric matrix prevented the HAP from coming into direct contact with the immersion solution. This finding also suggests that the ACP particles that converted to HAP were similarly isolated from the solution and possibly, as well, from those ACP particles that released Ca^{2+} and PO_4 ions by dissolution. The coupling agents could have further localized the conversion and dissolution events by effecting a finer dispersion of the ACP particles throughout the matrix. As stated above, it is unclear how this dispersion affected directly the ion releasing properties of the composite, but by reducing the clustering of ACP particles within the matrix, finer dispersion would more tightly confine HAP growth spatially, possibly making it more difficult for the HAP to interfere with the ion releasing process.

In conclusion, results from this study show that ACP-methacrylate composites containing the bifunctional, surface active monomers ZrM and MPTMS released Ca^{2+} and PO_4 ions into buffered pH 7.4 saline solutions at sustainable levels that were significantly higher that did composites without these coupling agents. This additional release occurred without significant deterioration in the strength of the composites. The added ions also made solution levels more than adequate to promote HAP formation in mineralized tissue such as teeth and bone. Therefore, the inclusion of the coupling agents in ACP/methacrylate composites should make these materials potentially even more effective in dental material applications where stabilization or remineralization of damaged tooth structure is desirable.

Dislaimer

Certain commercial materials and instruments are identified in this report to specify adequately the experimental procedure. In no instance does such identification imply recommendation or endorsement by the National Institute of Standards and Technology nor does it imply that the material or instrument identified is necessarily the best available for this purpose.

Literature Cited

1. LeGeros, R.Z.; *Adv. Dent. Res.* **1988**, *2(1)*, 164.
2. Alexander, H.; Parsons, J.R.; Ricci, J.L.; Bajpai, P.K.; Weis, A.B. In *Critical reviews on biocompatibility*, Williams, C., Ed.; CRC Press: Boca Raton, FL, 1987, pp.43.
3. Brown, W.E.; Chow, L.C. In *Cement research progress 1986*, Brown, P.W., Ed.; American Ceramic Society: Westerville, OH, 1986, pp.352.
4. LeGeros, R.Z. *Calcium phosphates in oral biology and medicine*, Karger: Basel, Switzerland, 1991, pp.82.
5. Driessens, F.M.C. In *Bioceramics of calcium phosphate*, deGroot, K., Ed.; CRC Press, Boca Raton, FL, 1983, pp.1.
6. Antonucci, J.M.; Misra, D.N.; Pecko, R.J. *J. Dent. Res.* **1981**, *60(7)*, 1332.
7. Sugawara, A.; Antonucci, J.M.; Takagi, S.; Chow, L.C.; Ohashi, M.J. *Nihon Univ. Sch. Dent.* **1989**, *31*, 372.
8. Antonucci, J.M.; Fowler, B.O.; Venz, S. *Dent. Mater.* **1991**, *7*, 124.
9. Miyazaki, K.; Horibe, T.; Antonucci, J.M.; Takagi, S.; Chow, L.C. *Dent. Mater.* **1993**, *9*, 41.
10. Miyazaki, K.; Horibe, T.; Antonucci, J.M.; Takagi, S.; Chow, L.C. *Dent. Mater.* **1993**, *9*, 46.
11. Dickens-Venz, S.; Takagi, S.; Chow, L.C.; Bowen, R.F., Johnson, A.D.; Dickens, B. *Dent. Mater.* **1994**, *10*, 100.
12. Meyer, J.L.; Eanes, E.D. *Calcif. Tissue Res.* **1979**, *25*, 59.
13. Eanes, E.D.; Gillesen, I.H.; Posner, A.S. *Nature* **1965**, *208(5008)*, 365.

14. Eanes, E.D.; Posner, A.S. *Calc. Tissue Res.* **1968**, *2*, 38.
15. Greenfield, D.J.; Eanes, E.D. *Calc. Tissue Res.* **1972**, *9*, 152.
16. Wachtman, J.B.Jr.; Capps, W.; Mandel, J. *J. Materials* **1972**, *7(2)*, 188.
17. Ban, S.; Anusavice, K.J. *J. Dent. Res.* **1990**, *69(12)*, 1791.
18. Council on Dental Materials, Instrument and Equipment, New American Dental Association Specification No. 27 for Direct Filling Resins. *J. Am. Dent. Assoc.* **1977**, *94*, 1191.

RECEIVED November 13, 1995

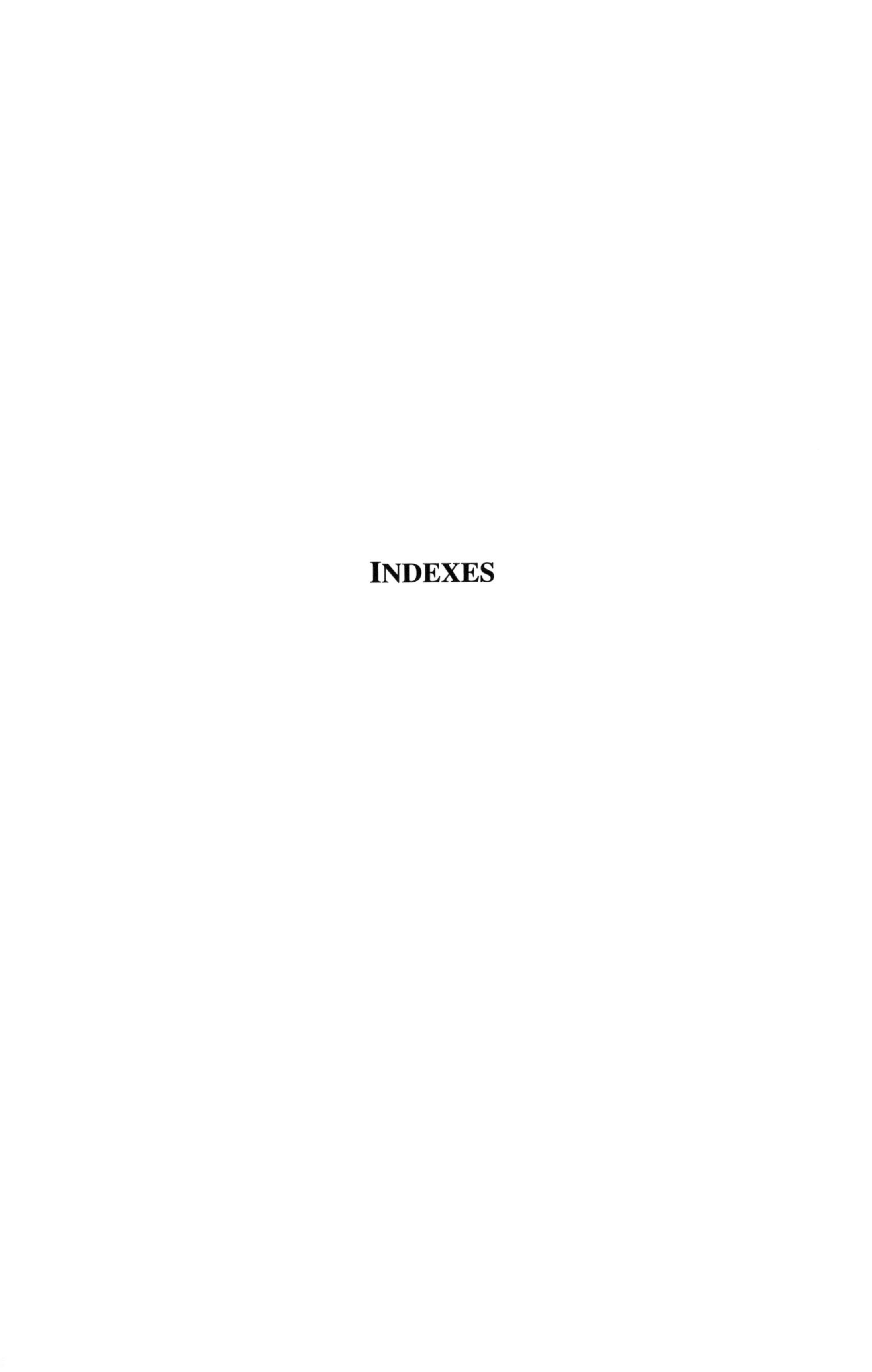

INDEXES

Author Index

Affiliation Index

Subject Index

Highlights from ACS Books

Good Laboratory Practice Standards: Applications for Field and Laboratory Studies
Edited by Willa Y. Garner, Maureen S. Barge, and James P. Ussary
ACS Professional Reference Book; 572 pp; clothbound ISBN 0–8412–2192–8

Silent Spring Revisited
Edited by Gino J. Marco, Robert M. Hollingworth, and William Durham
214 pp; clothbound ISBN 0–8412–0980–4; paperback ISBN 0–8412–0981–2

The Microkinetics of Heterogeneous Catalysis
By James A. Dumesic, Dale F. Rudd, Luis M. Aparicio, James E. Rekoske, and Andrés A. Treviño
ACS Professional Reference Book; 316 pp; clothbound ISBN 0–8412–2214–2

Helping Your Child Learn Science
By Nancy Paulu with Margery Martin; Illustrated by Margaret Scott
58 pp; paperback ISBN 0–8412–2626–1

Handbook of Chemical Property Estimation Methods
By Warren J. Lyman, William F. Reehl, and David H. Rosenblatt
960 pp; clothbound ISBN 0–8412–1761–0

Understanding Chemical Patents: A Guide for the Inventor
By John T. Maynard and Howard M. Peters
184 pp; clothbound ISBN 0–8412–1997–4; paperback ISBN 0–8412–1998–2

Spectroscopy of Polymers
By Jack L. Koenig
ACS Professional Reference Book; 328 pp;
clothbound ISBN 0–8412–1904–4; paperback ISBN 0–8412–1924–9

Harnessing Biotechnology for the 21st Century
Edited by Michael R. Ladisch and Arindam Bose
Conference Proceedings Series; 612 pp;
clothbound ISBN 0–8412–2477–3

From Caveman to Chemist: Circumstances and Achievements
By Hugh W. Salzberg
300 pp; clothbound ISBN 0–8412–1786–6; paperback ISBN 0–8412–1787–4

The Green Flame: Surviving Government Secrecy
By Andrew Dequasie
300 pp; clothbound ISBN 0–8412–1857–9
